Advances in
MICROBIAL ECOLOGY

Volume 6

ADVANCES IN MICROBIAL ECOLOGY

Sponsored by the International Committee on Microbial Ecology
(ICOME), a committee of the International Union of
Microbiological Societies (IUMS), and the International Union of
Biological Sciences (IUBS)

A Continuation Order Plan is available for this series. A continuation order will bring delivery of
each new volume immediately upon publication. Volumes are billed only upon actual shipment.
For further information please contact the publisher.

Advances in
MICROBIAL ECOLOGY

Volume 6

Edited by

K. C. Marshall
University of New South Wales
Kensington, New South Wales, Australia

PLENUM PRESS · NEW YORK AND LONDON

The Library of Congress cataloged the first volume of this title as follows:

Advances in microbial ecology. v. 1–
    New York, Plenum Press c1977 –
        v. ill. 24 cm.
    Key title: Advances in microbial ecology, ISSN 0147-4863

    1. Microbial ecology–Collected works.
QR100.A36   576′.15   77-649698

Library of Congress Catalog Card Number 77-649698
ISBN 0-306-41064-8

©1982 Plenum Press, New York
A Division of Plenum Publishing Corporation
233 Spring Street, New York, N.Y. 10013

Printed in the United States of America

Contributors

John W. Doran, Agricultural Research Service, United States Department of Agriculture, Lincoln, Nebraska 68583

Madilyn Fletcher, Department of Environmental Sciences, University of Warwick, Coventry CV4 7AL, England

K. C. Marshall, School of Microbiology, The University of New South Wales, Kensington, N.S.W. 2033, Australia

L.-A. Meyer-Reil, Institut für Meereskunde, Abteilung Marine Mikrobiologie, 2300 Kiel 1, Federal Republic of Germany

Richard Y. Morita, Department of Microbiology and School of Oceanography, Oregon State University, Corvallis, Oregon 97331

Hans W. Paerl, Institute of Marine Sciences, University of North Carolina, Morehead City, North Carolina 28557

G-Yull Rhee, Environmental Health Institute, Division of Laboratories and Research, New York Department of Health, Albany, New York 12201

F. B. van Es, Department of Microbiology, Kerklaan 30, 9751 NN, Haren, The Netherlands

Preface

This volume of *Advances in Microbial Ecology* marks a change in the editorship of the series. The Editorial Board wishes to take this opportunity to express its gratitude to Martin Alexander, the founding editor and editor of the first five volumes, for his enterprise in establishing the series and in ensuring that *Advances* has become an outstanding focal point for the identification of new developments in the rapidly expanding field of microbial ecology. With the publication of this volume, we welcome Howard Slater to the Editorial Board.

The policies of the Editorial Board remain the same as before. Most contributions to *Advances in Microbial Ecology* will be solicited by the Board. However, individuals are encouraged to submit outlines of unsolicited contributions to the Board for consideration for inclusion in the series. *Advances* is designed to serve an international audience and to provide critical reviews on basic and applied aspects of microbial ecology.

Contributions in the present volume are predominantly concerned with the ecology of aquatic microorganisms, but encompass a variety of approaches to this area. The exception is the chapter by J. W. Doran on the role of microorganisms in the cycling of selenium. G-Y. Rhee discusses the effects of environmental factors on phytoplankton growth. The factors limiting the productivity of freshwater microbial ecosystems are considered by H. W. Paerl. Problems in assessing biomass and metabolic activity of heterotrophic bacteria is the subject of the review by F. B. van Es and L.-A. Meyer-Reil. In considering nutrient-deficient marine habitats, R. Y. Morita presents evidence for mechanisms related to the survival of heterotrophic bacteria in such extreme conditions. M. Fletcher and K. C. Marshall discuss the role of solid surfaces in the ecology of aquatic bacteria.

K. C. Marshall, Editor
M. Alexander
T. Rosswall
J. H. Slater

Contents

Chapter 3

Factors Limiting Productivity of Freshwater Ecosystems

Hans W. Paerl

Chapter 4

Biomass and Metabolic Activity of Heterotrophic Marine Bacteria

F. B. van Es and L.-A Meyer-Reil

Chapter 5

Starvation-Survival of Heterotrophs in the Marine Environment

Richard Y. Morita

Chapter 6

Are Solid Surfaces of Ecological Significance to Aquatic Bacteria?

Madilyn Fletcher and K. C. Marshall

Microorganisms and the Biological Cycling of Selenium

JOHN W. DORAN

1. Introduction

Most studies on the microbial transformations of elements have emphasized nutrient cycling within the biosphere or the economics of agricultural or industrial processes. Cyclic transformations within the biosphere between soluble, insoluble, and gaseous forms of carbon, nitrogen, hydrogen, oxygen, and sulfur are well known. Recently, attention has been focused on the role of microorganisms in the production and degradation of chemicals containing toxic elements (Alexander, 1973; Wood, 1974). Measures to increase animal and food crop production or disposal of waste materials can result in the introduction of elements in amounts harmful to terrestrial and aquatic ecosystems. Many elements and their compounds vary widely in both toxicity and mobility. Consequently, their safe disposal or effective recycling requires an understanding of their potential toxicities and possible transformations in the environment.

Chemically, selenium (Se) is similar to sulfur (S), and both elements are placed in group VIA of the periodic table along with oxygen, tellurium, and polonium. The inorganic forms of Se (Table I) are structurally similar to their S analogs. Selenium and S exist in oxidation states of $+6$, $+4$, 0, and -2, and both have the six-electron system of valence orbitals. Because of its low reduction potential, free selenide is very unstable and is readily oxidized to elemental Se in the presence of O_2; thus, very little exists in soils with pH values below 9 10. The only stable selenides are those bound to metals or in organic

JOHN W. DORAN • Agricultural Research Service, U.S. Department of Agriculture, Lincoln, Nebraska 68583.

**Table I. Inorganic Forms of Selenium and Their Reduction
Potentials[a]**

Form	Oxidation state	E_0'	Aqueous form
Selenate	+6	+0.44	SeO_4^{2-}
Selenite	+4	+0.21	SeO_3^{2-}
Elemental selenium	0	−0.73	Se^0
Selenide	−2		HSe^-

[a]Reduction potentials (E_0') at pH 7.5 and unit activity for Se species were calculated
from values given by Pourbaix (1966).

combination. Like elemental S, elemental Se occurs in three allotropic forms:
"metallic," crystalline, and amorphous. Biological oxidation or reduction of
inorganic Se can often be detected by the orange-red color characteristic of the
freshly precipitated amorphous form of elemental Se. Except for elemental Se,
which is particulate and insoluble, the other forms of inorganic Se, shown in
Table I, are anionic and water soluble at pH 7.5.

Selenium is of considerable biological interest because it is required by
animals, but it can also be highly toxic. The toxicity of Se was first recognized
in the 1930s, when it was established that certain plants growing on high-Se
calcareous soils accumulated the element in amounts which could be harmful
to animals eating them (Franke, 1934). Moxon (1937) stated that "alkali disease," which plagued many stockmen in the western United States, was a form
of Se toxicity. Research reports from other countries have indicated that haz-
ards to the health of humans and domestic animals can result from production
of food crops containing toxic Se levels (Lakin, 1972; Shapiro, 1973).

The toxicity of Se compounds varies greatly, but soluble, inorganic forms
of Se are generally considered the most toxic of all Se compounds. Selenite, a
powerful oxidizing agent, readily denatures sulfhydryl enzymes and oxidizes
sulfhydryl groups to form disulfide and unstable selenotrisulfides (RS-Se-SR).
The toxicity of selenate is less than that of selenite and is apparently manifested only after it has first been reduced to selenite (Oehme, 1972). Hydrogen
selenide, an inhibitor of terminal cytochrome oxidase, is 100 times more toxic
than hydrogen cyanide and is considered the most toxic Se compound (Martin,
1973; Painter, 1941). One of the major mechanisms of Se toxicity may be its
substitution for S in many proteins and a resultant instability of the Se-substi-
tuted compounds (Shrift, 1972).

The toxicity of Se was considered its only role in animal nutrition, until
Schwarz and Foltz (1957) discovered that certain Se compounds could replace
a dietary factor that prevented severe necrotic degeneration of the liver in rats.
Selenium has since been shown to be essential in preventing animal diseases
involving tissue integrity and function (Muth *et al.,* 1967; Scott, 1973), to be

a necessary component of several bacterial and mammalian enzyme systems (Stadtman, 1980), and is required for the growth of at least two bacteria (Dürre et al., 1981; Jones and Stadtman, 1977). Although research data are limited, dietary Se has also been implicated in detoxification of heavy metals and prevention of some forms of cancer (Frost, 1972). Many soils do not supply Se to crop plants in amounts adequate to support proper animal nutrition. Today, Se deficiencies in domestic animals are often corrected by injection or supplementation of animal feeds with selenite and selenate. Attempts to eliminate Se deficiencies by prophylactic use of Se compounds were blocked for many years by federal law because inorganic Se had been labeled a carcinogen (Frost, 1960).

2. Selenium in the Environment

2.1. Biogeochemistry

Selenium is considered one of the least plentiful but most toxic elements in the earth's crust (Frost, 1972). The distributions of Se in terrestrial, aquatic, anthropogenic, and extraterrestrial substances are given in Table II. Since Se and S are closely related, their biogeochemistry is similar. The selenide ion substitutes for the sulfide ion, and Se is often concentrated in the major sulfide minerals and pyritic coal deposits (Lakin, 1972). The average Se content of most soils ranges from 0.1 to 2 ppm, but soils derived from parent material formed during the Cretaceous period can contain more than 100 ppm Se. Crop plants grown on high-Se soil can accumulate concentrations of Se (>4 ppm) that are toxic to livestock and humans; however, in many parts of the world, the Se content of vegetation is less than 0.05 ppm and is not sufficient for proper animal nutrition. In the United States, the majority of forage and crop plants contain less than 0.1 ppm Se (Kubota et al., 1967).

The Se content of both fresh and salt waters is much lower than terrestrial materials and usually ranges from 0.05 to 4 ppb. A cursory examination of the data in Table II indicates the marked propensity for food-chain magnification of Se in aquatic ecosystems. Sandholm et al. (1973) reported that some zooplankton can absorb Se as selenite, and phytoplankton can actively concentrate certain organic Se compounds, such as selenomethionine. The same authors commented that fish concentrate little inorganic or organic Se directly from water, but accumulate Se when feeding on Se-containing phytoplankton and zooplankton. Lindström and Rhode (1978) demonstrated a selenium requirement for growth of the dinoflagellate Peridinium cinctum fa. westii. Optimal growth of this photosynthetic alga occurred at a selenite concentration in lake water as low as 0.05 ppm.

Any meaningful discussion of the distribution and biological availability

Table II. Distribution of Selenium in Natural and Anthropogenic Materials

Material	Selenium concentration (ppm Se)	Reference
Terrestrial		
Earth's crust	0.09	Lakin (1972)
Limestone	0.1–14	Rosenfeld and Beath (1964)
Shales and phosphate rocks	<1–55	Lakin (1972)
Crude oil	0.06–0.39	Pillay et al. (1969)
Coal	0.5–11	Pillay et al. (1969)
Soils		
Nonseleniferous	<0.1–2.0	Swaine (1955)
Seleniferous	2–200	Natl. Acad. Sci. (1971)
Vegetation		
Primary selenium accumulators	51–4474	Rosenfeld and Beath (1964)
Grains (seleniferous areas)	0.1–30	Rosenfeld and Beath (1964)
Crop plants (low-Se areas)	<0.05	Allaway (1973)
Fruits and vegetables	<0.01	Allaway (1973)
Blood		
Animals with Se poisoning	1–27	Rosenfeld and Beath (1964)
Human	0.1–0.34	Allaway (1973)
Aquatic		
Ocean waters	0.0001–0.004	Schutz and Turekian (1965)
River waters	0.0001–0.0004	Kharkar et al. (1968)
Aquatic plants	0.02–0.14	Sandholm et al. (1973)
Plankton	1.1–2.4	Sandholm et al. (1973)
Fish	0.5–6.2	Lakin (1972)
Anthropogenic		
Petroleum products	0.15–1.65	Lakin (1972)
Fly ash	1.2–16.5	Gutenmann et al. (1976)
Sewage sludge	1.8–4.8	Furr et al. (1980)
Certain paper products	1.6–19	Lakin (1972)
Extraterrestrial		
Lunar basalts	0.14–0.25	Lakin (1972)
Meteorites	3–15	Rosenfeld and Beath (1964)

of Se in the environment must be directed towards specific compounds or chemical forms of the element (Allaway, 1973). The total Se content of soils is of little use in determining the availability of Se to plants or microorganisms because the chemical forms of Se present in soil vary widely in availability. The forms of Se present in soil include (1) metal selenides, (2) elemental Se, (3) selenite, (4) selenate, and (5) organic Se. The pyritic, heavy metal selenide, and elemental forms of Se are essentially insoluble or very slowly soluble and of limited biological availability. Selenite and selenate, both water-soluble anions, are potentially available for biological uptake. Selenate occurs in appreciable quantities only in the highly oxidizing soils of arid regions (Geering *et al.*, 1968). Selenite is readily sorbed by iron hydroxide complexes in soils and is less available in acid than basic soils (Cary *et al.*, 1967). Thus, acidic lateritic soils in Hawaii and Puerto Rico, containing up to 30 ppm total Se, usually do not produce vegetation containing Se levels toxic to animals (Bisbjerg, 1972). As in soil, free selenite in water is readily adsorbed by iron hydroxides and manganese oxides. From 30 to 50% of this "sorbed" Se, which is carried in the suspended load of streams, can be desorbed when these waters mix with salt water (Kharkar *et al.*, 1968).

Olson and Moxon (1939) presented data that indicated up to 40% of the total Se of some soils is in the organic form. These same authors also stated that the availability of Se in soils seems to be dependent upon the amount of water-soluble Se, which, in turn, is correlated with the organic fraction of soil Se. Thus, the cycling of Se in soils is apparently related to the mineralization of Se-containing organic matter and the continuous release of soluble Se.

2.2. Mobilization of Selenium

Selenium is mobilized in the environment through natural processes of weathering, disposal of wastes, ore processing, and gaseous emissions to the atmosphere. Significant quantities of Se are currently moving through the food chain in the United States (Allaway, 1973), and the disposal of animal and domestic wastes may represent a significant source of Se to aquatic and terrestrial ecosystems. The Se contained in such wastes originates from dietary amendments or injections, food and feeds, water, industrial wastes, and from commercial products, like some Se-containing medicated shampoos. The Se contained in sewage sludge is apparently in a form available for biological uptake. Furr *et al.* (1976, 1980) found that the Se contents of eight vegetable crops grown on sludge-amended soils were 2–11 times higher than those from nonamended soil.

Like S, Se is emitted to the atmosphere via volcanic activity and fossil fuel burning. However, unlike S, which is present predominantly as SO_2, the major Se form in these gaseous emissions, as determined from thermodynamic calculations and chemical characteristics, is elemental Se (Andren *et al.*, 1975; Suzuoki, 1965). The quantity of Se discharged to the earth's atmosphere from the burning of coal represents 6–11% of the Se mobilized through weathering processes and river flow (Bertine and Goldberg, 1971; Andren *et al.*, 1975). Fly ash, the residue produced at steam generation power plants during the burning of coal, can contain relatively high Se concentrations. Andren *et al.* (1975) estimated that 1.5–2.3 times as much Se is mobilized through disposal of fly ash and slag wastes than by natural weathering and erosion of crustal materials. The addition of fly ash to soil or aquatic habitats can result in increased Se contents in the plants, animals, and other organisms indigenous to these environments. Gutenmann *et al.* (1976) reported that cabbage plants grown on soils containing 10% fly ash absorbed Se (up to 3.7 ppm) in direct proportion to the Se content of the fly ash. Also, sweetclover plants growing directly on fly ash disposal piles contained over 200 ppm Se. Since most of the Se in fly ash is in the elemental form (Andren *et al.,* 1975), the finding that plants grown on fly-ash-amended soils can contain toxic Se concentrations indicates that the Se originally present was transformed to a more available form. The release of fly ash to the atmosphere through burning of fossil fuels has also been implicated in the food-chain magnification of Se in aquatic ecosystems. Copeland (1970) reported that Se levels in zooplankton in Lake Michigan increased with proximity to metropolitan areas. Selenium levels in zooplankton in most areas of the lake averaged 1 to 2 ppm, but increased to maxima of 3 and 6 ppm in areas directly downwind from Milwaukee and Chicago, respectively.

3. Microbial Transformations of Selenium

Selenium appears to be cycled predominantly via biological pathways (Shrift, 1973). The inorganic forms of Se are converted to reduced organic forms by plants, animals, and microorganisms. Microorganisms bring about the decomposition of biological residues and, thus, function as catalysts essential to Se recycling. The major microbial transformations of Se can be divided into three categories: oxidation and reduction, immobilization and mineralization, and methylation. These Se transformations have been categorized to facilitate discussion and, as shown in Fig. 1, are not mutually exclusive because

OXIDATION AND REDUCTION

$$Se^{2-} \rightleftharpoons Se^{\circ} \rightleftharpoons SeO_3^{2-} \rightleftharpoons SeO_4^{2-}$$

selenide   elemental   selenite   selenate

IMMOBILIZATION AND MINERALIZATION

INORGANIC Se $\underset{\text{min.}}{\overset{\text{immob.}}{\rightleftharpoons}}$ ORGANIC Se

METHYLATION

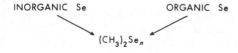

INORGANIC Se   ORGANIC Se

$(CH_3)_2Se_n$

Figure 1. Microbial transformations of selenium.

some overlap occurs. Many microorganisms have the capacity to reduce oxidized compounds, whereas a smaller number have been demonstrated to carry out the oxidation of inorganic Se. Because Se becomes incorporated into the cells of organisms in forms unavailable for uptake and metabolism by most plants and animals, the microbial mineralization of organic forms can regulate Se availability. The microbial methylation of Se represents both reduction and immobilization of Se and may be an important source of atmospheric Se.

3.1. Oxidation and Reduction of Inorganic Selenium

Because of the amount of Se that is assimilated into organic materials in plants, animals, and microorganisms, obviously a mechanism for recycling to the more-available oxidized forms of inorganic Se must exist. The energy available from the oxidation of inorganic Se would be sufficient to serve as an energy source for microorganisms; however, unlike S, the microbial oxidation of inorganic Se has received only limited investigation. Two microorganisms, an aerobic soil bacterium (Lipman and Waksman, 1923) and a purple sulfur bacterium (Saposnikov, 1937), were reported to have used the oxidation of elemental Se to selenic acid (H_2SeO_4) as a sole source of energy. Torma and Habashi (1972) reported that a strain of *Thiobacillus ferroxidans* derived its energy from the oxidation of copper selenide, with the resultant products being cupric ions and elemental Se. Limited research indicates that microorganisms are active in the oxidation of elemental Se and selenite in soils (Bisbjerg, 1972; Geering *et al.*, 1968). Recently, a strain of *Bacillus megaterium*, a heterotrophic bacterium isolated from soil, was found to oxidize elemental Se to selenite in laboratory cultures (Sarathchandra and Watkinson, 1981). Unfortunately, these brief accounts of microbial oxidation of Se were not followed by more extensive studies.

The microbial reduction of oxidized, inorganic Se compounds usually results in incorporation of selenide into organic Se compounds or the formation of elemental Se. The immobilization of Se will be discussed in the next section, and only transformations to more reduced inorganic forms will be discussed here. Many fungi, actinomycetes, and bacteria are capable of reducing inorganic Se salts. The reduced Se usually appears as red intracellular deposits, and several reports indicate that amorphous elemental Se is the final product in microbial reduction of selenite and selenate (Falcone and Nickerson, 1963; Levine, 1925; Zalokar, 1953). The reduction of selenite to elemental Se is reportedly a detoxification mechanism which enables *Salmonella* to tolerate higher concentrations of selenite than other microorganisms (McCready *et al.*, 1966).

Levine (1925) studied the reducing properties of several microorganisms and concluded that the reduction of selenate and selenite is an intracellular phenomenon and a function of growth. Fewer microorganisms could reduce selenate, and the formation of elemental Se was slower from selenate than from selenite. Thus, it was concluded the reduction of selenate is a two-step process in which selenite is an intermediate product. McCready *et al.* (1966) presented data indicating the presence of an intermediate Se compound in the reduction of selenite to elemental Se by *Salmonella heidelberg.* This intermediate Se form was reported to be an inorganic compound containing Se in the divalent state and was more toxic to the organism than selenite, its tetravalent precursor.

The reduction of both selenate and selenite is enzymatic and metabolically driven. Zalokar (1953) found that only living cells of *Neurospora crassa* could reduce selenite, and the reduction was inhibited by compounds that were known poisons of respiratory enzymes. Woolfolk and Whiteley (1962) studied the reduction of selenite by cell-free extracts of *Micrococcus lactilyticus* incubated in hydrogen atmosphere. Measurements of hydrogen uptake (manometric) and rate of formation of elemental Se and selenide suggested that the reduction involves two steps (equations 1 and 2):

$$\underset{\text{selenite}}{HSeO_3^-} + 2H_2 \rightarrow \underset{\text{elemental Se}}{Se^0} + 2H_2O + OH^- \tag{1}$$

and

$$\underset{\text{elemental Se}}{Se^0} + H_2 \rightarrow \underset{\text{selenide}}{HSe^-} + H^+ \tag{2}$$

The overall reduction of selenite to selenide was quantitative and involved a transfer of six electrons. The same authors demonstrated that hydrogenase-containing extracts of *Desulfovibrio desulfuricans* and *Clostridium pasteurianum* also utilized hydrogen to catalyze the reduction of selenite. Nickerson and Falcone (1963) also found that a specific dehydrogenase was involved in the reduction of selenite by yeast, and a NADPH-dependent reduction of selenite to elemental Se in the presence of glutathione reductase (isolated from yeast) has also been demonstrated (Ganther, 1974). Less information is available on the enzymatic reduction of selenate. In bacteria and yeasts, the enzyme responsible for the first step in the assimilative reduction of sulfate is also active on selenate. Wilson and Bandurski (1958) demonstrated that the ATP sulfurylase isolated from yeast also forms adenosine-5'-phosphoselenate (APSe) when incubated with selenate.

In soil, the reduction of oxidized forms of inorganic Se is often related to microbial activity. Some fungi and a large proportion of actinomycetes and bacteria isolated from soil can reduce selenate and selenite to elemental Se

**Table III. Proportion of Soil Microbial Groups Capable of
Reducing Selenate and Selenite to Elemental Selenium**[a]

Se compound reduced	Se reducers (% of total numbers)		
	Fungi	Actinomycetes	Bacteria
SeO_4^{2-}	11	48	17
SeO_3^{2-}	3	71	43

[a]Data from Bautista and Alexander (1972).

(Table III). Kovalski *et al.* (1968) reported that the mechanism of adaptation and resistance to high Se concentrations of microorganisms, isolated from high-Se soils, was the ability of these organisms to reduce Se to the elemental state. Selenate and selenite added to soil are transformed to organic Se, elemental Se, or metallic selenides within a relatively short time (Bisbjerg, 1972). Doran and Alexander (1977a) found that, under anaerobic conditions, hydrogen selenide (H_2Se) is evolved from soils amended with either selenate, selenite, or elemental Se. Reduction of the Se compounds was stimulated by the addition of an available C source, and no activity was noted in steam-sterilized soils. It is likely that most of the H_2Se produced in anaerobic zones of soils and aquatic habitats is either precipitated with heavy metals or oxidized to elemental Se when it diffuses to O_2-containing zones. The formation of reduced insoluble forms of Se may explain why lake sediments seem to be reservoirs for Se (Wiersma and Lee, 1971).

3.2. Immobilization and Mineralization

Immobilization and mineralization, terms often used synonymously with assimilation and metabolism, are microbial processes that influence transformations between inorganic and organic forms of an element. In soil microbiology, immobilization refers almost exclusively to the loss in availability of an essential element to plants as a result of microbial assimilation. Within the realm of environmental microbiology, however, immobilization can be defined as any biological process that effectively reduces the availability of a toxic or essential element to plants, animals, or microorganisms in terrestrial or aquatic ecosystems. The opposing process, mineralization, is the conversion of combined organic forms of an element to less complex inorganic substances. The microbial decomposition of plant and animal residues results in the release of CO_2, hydrogen, simple organics, and inorganic (mineral) elements, i.e., mineralization.

3.2.1. Immobilization of Selenium

Sulfur is an essential element for life and is present in the cellular components of all organisms, predominantly in amino acids and proteins. Since Se is chemically similar to S and can replace it in many biological systems (Shrift, 1961), it is not surprising that the Se analogs of organic S compounds occur in living organisms. Many microorganisms, plants, and animals are able to convert inorganic Se into the more-reduced organic forms. The major groups of organic Se compounds that have been isolated from living systems (Table IV)

Table IV. Organic Selenium Compounds Isolated from Plants, Animals, and Microorganisms[a]

Compound	Formula	Source
Amino acids		
Selenomethionine	$CH_3SeCH_2CH_2CHNH_2COOH$	Proteins, plants, yeasts, and bacteria
Selenocystine	$SeCH_2CHNH_2COOH$ $\mid$ $SeCH_2CHNH_2COOH$	Proteins, plants, yeasts, and bacteria
Selenohomocystine	$SeCH_2CH_2CHNH_2COOH$ $\mid$ $SeCH_2CH_2CHNH_2COOH$	Se accumulator plants
Selenocystathionine	$Se{\diagup}^{CH_2CH_2CHNH_2COOH}_{\diagdown CH_2CHNH_2COOH}$	Se accumulator plants
Se-methylselenocysteine	$CH_3SeCH_2CHNH_2COOH$	Se accumulator plants
Selenonium compounds		
Trimethylselenonium	$(CH_3)_3Se^+$	Animals
Se-methylselenomethionine	$(CH_3)_2\overset{+}{Se}CH_2CH_2CHNH_2COOH$	Plants
Se-adenosyl selenomethionine	$Adenosine\text{-}Se^+CH_2CH_2CHNH_2COOH$ $\mid$ CH_3	Yeast
Selenides		
Dimethyl selenide	CH_3SeCH_3	Animals, fungi, bacteria
Dimethyl diselenide	$CH_3SeSeCH_3$	Plants (Se accumulators and nonaccumulators), microorganisms
Miscellaneous compounds		
Selenocysteineseleninic acid	$HOOSeCH_2CHNH_2COOH$	Plants
Selenomethionine selenoxide	$CH_3SeOCH_2CH_2CHNH_2COOH$	Plants

[a]From Ganther (1974) and Shrift (1973).

include amino acids, proteins, selenonium compounds, and selenides. Not included in this list are the Se analogs of some of the S-containing vitamins, coenzymes, and purine or pyrimidine bases, nor the specific Se-containing enzymes (Stadtman, 1974, 1980).

Much of the organic Se in plants, animals, and microorganisms is found in amino acids and proteins. Selenomethionine and selenocystine are two amino acids most commonly isolated from plants that do not accumulate Se. Many of the less-common selenoamino acids are present in the Se-accumulator plants of the genera *Astragalus, Stanleya,* and *Neptunia,* for example. These Se-accumulating plants are often used as indicators in geoprospecting for high Se areas. Microorganisms also incorporate inorganic Se into amino acids, and both selenocystine and selenomethionine are reportedly synthesized from selenite by several bacteria and yeast (Blau, 1961; Weiss *et al.,* 1965). However, as pointed out by Huber and Criddle (1967), extreme care must be used in interpreting data indicating the presence of selenocystine in acid hydrolysates of protein. The synthesis of selenomethionine from selenite by *Escherichia coli* (Tuve and Williams, 1961) and *Candida albicans* (Falcone and Giambanco, 1967) probably occurs through pathways by which S is incorporated into amino acids. In *Escherichia coli,* selenomethionine is active in the first step of protein synthesis (Hoffman *et al.,* 1970) and is known to be incorporated into the proteins of this and other bacteria (Coch and Greene, 1971; Hidiroglou *et al.,* 1968).

3.2.2. Metabolism of Organic Se Compounds

Organic Se compounds are metabolized along many of the same routes as S, and many isolated enzyme systems can utilize S and Se analogs interchangeably (Shrift, 1973). In the cycling of S, the complex organic compounds present in plant and animal residues are transformed by microorganisms to simpler materials and, ultimately, to inorganic S compounds. Shrift (1973) has indicated that Se is cycled by predominantly biological pathways, which are strikingly similar to the S cycle. However, the evidence for such a natural cycle for Se is weak because of the paucity of information available concerning the microbial metabolism (mineralization) of organic Se compounds. Comparative biochemistry and our understanding of S metabolism must be relied upon rather heavily to explain or suggest pathways by which Se is metabolized.

The major S products formed during the decomposition of methionine by aerobic and anaerobic microorganisms in both soil and axenic culture are methyl mercaptan and dimethyl disulfide (Freney, 1967; Segal and Starkey, 1969). Methionine degradation frequently leads to the formation of α-ketobutyric acid, ammonia, and methyl mercaptan (equation 3), and in the presence of O_2, methyl mercaptan is oxidized to dimethyl disulfide (equation 4):

$$CH_3SCH_2CH_2CHNH_2COOH \rightarrow CH_3CH_2COCOOH + NH_3$$
methionine   α-ketobutyric   ammonia
                                               acid

$$+ \quad CH_3SH \quad (3)$$
methyl
mercaptan

$$\overset{+O_2}{CH_3SH \quad \rightarrow \quad CH_3SSCH_3} \quad (4)$$
methyl mercaptan   dimethyl disulfide

The microbial metabolism of selenomethionine seems to parallel that of its S analog. Cowie and Cohen (1957) and Coch and Greene (1971) have demonstrated that selenomethionine (1×10^{-4} M) can support exponential growth of methionine-requiring mutants of *Escherichia coli*. The growth of *Escherichia coli* K_{12}, however, was markedly inhibited by selenomethionine concentrations of 1×10^{-4} M (Coch and Greene, 1971). Doran and Alexander (1977b) found that four methionine-utilizing bacteria, isolated from sewage and soil, were unable to utilize selenomethionine as a sole C source, and the selenoamino acid (6×10^{-4} M) inhibited growth of the organisms in a methionine-containing growth medium. When the cultures grown on methionine were incubated with selenomethionine, an unpleasant odor was evolved from the culture, which was later identified as dimethyl diselenide. Resting-cell studies with one of the methionine-utilizing bacteria, identified as *Pseudomonas* sp., revealed that selenomethionine was quantitatively transformed to dimethyl diselenide. Thus, although selenomethionine at substrate concentrations is toxic to some microorganisms, it is apparently metabolized in the same manner as its S analog.

The microbial decomposition of selenocystine also appears to occur by one of the same pathways as its S analog. The mineralization of cystine by fungi and bacteria occurs by either of two general pathways and results in the release of either H_2S or sulfate (Fig. 2). In one pathway, cystine is reduced to cysteine, from which hydrogen sulfide is cleaved by either cysteine desulfhydrase (Collins *et al.*, 1973; Delwiche, 1951; Singer and Kearney, 1955) or serine sulfhydrase (Freney, 1967). In the other pathway, cystine is oxidized through a series of organic S compounds to sulfate.

Some cystine-utilizing microorganisms can also utilize selenocystine for growth. Doran and Alexander (1977b) found that, of 10 microorganisms able to use cystine as a sole C source, three fungi and three bacteria could also use

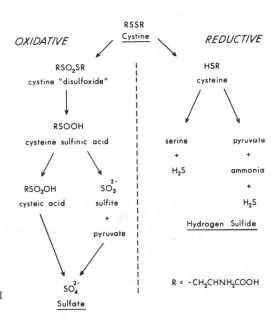

Figure 2. Pathways for the microbial metabolism of cystine.

selenocystine as their only C source. The Se products identified from seleno-cystine metabolism were hydrogen selenide and elemental Se. No volatile organic Se compounds were evolved from the cultures, and no soluble organic Se forms, other than selenocystine, were detected in the culture solutions. Consequently, selenocystine seems to be metabolized by a reductive pathway that results in the production of hydrogen selenide, which is oxidized to elemental Se in air. This study of the microbial formation of inorganic Se from seleno-cystine is probably the first report of the mineralization of organic Se. Hidi-roglou *et al.* (1974) reported that selenocystine was a product of selenome-thionine metabolism in rumen microorganisms. The identification of elemental Se in protein enzymatic hydrolysates of the same organisms might suggest that the selenocystine formed was further metabolized.

Analogies also exist between the metabolism of selenonium and sulfonium compounds. The metabolism of trimethylsulfonium $[(CH_3)_3S^+]$ by *Pseudomonas* sp. involves the cleavage of the C-S bond, resulting in the formation of dimethyl sulfide and a methyl group that can be used for growth (Wagner *et al.,* 1967). Trimethylselenonium, a common excretion product of animals metabolizing Se, is apparently decomposed by a similar mechanism. A pseu-domonad that has been isolated from soil can utilize trimethylselenonium as its sole source of C, one of the resultant products being dimethyl selenide (Doran and Alexander, 1977b). The bacterial degradation of methyl methionine (Mazelis *et al.,* 1965) and other sulfonium compounds (Kadota and Ish-ida, 1972) also results in the formation of dimethyl sulfide. Although there are no reports in the literature of the microbial metabolism of Se-methylseleno-methionine to dimethyl selenide, such a cleavage has been noted in plant leaves (Lewis *et al.,* 1974).

3.3. Methylation and Volatilization

The methylation of toxic elements is an important process because it often leads to a change in both mobility and toxicity. Microorganisms are essential catalysts in the methylation of mercury (Wood, 1971), arsenic (Wood, 1974; Woolson and Kearney, 1973), tellurium (Fleming and Alexander, 1972), selenium (Chau *et al.,* 1976; Doran and Alexander, 1977a), tin (Huey *et al.,* 1974), and lead (Wong *et al.,* 1975) in soil and aquatic environments. With Se, methylation represents the conversion of a nonvolatile precursor to a volatile product, but dimethyl selenide, the major product formed, seems much less toxic than other Se forms (McConnell and Portman, 1952).

3.3.1. Soils, Sediments, and Sewage

The microbial formation of methylated Se compounds occurs in soils, sediments, and sewage. As shown in Table V, both dimethyl selenide and dimethyl diselenide are evolved from Se-amended soil under both aerobic and anaerobic conditions (Doran and Alexander, 1977a). In this study, dimethyl selenide was the predominant product, and organic Se compounds were more rapidly converted to volatile methylated Se compounds than were inorganic Se com-

Table V. Evolution of Methylated Selenium Compounds from Soil Incubated in Air for 17 Days and Under Anaerobiosis for 27 Days[a]

Se added	μg Se/g soil	% Se recovered as dimethyl selenide	
		In air	Under anaerobiosis
Elemental Se	1000	0.01	0.00
Selenite	100	0.59	0.00
Selenate	100	0.10	0.005
Selenocystine	10	1.0	2.3
Selenomethionine	10	15.8[b]	8.8[b]
Trimethylselenonium	10	31.5	8.2

[a]From Doran and Alexander (1977a).
[b]Dimethyl diselenide was also formed in soil amended with selenomethionine, the Se recovery being 3.5% in air and 11.1% under anaerobiosis.

pounds. The decrease in dimethyl selenide production from soil under anaerobic conditions probably resulted from decreased microbial activity and reduction of the inorganic Se compounds to elemental Se or H_2Se gas. Reamer and Zoller (1980) reported a similar decrease in evolution of methylated Se compounds when soil amended with inorganic Se was incubated with N_2 instead of air. In their study, recoveries of methylated products from inorganic Se were low (0.005–0.034% of Se added) and dimethyl selenide, dimethyl diselenide, and dimethyl selenone [$(CH_3)_2SeO_2$] were the methylated Se compounds identified. The formation of methylated Se compounds also occurs in aquatic environments. Chau *et al.* (1976) have observed that the addition of selenate, selenite, selenocystine, selenomethionine, or selenourea to lake sediments results in the evolution of dimethyl selenide and dimethyl diselenide. As in other studies, the total quantity of methylated products evolved from sediments amended with selenite or selenate was low and represented only 0.001–0.01% of the Se added.

It is apparent that organic Se compounds are readily metabolized in soil, and relatively large quantities of methylated Se can be released from soils amended with organic Se. Doran and Alexander (1977a) found that, in a loam soil incubated in air, 87, 28, and 7% of the Se added as trimethylselenonium, and selenocystine, respectively, was recovered as a volatile product within 40 days. Dimethyl selenide was identified as the major product evolved, but, for selenomethionine, some dimethyl diselenide was also identified. The total recovery of volatile Se from anaerobically incubated soils receiving trimethylselenonium, selenomethionine, and selenocystine was similar to soils incubated in air. The formation of dimethyl selenide and dimethyl diselenide from trimethylselenonium and selenomethionine in soil apparently results from the microbial metabolism of these Se compounds. The formation of dimethyl selenide from selenocystine could also result from degradation of this amino acid, since elemental Se, the demonstrated product of selenocystine metabolism by soil microorganisms, is methylated in soil to form dimethyl selenide.

Inorganic Se forms are less readily transformed to dimethyl selenide than are organic Se compounds. Results of published research indicate that 0.003–8.1% of the inorganic Se added to soils, sediments, and sewage is evolved as volatile Se products (Table VI). However, where environmentally relevant concentrations of soluble Se were used (< 100–200 ppm for selenite or selenate), volatile Se recovered generally averaged < 1% of the added Se, and the major volatile product identified was dimethyl selenide. Significantly more Se was evolved from soils, sediments, and sewage incubated aerobically than under anaerobic conditions. Selenite (SeO_3^{2-}) and selenate (SeO_4^{2-}), both soluble forms of Se, were more readily methylated than elemental Se. Doran and Alexander (1977a) reported 20–300 times as much volatile Se was evolved from soils amended with selenite or selenate than with elemental Se.

The evolution of volatile Se from unamended-seleniferous or selenite-

Table VI. Influence of Inorganic Selenium Form and Aeration Status on the Evolution of Volatile Selenium from Soil, Sewage, and Lake Sediments

Se form added, biological medium	Se amount (ppm)	Duration of incubation (da)	Volatile Se evolved (% of Se added)		References
			Aerobic	Anaerobic	
SeO_4^{2-}					
Soils	100	17–40	0.1–0.24	0.005–0.12	Doran and Alexander (1977a)
Lake sediments	5	7	—	0.003–0.034[a]	Chau et al. (1976)
SeO_3^{2-}					
Soil	1000	30	0.034	0.024	Reamer and Zoller (1980)
Soils	473	45	0.4–3.5	—	Francis et al. (1974)
Soils	100	17–40	0.2–0.59	0.0–1.4	Doran and Alexander (1977a)
Soils	4	121	0.5–8.1	—	Hamdy and Gissel-Nielsen (1976)
Soil	0.1	20	0.14–0.25	—	Zieve and Peterson (1981)
Lake sediments	5	7	—	0.003–0.034[a]	Chau et al. (1976)
Sewage	1000	30	7.9	0.09	Reamer and Zoller (1980)
Sewage	1–100	30	0.55–1.3	—	Reamer and Zoller (1980)
Se^0					
Soils	1000	17–40	0.009–0.01	0–0.004	Doran and Alexander (1977a)
Soil	200	30	0.004	0.003	Reamer and Zoller (1980)
Sewage	500	30	0.005	0.0005	Reamer and Zoller (1980)

[a]Assumed to be anaerobic (50 g sediments + 150 ml water).

amended soils is directly related to the content of soluble Se (Abu-Erreish et al., 1968; Zieve and Peterson, 1981). Hamdy and Gissel-Nielsen (1976) concluded that the higher recoveries of volatile Se from a selenite-amended sand (8.1%), as compared with clay or organic soils (0.5–3%), resulted from less fixation of selenite in the sand. Thus, in soils that fix large amounts of soluble Se, the evolution of methylated Se compounds from inorganic Se may be greatly reduced. This may be one reason why the Se content of soils is not always a good index of the potential for microbial conversion of inorganic Se into dimethyl selenide (Francis et al., 1974).

The formation of methylated Se products from inorganic Se, like that for organic Se, is largely mediated by microbial activity. Consequently, the evolution of volatile Se from soils, sediments, and sewage is greatly influenced by availability of carbon, temperature, and water content, and none occurs when these biological media are steam sterilized. The addition of available carbon sources markedly stimulates the evolution of volatile Se from both seleniferous soils and soils amended with selenate or selenite (Abu-Erreish *et al.*, 1968; Doran and Alexander, 1977a). Methylation of Se is also influenced by temperature and, in soil and lakes amended with selenite or selenate, evolution of volatile Se at 20°C was 1.3–7.5 and 10–15 times greater than at 10°C and 4°C, respectively (Chau *et al.*, 1976; Zieve and Peterson, 1981).

The effect of soil water content on the evolution of volatile Se is indirect and apparently the result of changes in microbial activity. Selenium volatilization is greatest at water contents that permit maximum aerobic microbial activity (Hamdy and Gissel-Nielsen, 1976; Zieve and Peterson, 1981).

3.3.2. Organisms Involved

The specific role of microorganisms in the methylation of inorganic Se was first discovered by Challenger and North (1934) during their studies of the action of fungi upon inorganic Se. They found that the fungus *Scopulariopsis brevicaulis* could convert both selenite and selenate into dimethyl selenide. Since then, many other fungi and bacteria have been shown to have the capacity to form dimethyl selenide from inorganic Se (Table VII). The number of microorganisms forming dimethyl selenide and the habitats from which they can be isolated are quite numerous. Fungi have been the predominant group of microorganisms demonstrating a capacity to form dimethyl selenide from

Table VII. Microbial Formation of Dimethyl Selenide

Microorganism	Source of isolate	Reference
Scopulariopsis brevicaulis	Bread	Challenger and North (1934)
Schizophyllum commune	Bread	Challenger and Charlton (1947)
Aspergillus niger	Bread	Challenger *et al.* (1954)
Penicillium sp.	Sewage	Fleming and Alexander (1972)
Candida humicola	Sewage	Cox and Alexander (1974)
Cephalosporium spp., *Fusarium* spp., *Scopulariopsis* sp., *Penicillium* sp.	Soil	Barkes and Fleming (1974)
Corynebacterium sp.	Soil	Doran and Alexander (1975)
Aeromonas sp., *Flavobacterium* sp., *Pseudomonas* sp.	Lake sediment	Chau *et al.* (1976)

selenite; however, in most past research, the levels of selenite-Se used in media from which Se-methylating organisms were isolated ranged from 460 to 2900 ppm. Selenite-Se concentrations greater than 220 ppm inhibit many bacteria and, at 1200 ppm, growth of most bacteria is inhibited (Smith, 1959).

The microbial production of volatile Se compounds appears to be a general phenomenon and not restricted to any one group of microorganisms. Abu-Erreish *et al.* (1968) concluded, based on observations of changes in Se evolution under different aeration conditions, that soil fungi are the predominant group of microorganisms volatilizing Se. However, Zieve and Peterson (1981) reported that the volatilization of selenite from soil was reduced 50% by the addition of chloramphenicol—thus indicating the importance of bacteria in this process. Undoubtedly, the predominance of microbial groups and their relative activity in Se volatilization are determined as much by the physical, chemical, and biological characteristics of each ecosystem as by innate differences between microbial groups.

In addition to microorganisms, plants and animals can also methylate inorganic Se (Byard, 1969; Ganther, 1966; Lewis *et al.*, 1974). The methylation of inorganic Se is an energy-requiring process, and how organisms benefit from this transformation is not immediately clear. However, the methylation of Se compounds greatly decreases their toxicity, and the formation of selenonium compounds and methylation of selenide may be important mechanisms by which plants, animals, and microorganisms detoxify and dispose of Se (Byard, 1969; Chau *et al.*, 1976; Ganther, 1966; Shrift, 1973).

3.3.3. Mechanisms

An understanding of the mechanisms involved in formation of methylated Se compounds may be important in understanding Se cycling in the environment. In plants, animals, and microorganisms, methylated Se compounds are formed through either enzymatic cleavage of preformed organic Se compounds or by direct methylation of the Se atom itself. The formation of methylated Se compounds from several organic precursors is illustrated in Fig. 3. Since the mechanisms for the formation of methylated Se compounds through microbial metabolism of selenoamino acids and selenonium compounds have already been discussed in Section 3.2.2., only the methylation of inorganic Se will be discussed here.

Because methylation and transmethylation are general processes essential to all living organisms, it is not surprising that many different elements appear to be methylated through similar pathways. Methylation of most elements or compounds usually involves the enzymatic transfer of a methyl group from a donor compound. The three cofactors known to be involved in the transfer of

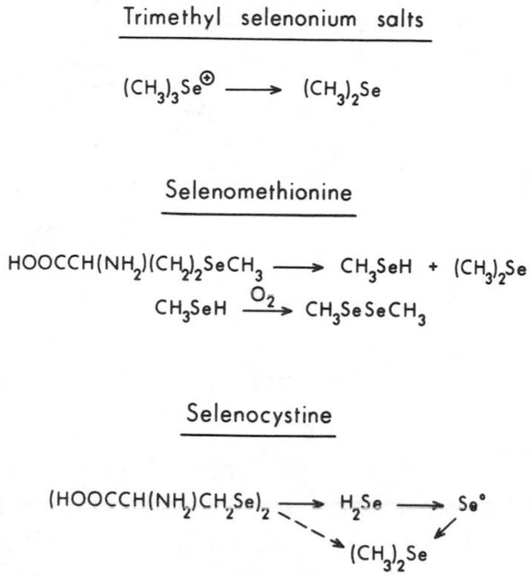

Trimethyl selenonium salts

$$(CH_3)_3Se^{\oplus} \longrightarrow (CH_3)_2Se$$

Selenomethionine

$$HOOCCH(NH_2)(CH_2)_2SeCH_3 \longrightarrow CH_3SeH + (CH_3)_2Se$$
$$CH_3SeH \xrightarrow{O_2} CH_3SeSeCH_3$$

Selenocystine

Figure 3. Methylated products formed from microbial metabolism of organic Se compounds.

$$(HOOCCH(NH_2)CH_2Se)_2 \longrightarrow H_2Se \longrightarrow Se^{\circ}$$
$$\searrow (CH_3)_2Se \nearrow$$

methyl groups in biological systems are (1) N^5-methyl-tetrahydrofolate derivatives, (2) cobalamin derivatives, and (3) S-adenosylmethionine. Each of these methyl carriers serves a unique and distinctive role in transmethylation: Tetrahydrofolate has the capacity to accept one-C fragments from many donors (Mudd and Catoni, 1964); methyl-cobalamine can transfer methyl groups to either cations, anions, or nonionized species (Stadtman, 1971); and S-adenosylmethionine, often called "the" biological methyl donor, is involved in many biosynthetic methylations, including the synthesis of its own precursor, methionine.

Both methyl cobalamine and S-adenosylmethionine function as methyl donors in the microbial methylation of inorganic Se (Doran and Alexander, 1977b; McBride and Wolfe, 1971). While studying the production of dimethylarsine by *Methanobacterium* sp., McBride and Wolfe (1971) found that cell-free extracts of this organism readily reduced and methylated selenite when methyl cobalamine was added to the reaction mixture. Doran and Alexander (1977b) reported that the production of dimethyl selenide from selenite and elemental Se, as catalyzed by cell-free extracts of *Corynebacterium* sp., was enhanced by S-adenosylmethionine. S-adenosylmethionine is also a cofactor in the enzymatic formation of dimethyl selenide in mammals (Bremer and Natori, 1960; Hsieh and Ganther, 1975).

Although the methylation of inorganic Se is known to involve both reduction and methylation of the Se atom, the specific pathway for the methylation is somewhat uncertain. Challenger (1945) suggested that methylation of selenite by fungi involves a series of four steps in which the Se atom is successively methylated and reduced, with dimethyl selenide as the final product (equation 5):

$$\underset{\text{selenite}}{\text{HSeO}_3^-} \xrightarrow{\text{CH}_3^+} \underset{\substack{\text{methane selenonic} \\ \text{acid}}}{\text{CH}_3\text{SeO}_3\text{H}} \xrightarrow[\text{and reduction}]{\text{ionization}} \tag{5}$$

$$\underset{\substack{\text{ion of} \\ \text{methaneseleninic} \\ \text{acid}}}{\text{CH}_3\text{SeO}_2^-} \xrightarrow{\text{CH}_3^+} \underset{\substack{\text{dimethyl} \\ \text{selenone}}}{(\text{CH}_3)_2\text{SeO}_2} \xrightarrow{\text{reduction}} \underset{\substack{\text{dimethyl} \\ \text{selenide}}}{(\text{CH}_3)_2\text{Se}}$$

Originally, the steps of this pathway were based on the knowledge that all compounds known to undergo methylation by fungi were negative ions ($H_2AsO_3^-$, $HTeO_3^-$, etc.) and that cultures of *Scopulariopsis brevicaulis* and *Penicillium* spp. convert methaneselenonic and methaneseleninic acids to dimethyl selenide (Bird and Challenger, 1942; Challenger, 1945). Uncertainty concerning the proposed pathway arises from the following facts: None of the postulated, intermediate Se compounds were detected in culture solution; dimethyl selenone was never tested; and the potassium salt of methane selenonic acid used in these studies forms selenite upon hydrolysis.

Recently, Reamer and Zoller (1980) identified dimethyl selenide, dimethyl diselenide, and dimethyl selenone as products from soil and sewage amended with selenite or elemental Se. Dimethyl selenide was the predominant product at low concentrations of selenite (1–10 ppm Se), and relatively large proportions of dimethyl diselenide or dimethyl selenone were only produced where large amounts of selenite (100–1000 ppm) were added to sewage (Table VIII). Dimethyl selenide was the only product recovered when sewage sludge was amended with elemental Se. The authors proposed a pathway for formation of dimethyl selenide and dimethyl diselenide, which included dimethyl selenone (or methyl methylselenite) as an intermediate. However, as in the studies by Challenger (1945), the possibility that methylated Se compounds were formed through microbial metabolism of organic Se compounds (synthesized from inorganic Se in the test medium) was not evaluated. Blau (1961) and Tuve and Williams (1961) reported that, within hours, 7.5–40% of the selenite added to biological media can be synthesized as selenomethionine or selenocystine by *Saccharomyces cerevisiae* or *Escherichia coli*. Therefore, the dimethyl diselenide and, possibly, dimethyl selenone identified by Reamer and Zoller (1980) were likely products of the metabolism of selenoamino acids which were biologically produced in the test medium.

Our research studies suggest that the methylation of inorganic Se may first involve reduction of Se to the selenide form, which is subsequently methylated to form dimethyl selenide. From data obtained in methylation studies

Table VIII. Effect of Form and Concentration of Inorganic Se on the Evolution of Methylated Selenium Compounds from Sewage after 30 Days[a]

Se form	Concentration added (ppm)	% Se recovered as		
		$(CH_3)_2Se$	$(CH_3)_2Se_2$	$(CH_3)_2SeO_2$
SeO_3^{2-}	1000	0.01	4.9	3.0
	100	0.17	0.07	0.31
	10	0.54	0.03	0.05
	1	1.3	0	0
Se^0	500	0.005	0	0

[a]Data from Reamer and Zoller (1980).

with microorganisms and from the literature, one pathway for methylation of inorganic Se can be given as equation (6):

$$SeO_3^{2-} \rightarrow Se^0 \rightarrow HSeX \rightarrow CH_3SeH \rightarrow (CH_3)_2Se \qquad (6)$$

$$\text{selenite} \quad \text{elemental Se} \quad \text{selenide} \quad \text{methane selenol} \quad \text{dimethyl selenide}$$

In the major research study from which this pathway was developed (Doran, 1976), the formation of dimethyl selenide from selenite by a soil corynebacterium was found to be preceded by formation of elemental Se. Also, cell-free extracts of the same organism readily methylated both selenite and elemental Se, when S-adenosylmethionine was present in the reaction mixture. Selenide and methane selenol were not tested as intermediate compounds in this study. Organic intermediates were probably not involved because neither selenomethionine, selenocystine, trimethylselenonium, nor methaneselenonate were converted enzymatically to dimethyl selenide under the same conditions in which selenite and elemental Se were methylated. Methane seleninate was converted to dimethyl selenide under the conditions of the above experiments, but this transformation also occurred nonenzymatically and, apparently, by a pathway different from the one by which selenite and selenate were methylated.

The role of selenide and methane selenol as intermediates in the methylation of inorganic Se has been suggested by other research studies. Bird and Challenger (1942) detected traces of methane selenol in fungal cultures which were actively producing dimethyl selenide, and Diplock et al. (1973) proposed selenol as an intermediate in the formation of dimethyl selenide from selenite in rats. Bremer and Natori (1960), also working with rats, observed that S-adenosylmethionine functioned as a methyl donor in the enzymatic formation of methane selenol and dimethyl selenide from sodium selenide. Based on their observations from studies on glutathione-reductase-mediated reduction of selenite and formation of dimethyl selenide in mammalian systems, Hsieh and Ganther (1975) presented a pathway for dimethyl selenide formation from selenite that involved hydrogen selenide and methane selenol, as intermediates or products, and S-adenosylmethionine, as the specific methyl donor.

4. Biological Cycling of Selenium

4.1. Soils, Plants, and Animals

Research into the biological cycling of Se has been stimulated by the knowledge that this element is both essential and toxic to warm-blooded animals. For proper nutrition of both humans and domestic animals, it is desirable to keep the Se concentrations in food and feed between 0.1 and 1.0 ppm (Allaway, 1968). The Se content of crop plants, within the United States, varies according to the areas in which these plants are grown (Kubota et al., 1967). Areas where the Se content of plants is low correspond fairly well with areas where Se-responsive diseases of livestock occur (Allaway, 1973). Within the United States, the high-Se areas, which support growth of plants containing toxic Se concentrations, are located predominantly in the dry, nonagricultural regions (Rosenfeld and Beath, 1964). The addition of Se, by disposal or recycling of waste materials, to areas where agriculture is more intensive could effectively alter the biological balance of Se in the food chain.

4.1.1. Pathways Involved

Several pathways have been suggested for the cycling of Se through soil, water, plants, and animals (Allaway, 1973; Doran, 1976; Frost, 1972; Olson, 1967; Shrift, 1973). Although presented from different points of view, these pathways have two points in common. First, the transformations within and between inorganic and organic Se forms, which are usually catalyzed by microorganisms, are essential to cycling of Se. Second, the formation of inorganic selenides, elemental Se, and complexed forms of selenite lead to a reduction in the biological availability of Se. The importance of volatile forms in the cycling of Se has been noted by many researchers. Some have incorporated an atmospheric component into their representations of the Se cycle.

4.2. Atmospheric Selenium

Investigations into the sources of atmospheric Se have indicated that much of this Se is produced by natural processes (Duce et al., 1975; Weiss et al., 1971). According to Weiss et al. (1971), measurements of S and Se in glacial ice of varying ages indicated that, in the combustion of fossil fuels, Se is mobilized in the atmosphere to a much lesser extent than is S. The formation of volatile compounds in biological processes was suggested as the predominant source of atmospheric Se in remote regions. In a survey of trace elements in Norway, Låg and Steinnes (1974) discovered that the Se concentration in humus layers of forest soils was highly correlated with rainfall distribution. They also concluded that natural processes, such as volcanism or plant decomposition, were more important than anthropogenic activity as sources of Se in rainfall.

Mackenzie *et al.* (1979) calculated the relative inputs of Se to the atmosphere from all known sources. Anthropogenic activities, such as fossil fuel burning or mining of sulfide ores, were the largest single input (120×10^8 g/ yr), and biological methylation from land (30×10^8 g/yr) accounted for 15% of the total inputs to atmospheric Se. However, their estimates for rates of biological methylation from soil were approximate and based upon results of *in vitro* studies, not direct measurements from natural ecosystems. The geographic pattern of high Se in rainwater from Denmark and regions of the United States indicates that industrial activity and the use of fossil fuels are major sources of atmospheric Se in many regions (Kubota *et al.*, 1975). The importance of biologically produced Se should not be underestimated, however, because contents of Se in rainfall and the atmosphere are higher for samples taken in spring and summer than in fall or winter (Kubota *et al.*, 1975; McDonald and Duncan, 1979). Biological inputs to atmospheric Se would follow this pattern, but anthropogenic inputs should peak in fall and winter when combustion of fossil fuel is greatest.

As mentioned earlier, microorganisms play an important role in the production of methylated Se compounds that are released into the atmosphere. As early as 1942, it was recognized that volatilization of Se was correlated with lower soluble Se contents in the surface of seleniferous soils as compared with deeper layers (Olson *et al.*, 1942). Later, it was discovered that volatilization of Se from these soils was microbially mediated and stimulated by the addition of organic matter (Abu-Erreish *et al.*, 1968). Measurements of air over soil and sediments in New York State reveal that volatile Se is being released to the atmosphere even in low-Se regions (Doran, 1976).

Since the biological conversions of Se compounds closely parallel the transformations of their S analogs, volatile Se compounds probably play as important a role in the cycling of Se as do the gaseous S compounds in the S cycle (Kellogg *et al.*, 1972). Emissions of dimethyl sulfide and dimethyl disulfide by microorganisms, together with hydrogen sulfide, may be major natural sources of atmospheric S (Lovelock *et al.*, 1972; Rasmussen, 1974). Formation of H_2Se, in contrast with H_2S, would not contribute much of the element to the atmosphere, since it is readily oxidized to elemental Se in the presence of O_2; however, microbially derived dimethyl selenide, and possibly dimethyl diselenide, might play an important role in the atmospheric phase of the Se cycle.

The microbial formation of dimethyl selenide is widespread, and a large quantity of the organic Se added to soil, as plant and animal residues, may be converted to this volatile metabolite. Figure 4, which is a schematic representation of biological Se cycling, summarizes the importance of methylated Se compounds (predominantly dimethyl selenide and dimethyl diselenide). Plants and animals also produce volatile Se, but the quantity released to the environment would not be great, since this loss is usually only associated with metabolism of high concentrations of Se (Allaway, 1973). In soil, sewage, and sediments, dimethyl selenide and some dimethyl diselenide are produced from many forms of Se, but production from organic precursors is much greater than from inorganic Se compounds. Since dimethyl selenide and dimethyl diselenide are the major products of microbial degradation of organic Se compounds, some mechanism must exist by which at least some of these volatile Se forms are returned to soil or aquatic environments. Plants can absorb Se from the air (Lewis, 1976), and soils are capable of reabsorbing volatile Se, although the process is largely a chemical reaction (Peterson *et al.*, 1981). Bacteria have been isolated that can utilize dimethyl selenide and dimethyl diselenide as C sources for growth, but the products of metabolism were not identified (Doran and Alexander, 1977b).

Dimethyl selenide released into the atmosphere may be photochemically oxidized to available, water-soluble forms. Bently *et al.* (1972) indicated that atmospheric dimethyl sulfide is oxidized to water-soluble, inorganic, and organic S compounds, which return to the earth's surface in rainfall. A considerable quantity of Se is added to soils in precipitation, and as much as 50% of

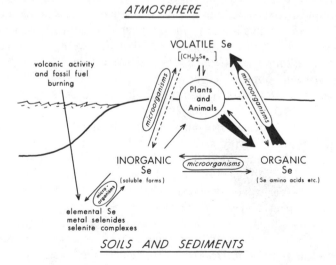

Figure 4. The role of dimethyl selenide in the cycling of selenium.

the Se contained in rainfall is water-soluble, presumably selenate or selenite (Kubota *et al.,* 1975).

Research is needed to determine the rates at which Se is being released from and/or returned to aquatic and terrestrial environments. The methylation of Se compounds is dependent on biological activity, which, in turn, is determined by ecosystem characteristics such as water and organic-matter availability, oxygen status, temperature, and presence of toxic materials. Thus, accurate determinations of evolution rates can only be made using methodology and measurements relevant to natural ecosystems and not from extrapolations of laboratory findings. It is also necessary to determine if dimethyl selenide production from terrestrial and aquatic ecosystems represents a danger to public health. Although dimethyl selenide is considered one of the least toxic Se forms, its toxicity can be greatly increased when animals are also exposed to subtoxic levels of arsenic or mercury (Obermeyer *et al.,* 1971; Parizek *et al.,* 1971). Ganther *et al.* (1972) found that dietary Se was effective in reducing methyl mercury toxicity in rats. The potential for transmethylation reactions with other toxic elements must be considered in any evaluation of dimethyl selenide release to the environment.

5. Conclusions and Remarks

Selenium, an element that is both essential and toxic to warm-blooded animals and certain microorganisms, is cycled predominantly via biological pathways. Soils producing foodstuffs containing toxic Se levels are predominantly confined to dry, basic, nonagricultural soils of the United States. The disposal or release of animal and industrial residues containing relatively high concentrations of Se results in bioaccumulation of Se in aquatic and terrestrial habitats in low-Se areas. At present, the burning of fossil fuels and disposal of fly ash and sewage sludges are important sources of environmental Se. These sources will undoubtedly increase with future population growth.

The availability of Se compounds for biological uptake and cycling is largely influenced by microbial transformations of this element, which closely resemble those of S. The oxidation of inorganic Se can serve as an energy source, and many microorganisms reduce oxidized forms of inorganic Se. The behavior of inorganic Se in terrestrial and aquatic ecosystems differs from that of S because the availability of selenite is less than sulfite, and selenate, unlike sulfate, is usually only formed under strongly oxidizing alkaline conditions. Thus, the availabilities of soluble forms of inorganic Se for biological uptake are more limited than those of their S analogs.

Organic compounds of Se and S are apparently metabolized by similar mechanisms; however, the assimilation of analogs of these elements can differ. Metabolism of selenoamino acids and selenonium compounds by most microorganisms and plants is apparently related to normal biochemical reactions and occurs along many of the same pathways that metabolize organic S. The microbial cometabolism of organic Se compounds can be limited by the higher toxicity of Se as compared to their S analogs. The mineralization of organic Se compounds has been demonstrated in only one instance, but methylated Se compounds are apparently converted to soluble inorganic forms in the earth's atmosphere.

The microbial formation of methylated Se compounds from organic and inorganic Se is widespread and apparently plays an important role in the cycling of Se. Rates of formation of methylated Se products in soil, sewage, and sediments are largely controlled by the quantity and forms of Se present and by environmental factors which regulate microbial activity. Dimethyl selenide is the predominant product of microbial methylation of Se, but dimethyl diselenide is also produced from metabolism of organic Se compounds such as selenomethionine. The relative quantities of methylated Se compounds released to the atmosphere from microbial metabolism of organic Se compounds are much greater than those formed from inorganic Se.

Future research endeavors should focus on determining the magnitude and rates at which methylated Se compounds are released to the atmosphere. An understanding of how these compounds are reabsorbed into terrestrial and aquatic ecosystems is also needed. These investigations, however, should include measurements and methodologies that are relevant to the forms and concentrations of Se and environmental conditions that are characteristic of terrestrial and aquatic ecosystems. A phenomenon that can only be demonstrated in a test tube is not relevant to understanding the intricacies of the total environment in which we live.

ACKNOWLEDGMENTS. The author is indebted to M. Alexander and W. H. Allaway for helpful guidance and suggestions during early phases of research. Appreciation is extended to O. E. Olson, E. E. Cary, and A. Shrift for careful and critical review of the manuscript. A special thanks to my wife, Janet, for her support of research efforts and patience during times when the unique olfactory nature of methylated Se compounds would have discouraged even the most avid researcher.

References

Abu-Erreish, G. M., Whitehead, E. I., and Olson, O. E., 1968, Evolution of volatile selenium from soils, *Soil Sci.* **106:**415–420.
Alexander, M., 1973, Microorganisms and chemical pollution, *Bioscience* **23:**509–515.
Allaway, W. H., 1968, Agronomic controls over the environmental cycling of trace elements, in:

Advances in Agronomy (A. G. Norman, ed.), Vol. 20, pp. 235–274, Academic Press, New York.

Allaway, W. H., 1973, Selenium in the food chain, *Cornell Vet.* **63**:151–170.

Andren, A. W., Klein, D. H., and Talmi, Y., 1975, Selenium in coal-fired steam plant emissions, *Environ. Sci. Tech.* **9**:856–858.

Barkes, L., and Fleming, R. W., 1974, Production of dimethylselenide gas from inorganic selenium by eleven soil fungi, *Bull. Environ. Contam. Toxicol.* **12**:308–311.

Bautista, E. M., and Alexander, M., 1972, Reduction of inorganic compounds by soil microorganisms, *Soil. Sci. Soc. Am. Proc.* **36**:918–920.

Bently, M. O., Douglass, I. B., Lacadie, J. A., and Whittier, R. R., 1972, The photolysis of DMS in air, *J. Air Pollut. Contr. Assoc.* **22**:359–363.

Bertine, K. K., and Goldberg, E. D., 1971, Fossil fuel combustion and the major sedimentary cycle, *Science* **173**:233–235.

Bird, M. L., and Challenger, F., 1942, Studies in biological methylation. Part IX. The action of *Scopulariopsis brevicaulis* and certain penicillia on salts of aliphatic seleninic and selenonic acids, *J. Chem. Soc.* **1942**:574–577.

Bisbjerg, B., 1972, Studies on selenium in plants and soils, Danish Atomic Energy Comm. Res. Estab. Risø Rep. No. 200.

Blau, M., 1961, Biosynthesis of (^{75}Se) selenomethionine and (^{75}Se) selenocystine, *Biochim. Biophys. Acta* **49**:389–390.

Bremer, J., and Natori, Y., 1960, Behavior of some selenium compounds in transmethylation, *Biochim. Biophys. Acta* **44**:367–370.

Byard, J. L., 1969, Trimethyl selenide. A urinary metabolite of selenite, *Arch. Biochem. Biophys.* **130**:556–560.

Cary, E. E., Wieczorek, G. A., and Allaway, W. H., 1967, Reactions of selenite-Se added to soils that produce low selenium forages, *Soil. Sci. Soc. Am. Proc.* **31**:21–26.

Challenger, F., 1945, Biological methylation, *Chem. Rev.* **36**:315–361.

Challenger, F., and Charlton, P. T., 1947, Studies on biological methylation. Part X. The fission of the mono- and di-sulfide links by moulds, *J. Chem. Soc.* **1947**:424–429.

Challenger, F., and North H. E., 1934, The production of organometal-loidal compounds by microorganisms. II. Dimethyl selenide, *J. Chem. Soc.* **1934**:68–71.

Challenger, F., Lisle, D. B., and Dransfield, P. B., 1954, Studies on biological methylation. Part XIV. The formation of trimethylarsine and dimethyl selenide in mould cultures from methyl sources containing 14C, *J. Chem. Soc.* **1954**:1760–1771.

Chau, Y. K., Wong, P. T. S., Silverberg, B. A., Luxon, P. L., and Bengert, G. A., 1976, Methylation of selenium in the aquatic environment, *Science* **192**:1130–1131.

Coch, E. H., and Greene, R. C., 1971, The utilization of selenomethionine by *Escherichia coli*, *Biochim. Biophys. Acta* **230**:223–236.

Collins, J. M., Wallenstein, A., and Monty, K. J., 1973, Regulatory features of the cysteine desulfhydrase of *Salmonella typhimurium*, *Biochim. Biophys. Acta* **313**:156–162.

Copeland, R., 1970, Selenium: The unknown pollutant, *Limnos* **3**:7–9.

Cowie, D. B., and Cohen, G. N., 1957, Biosynthesis by *Escherichia coli* of active altered proteins containing selenium instead of sulfur, *Biochim. Biophys. Acta* **26**:252–261.

Cox, D. P., and Alexander, M., 1974, Factors affecting trimethylarsine and dimethylselenide formation by *Candida humicola*, *Microb. Ecol.* **1**:136–144.

Delwiche, E. A., 1951, Activators for the cysteine desulfhydrase system of an *Escherichia coli* mutant, *J. Bacteriol.* **62**:717–722.

Diplock, A. T., Caygill, C. P. J., Jeffery, E. H., and Thomas, C., 1973, The nature of the acid-volatile selenium in the liver of the male rat, *Biochem. J.* **134**:283–293.

Doran, J. W., 1976, Microbial transformations of selenium in soil and culture, Ph.D. dissertation, Cornell University, Ithaca, N.Y.

Doran, J. W., and Alexander, M., 1975, Microbial formation of dimethyl selenide, Abstracts of the American Society for Microbiology, Annual Meeting, New York, p. 188 (N22).

Doran, J. W., and Alexander, M., 1977a, Microbial formation of volatile Se compounds in soil, *Soil Sci. Soc. Am. J.* **40**:687–690.

Doran, J. W., and Alexander, M., 1977b, Microbial transformations of selenium, *Appl. Environ. Microbiol.* **33**:31–37.

Duce, R. A., Hoffman, G. L., and Zoller, W. H., 1975, Atmospheric trace metals at remote northern and southern hemisphere sites: pollution or natural?, *Science* **187**:59–61.

Dürre, P., Andersch, W., and Andreesen, J. R., 1981, Isolation and characterization of an adenine-utilizing anaerobic sporeformer, *Clostridium purinolyticum* sp. nov. *Int. J. System. Bacteriol.* **31**:184–194.

Falcone, G., and Giambanco, V., 1967, Synthesis of seleno-amino acids in cell free extracts of *Candida albicans, Nature* **213**:396–398.

Falcone, G., and Nickerson, W. J., 1963, Reduction of selenite by intact yeast cells and cell-free preparations, *J. Bacteriol.* **85**:754–762.

Fleming, R. W., and Alexander, M., 1972, Dimethylselenide and dimethyltelluride formation by a strain of *Penicillium, Appl. Microbiol.* **24**:424–429.

Francis, A. J., Duxbury, J. M., and Alexander, M., 1974, Evolution of dimethylselenide from soils, *Appl. Microbiol.* **28**:248–250.

Franke, K. W., 1934, A new toxicant occurring naturally in certain samples of plant foodstuffs. I. Results obtained in preliminary feeding trials, *J. Nutr.* **8**:597–608.

Freney, J. R., 1967, Sulfur-containing organics, in: *Soil Biochemistry* (A. D. McLaren and G. H. Peterson, eds.), pp. 229–259, Marcel Dekker, New York.

Frost, D. V., 1960, Arsenic and selenium in relation to the food additive law of 1958, *Nutr. Rev.* **18**:129–132.

Frost, D. V., 1972, The two faces of selenium—can selenophobia be cured?, *Crit. Rev. Toxicol.* **1**:467–514.

Furr, A. K., Kelly, W. C., Bache, C. A., Gutenmann, W. H., and Lisk, D. J., 1976, Multielement absorption by crops grown in pots on municipal sludge-amended soil, *J. Agric. Food Chem.* **24**:889–892.

Furr, A. K., Parkinson, T. F., Bache, C. A., Gutenmann, W. H., Pakkala, I. S., and Lisk, D. J., 1980, Multielement absorption by crops grown on soils amended with municipal sludge ashes, *J. Agric. Food Chem.* **28**:660–662.

Ganther, H. E., 1966, Enzymic synthesis of dimethyl selenide from sodium selenite in mouse liver extracts, *Biochemistry* **5**:1089–1098.

Ganther, H. E., 1974, Biochemistry of selenium, in: *Selenium* (R. A. Zingaro and W. C. Cooper, eds.), pp. 546–614, Van Nostrand Reinhold, New York.

Ganther, H. E., Goudie, C., Sunde, M. L., Kopecky, M. J., Wagner, P., Ou, S-H, and Hoekstra, W. G., 1972, Selenium: relation to decreased toxicity of methylmercury added to diets containing tuna, *Science* **173**:1122–1124.

Geering, H. R., Cary, E. E., Jones, L. H. P., and Allaway, W. H., 1968, Solubility and redox criteria for the possible forms of selenium in soils, *Soil Sci. Soc. Am. Proc.* **32**:35–40.

Gutenmann, W. H., Bache, C. A., Youngs, W. D., and Lisk, D. J., 1976, Selenium in fly ash, *Science* **191**:966–967.

Hamdy, A. A., and Gissel-Nielsen, G., 1976, Volatization of selenium from soils, *Z. Pflanzenernaehr. Bodenkd.* **6**:671–678.

Hidiroglou, M., Heaney, D. P., and Jenkins, K. J., 1968, Metabolism of inorganic selenium in rumen bacteria, *Can. J. Physiol. Pharmacol.* **46**:229–232.

Hidiroglou, M., Jenkins, K. J., and Knipfel, J. E., 1974, Metabolism of selenomethionine in the rumen, *Can. J. Animal Sci.* **54**:325–330.

Hoffman, J. L., McConnel, K. P., and Carpenter, D. R., 1970, Aminoacylation of *Escherichia coli* methionine tRNA by selenomethionine, *Biochim. Biophys. Acta* **199**:531–534.

Hsieh, H. S., and Ganther, H. E., 1975, Acid-volatile selenium formation catalyzed by glutathione reductase, *Biochemistry* 14:1632–1636.

Huber, R. E., and Criddle, R. S., 1967, Comparison of the chemical properties of selenocysteine and selenocystine with their sulfur analogs, *Arch. Biochem. Biophys.* 122:164–173.

Huey, C., Brinkman, F. E., Grim, S., and Iverson, W. P., 1974, in: *Proceedings of the Intl. Conf. on Transport of Persistent Chemicals in Aquatic Ecosystems* (Q. N. Laltam, ed.), pp. II-73–II-78, Natl. Res. Council of Canada, Ottowa.

Jones, J. B., and Stadtman, T. C., 1977. *Methanococcus vannielii:* culture and effects of selenium and tungstate on growth. *J. Bacteriol.* 130:1404–1406.

Kadota, H., and Ishida, Y., 1972, Production of volatile sulfur compounds by microorganisms, *Annu. Rev. Microbiol.* 26:127–138.

Kellogg, W. W., Cadle, R. D., Allen, E. R., Lazarus, A. L., and Martell, E. A., 1972, The sulfur cycle, *Science* 175:587–596.

Kharkar, D. P., Turekian, K. K., and Bertine, K. K., 1968, Stream supply of dissolved silver, molybdenum, antimony, selenium, chromium, cobalt, rubidium, and cesium to the oceans, *Geochim. Cosmochim. Acta* 32:285–298.

Kovalski, V. V., Ermakov, V. V., and Lctunova, S. V., 1968, Geochemical ecology of microorganisms in soils with different selenium content, *Microbiology* 37:103–109.

Kubota, J., Allaway, W. H., Carter, D. L., Cary, E. E., and Lazar, V. A., 1967, Selenium in crops in the United States in relation to selenium-responsive diseases of animals, *J. Agric. Food. Chem.* 15:448–453.

Kubota, J., Cary, E. E., and Gissel-Nielsen, G., 1975, Selenium in the rainwater of the United States and Denmark, in: *Symposium on Trace Substances in Environmental Health* (D. D. Hemphill, ed.), pp. 123–130, University of Missouri, Columbia.

Låg, J., and Steinnes, E., 1974, Soil selenium in relation to precipitation, *Ambio* 3:237–238.

Lakin, H. W., 1972, Selenium accumulation in soils and its absorption by plants and animals, *Geol. Soc. Am. Bull.* 83:181–189.

Levine, V. E., 1925, The reducing properties of microorganisms with special reference to selenium compounds, *J. Bacteriol.* 10:217–263.

Lewis, B. G., 1976, Selenium in biological systems, and pathways for its volatilization in higher plants, in: *Environmental Biogeochemistry* (Vol. 1), *Carbon, Nitrogen, Phosphorus, Sulfur and Selenium Cycles* (J. O. Nriagu, ed.), pp. 389–409, Ann Arbor Science Publishers, Ann Arbor, Mich.

Lewis, B. G., Johnson, C. M., and Broyer, T. C., 1974, Volatile selenium in higher plants. The production of dimethyl selenide in cabbage leaves by enzymatic cleavage of Se-methyl selenomethionine selenonium salt, *Plant Soil* 40:107–118.

Lindström, K., and Rhode, W., 1978, Selenium as a micronutrient for the dinoflagellate *Peridinium cinctum* fa. *westii, Mitt. Int. Verein Limnol.* 21:168–173.

Lipman, L. G., and Waksman, S. A., 1923, The oxidation of selenium by a new group of autotrophic microorganisms, *Science* 57:60.

Lovelock, J. E., Maggs, R. J., and Rasmussen, R. A., 1972, Atmospheric dimethyl sulfide and the natural sulfur cycle, *Nature* 27:452–453.

Mackenzie, F. T., Lantzy, R. J., and Paterson, V., 1979, Global trace metal cycles and predictions, *J. Int. Assoc. Math. Geol.* 11:99–142.

Martin, J. L., 1973, Selenium assimilation in animals, in: *Organic Selenium Compounds: Their Chemistry and Biology* (D. L. Klayman and W. H. H. Gunther, eds.), pp. 663–691, John Wiley & Sons, New York.

Mazelis, M., Levin, B., and Mallinson, N., 1965, Decomposition of methyl methionine sulfonium salts by a bacterial enzyme, *Biochim. Biophys. Acta* 105:106–114.

McBride, B. C., and Wolfe, R. S., 1971, Biosynthesis of dimethylarsine by methanobacterium, *Biochemistry* 10:4312–4317.

McConnell, K. P., and Portman, O. W., 1952, Toxicity of dimethyl selenide in the rat and mouse, *Proc. Soc. Exp. Biol. Med.* **79:**230–231.

McCready, R. G. L., Campbell, J. N., and Payne, J. I., 1966, Selenite reduction by *Salmonella heidelberg, Can. J. Microbiol.* **12:**703–714.

McDonald, C., and Duncan, H. J., 1979, Atmospheric levels of trace elements in Glasgow, *Atmos. Environ.* **13:**413–417.

Moxon, A. L., 1937, Alkali disease or selenium poisoning?, *S. Dak. Agr. Exp. Sta. Bull.,* 311.

Mudd, S. H., and Catoni, G. L., 1964, Biological transmethylation, methyl group transfer and other "one-carbon" metabolic reactions dependent upon tetrahydrofolic acid, in: *Comprehensive Biochemistry,* Vol. 15 (M. Florkin and E. Stotze, eds.), pp. 1–47, Elsevier, Amsterdam.

Muth, O. H., Oldfield, J. E., and Weswig, P. H., eds., 1967, *Symposium: Selenium in Biomedicine,* AVI Publ. Co., Westport, Conn.

Natl. Acad. Sci., Nat. Res. Council, U.S.A., 1971, Selenium in nutrition, Subcommittee on Selenium, Committee on Animal Nutrition, Agricultural Board, NAS-NRC, Washington, D.C.

Nickerson, W. J., and Falcone, G., 1963, Enzymatic reduction of selenite, *J. Bacteriol.* **85:**763–771.

Obermeyer, B. D., Palmer, I. S., Olson, O. E., and Halverson, A. W., 1971, Toxicity of trimethylselenonium chloride in the rat with and without arsenite, *Toxicol. Appl. Pharmacol.* **20:**135–146.

Oehme, F. W., 1972, Mechanisms of heavy metal toxicities, *Clin. Toxicol.* **5:**131–167.

Olson, O. E., 1967, Soil, plant, animal cycling of excessive levels of selenium, in: *Symposium: Selenium in Biomedicine* (O. H. Muth, J. E. Oldfield, and P. H. Weswig, eds.), pp. 297–312, AVI Publ. Co., Westport, Conn.

Olson, O. E., and Moxon, A. L., 1939, Availability of selenium in soil. *Soil Sci.* **47:**305–311.

Olson, O. E., Whitehead, E. I., and Moxon, A. L., 1942, Occurrence of soluble selenium in soils and its availability to plants, *Soil Sci.* **54:**47–53.

Painter, E. P., 1941, The chemistry and toxicity of selenium compounds with special reference to the selenium problem, *Chem. Rev.* **28:**179–213.

Parizek, J., Ostadalova, J., Kalouskova, J., Babicky, A., and Benes, J., 1971, The detoxifying effects of selenium interrelations between compounds of selenium and certain metals, in: *Newer Trace Elements in Nutrition* (W. Mertz and W. E. Cornatzer, eds.), pp. 85–122, Marcel Dekker, New York.

Peterson, P. J., Benson, L. M., and Zieve, R., 1981, The metalloids, in: *Plant/Trace Metal Interactions* (N. W. Lepp, ed.), pp. 279–342, Applied Science, London.

Pillay, K. K. S., Thomas C. C. Jr., and Kaminski, J. W., 1969, Neutron activation analysis of the selenium content of fossil fuels, *Nucl. Applic. Technol.* **7:**478–483.

Pourbaix, Marcel, 1966, *Atlas of Electrochemical Equilibria in Aqueous Solutions,* Pergamon Press, Cebelcor, Brussels.

Rasmussen, R. A., 1974, Emission of biogenic hydrogen sulfide, *Tellus* **26:**254–260.

Reamer, D. C., and Zoller, W. H., 1980, Selenium biomethylation products from soil and sewage sludge, *Science* **208:**500–502.

Rosenfeld, I., and Beath, O. A., 1964, *Selenium—Geobotany, Biochemistry, Toxicity, and Nutrition,* Academic Press, New York.

Sandholm, M., Oksanen, H. E., and Pesonen, L., 1973, Uptake of selenium by aquatic organisms, *Limnol. Oceanog.* **18:**496–498.

Saposnikov, D. I., 1937, Substitution of sulfur with selenium in photoreduction of carbonic acid by purple sulfur bacteria, *Microbiologiya* **6:**643–644 (in Russian).

Sarathchandra, S. U., and Watkinson, J. H., 1981, Oxidation of elemental selenium to selenite by *Bacillus megaterium, Science* **211:**600–601.

Schutz, D. F., and Turekian, K. K., 1965, The investigation of the geographical and vertical distribution of several trace elements in sea water using neutron activation analysis, *Geochim. Cosmochim. Acta* **29**:259–313.

Schwarz, K., and Foltz, C. M., 1957, Selenium as an integral part of factor 3 against dietary necrotic liver degeneration, *J. Am. Chem. Soc.* **79**:3292–3293.

Scott, M. L., 1973, Nutritional importance of selenium, in: *Organic Se Compounds: Their Chemistry and Biology* (D. L. Klayman and W. H. H. Gunther, eds.), pp. 629–661, John Wiley & Sons, New York.

Segal, W., and Starkey, R., 1969, Microbial decomposition of methionine and identity of the resulting sulfur products, *J. Bacteriol.* **98**:908–913.

Shapiro, J. R., 1973, Selenium and human biology, in: *Organic Se Compounds: Their Chemistry and Biology* (D. L. Klayman and W. H. H. Gunther, eds.), pp. 693–726, John Wiley & Sons, New York.

Shrift, A., 1961, Biochemical interrelations between selenium and sulfur in plants and microorganisms, *Fed. Proc.* **20**:695–702.

Shrift, A., 1972, Selenium toxicity, in: *Phytochemical Ecology, Proc. Phytochem. Soc. Sym. 1971* (J. B. Harborne, ed.), pp. 145–161, Academic Press, New York.

Shrift, A., 1973, Metabolism of selenium by plants and microorganisms, in: *Organic Selenium Compounds: Their Chemistry and Biology* (D. L. Klayman and W. H. H. Gunther, eds.), pp. 763–814, John Wiley & Sons, New York.

Singer, T. B., and Kearney, E. B., 1955, Enzymatic pathways in the degradation of sulfur containing amino acids, in: *Amino Acid Metabolism* (W. D. McElroy and H. B. Glass, eds.), pp. 558–590, Johns Hopkins Press, Baltimore.

Smith, H. G., 1959, On the nature of the selective action of selenite broth, *J. Gen. Microbiol.* **21**:61–71.

Stadtman, T. C., 1971, Vitamin B_{12}, *Science* **171**:859–867.

Stadtman, T. C., 1974, Selenium biochemistry, *Science* **183**:915–922.

Stadtman, T. C., 1980, Selenium dependent enzymes, *Annu. Rev. Biochem.* **49**:93–110.

Suzuoki, T., 1965, A geochemical study of selenium in volcanic exhalation and sulfur deposits. II. On the behavior of selenium and sulfur in volcanic exhalation and sulfur deposits, *Chem. Soc. Japan Bull.* **38**:1940–1946.

Swaine, D. F., 1955, The trace-element content of soils, in: *Tech. Comm. No. 48 of the Commonwealth Bur. Soil Sci.,* pp. 91–99, Rothamsted Exp. Sta., Harpenden, England.

Torma, A. E., and Habashi, F., 1972, Oxidation of copper (II) selenide by *Thiobacillus ferroxidans, Can. J. Microbiol.* **18**:1780–1781.

Tuve, T., and Williams, H. H., 1961, Metabolism of selenium by *Escherichia coli:* biosynthesis of selenomethionine, *J. Biol. Chem.* **236**:597–601.

Wagner, C., Lusty, S. M., Jr., Kung, H. F., and Rogers, N. L., 1967, Preparation and properties of trimethylsulfonium-tetrahydrofolate methyltransferase, *J. Biol. Chem.* **242**:1287–1293.

Weiss, K. F., Ayres, J. C., and Kraft, A. A., 1965, Inhibitory action of selenite on *Escherichia coli, Proteus vulgaris,* and *Salmonella thompson, J. Bacteriol.* **90**:857–862.

Weiss, H. V., Koide, M., and Goldberg, E. D., 1971, Selenium and sulfur in a Greenland ice sheet: relation to fossil fuel combustion, *Science* **172**:261–263.

Wiersma, J. H., and Lee, G. F., 1971, Selenium in lake sediments—analytical procedure and preliminary results, *Environ. Sci. Technol.* **5**:1203–1206.

Wilson, L. G., and Bandurski, R. S., 1958, Enzymatic reactions involving sulfate, sulfite, selenate, and molybdate, *J. Biol. Chem.* **233**:975–981.

Wong, P. T. S., Chau, Y. K., and Luxon, P. L., 1975, Methylation of lead in the environment, *Nature* **253**:263–264.

Wood, J. M., 1971, Environmental pollution by mercury, *Adv. Environ. Sci. Technol.* **2**:36–56.

Wood, J. M., 1974, Biological cycles for toxic elements in the environment, *Science* **183**:1049–1052.

Woolfolk, C. A., and Whiteley, H. R., 1962, Reduction of inorganic compounds with molecular hydrogen by *Micrococcus lactilyticus, J. Bacteriol.* **84:**647–658.

Woolson, E. A., and Kearney, P. C., 1973, Persistence and reactions of (^{14}C)-cacodylic acid in soils, *Environ. Sci. Technol.* **7:**47–50.

Zalokar, M., 1953, Reduction of selenite by Neurospora, *Arch. Biochem. Biophys.* **44:**330–337.

Zieve, R., and Peterson, P. J., 1981, Factors influencing the volatilization of selenium from soil, *Sci. Total Environ.* **19:**277–284.

2

Effects of Environmental Factors and Their Interactions on Phytoplankton Growth

G-YULL RHEE

1. Introduction

The abundance and distribution of phytoplankton in the natural environment are regulated by various environmental factors, such as nutrient, light, and temperature. The effect, or stress, of each factor changes with time and space, and the relative importance of each factor also varies. Sometimes it is not difficult to identify the key factor controlling primary productivity, but quite frequently, it takes laborious efforts to discern the responsible environmental stress. This difficulty is due in good measure to the simultaneous action of more than one stress and is compounded by our general lack of understanding of the way one factor interplays with others in controlling growth.

Species also differ in their responses to environmental stresses, since they have different optimum conditions for growth. These differences are the driving forces responsible for species distribution and succession.

The present chapter will examine the effects of three principal factors—nutrient, light, and temperature—and their interactions on phyloplankton growth. The findings from continuous culture studies have recently been reviewed elsewhere (Rhee 1980; Rhee *et al.,* 1981).

2. Nutrients

Limiting nutrients in natural environments can be identified by various approaches, including the chemical analysis of the water and particulate mat-

G-YULL RHEE ● Environmental Health Institute, Division of Laboratories and Research, New York State Department of Health, Albany, New York 12201.

ter and bioassay methods. In the euphotic zone of marine environments nitrogen (N) is the most frequently limiting nutrient. This has been shown by, among other methods, comparing the elemental compositions of natural populations and their surrounding waters to the "average" composition of phytoplankton in the ocean. Redfield (1958) reported that the C:N:P atomic ratio of particulate matter in seawater is 106:16:1 and that these nutrients appear to be depleted in a similar proportion to this ratio during phytoplankton growth. Therefore, the ratio seems to reflect the relative requirements of these nutrients, and growth limitation by any one of these elements in a given body of water has often been defined by this ratio. For the photoautotrophs, carbon is rarely limiting (Schindler, 1977; Parsons et al., 1978); thus only the ratio of N to phosphorus (P) is of concern. If the N:P ratio is less than 16, N is considered to be limiting; if the ratio is larger, P is limiting.

The Redfield ratio is, however, a general value for the ocean as a whole, of which the euphotic zone comprises only 2% of its volume. Nevertheless, the average of relative requirements for N and P in various species has turned out to be close to 16, although there is wide variation among species (see below). Menzel and Ryther (1964) found N:P ratios of 5.4–17 in N-limited populations in the open sea; these also appear to be consistent with the Redfield ratio.

In euphotic zones, it is a rarity to find an N:P ratio of 16:1, except in areas of upwelling, where nutrient-rich deep water mixes with surface water. When the utilization of N and P was closely examined in surface waters, N was found to be depleted first with a significant amount of P always remaining in solution (Ryther and Dunstan, 1971). Therefore, N is generally considered as the limiting nutrient in seawaters.

Bioassay experiments in freshwater environments (e.g., Thomas, 1969; Fuhs et al., 1972; Lin and Schelske, 1979) show that P is generally the major limiting nutrient. A comparison of the elemental ratios of natural populations to the Redfield ratio also indicates P limitation. For example, Fuhs et al. (1972) found N:P ratios of natural populations ranging from 17 to 24 in Canadarago Lake, New York.

Occasional Si limitations for diatoms in both sea- and freshwaters have been reported (Smayda, 1974; Kilham, 1975). Although there have been no concerted efforts to determine the requirement for Si relative to other nutrients, the Si:P ratio ranges from about 8:1 to 12:1 (D'Elia et al., 1979; J. P. Barlow, personal communication). Ratios of 1:1–110:1 have been found in some species (S. S. Kilham, personal communication).

There is no single satisfactory method to identify the limiting nutrient. The average elemental ratio may be unreliable if detrital matter constitutes a significant part of the total particulates. The average ratio may also be unreliable if a few species dominate a community, because the relative requirement for N and P is quite variable between species. For example, when the dominant spccics has an N:P requirement of less than 7:1, a ratio of 10:1 for the community should be a sign of P limitation, rather than N limitation, as the Redfield ratio would suggest. The results of bioassays also cannot be accepted uncritically. Whether one uses the response of a single organism or of natural populations to nutrient additions as a measure of limitation, problems can arise from, among others, species differences in nutrient requirements and artificial successions (Rhee, 1980). Therefore, a combination of methods should be used to identify the limiting nutrient.

2.1. Limiting Nutrients: Growth and Uptake

2.1.1. External Concentrations and Growth

Many planktonic algae reproduce by binary fission. Under nutrient-unlimited conditions, therefore, the population growth after time t (N_t) is related to the original population (N_0) by

$$N_t = N_0 e^{t/t_d} \tag{1}$$

where t_d is the doubling time. Therefore the specific growth rate (μ), or the rate of increase in biomass per unit of biomass, is expressed as

$$\mu = d(\ln N)/dt = \ln 2/t_d \tag{2}$$

and μ has a unit of time^{-1}.

Under nutrient-limited conditions, algal growth rate increases hyperbolically as the concentration of limiting nutrient increases. This relationship is frequently described by the empirical equation of Monod (1942) for bacterial growth

$$\mu = \mu_m S/(K_s + S) \tag{3}$$

where μ_m is the maximum growth rate; S is the concentration of limiting nutrient, and K_s is the half-saturation constant or the value of S when $\mu = \mu_m/2$.

Some species appear to have a threshold concentration of nutrients (S_0) below which no growth can take place. The S_0 level of Si is 0.7 μM for *Thalassiosira pseudonana* when extrapolated from continuous culture data (Paasche, 1973a) and is about 0.43 μM for two strains of *Asterionella formosa* (Kilham, 1975). A threshold concentration is obviously an ecological disadvantage when nutrient concentrations are low. It seems to be influenced in part by growth conditions, since Guillard *et al.* (1973) did not find a threshold level for the same strain of *T. pseudonana* in batch cultures.

The Monod-type relationship between growth rate and limiting nutrient concentrations has been observed with many nutrients, including P, N, Si, CO_2, vitamin B_{12}, and iron in various species in both batch and continuous cultures (e.g., Davis, 1970; Müller, 1972; Guillard *et al.*, 1973; Swift and Taylor, 1974; Kilham, 1975; Picket, 1975; Ahlgren, 1977; Steeman-Nielsen, 1978; Goldman and Graham, 1981). For many species, K_s values for these nutrients are near or below the analytical limits of detection, indicating the high efficiency of utilization. For example, the values for P are 0.03 μM for *Oscillatoria agardhii,* and 0.014 μM for *Nitzschia actinastroides;* those for NH_4^+ are <0.03 μM for *T. pseudonana* and <0.02 μM for *Skeletonema costatum;* those for NO_3 are 0.04 μM for *O. agardhii* and 0.05 μM for *S. costatum;* those for CO_2 are 4.6 μM for *Chlorella vulgaris* and 0.16 μM for *Scenedesmus obliquus.*

The half-saturation constant has often been used to explain the success or failure of an organism in competitive interactions in nutrient-scarce environments; an organism with a lower K_s for limiting nutrients has been considered more competitive. However, comparisons of this constant alone are valid only when the values of μ_m for competing species are comparable, since the disadvantage of a high K_s can be offset by a high μ_m. Healey (1980) suggested that the μ_m/K_s ratio should be employed, but the utility of this ratio is limited to the concentration range where growth rate varies linearly with concentration. In the Monod-type plot, there is a 10% error from linearity when $S = 0.1K_s$, and the error increases with increasing S (Plowman, 1972). Graphic comparisons may be more useful. At a given limiting nutrient concentration, the species with the higher growth rate is likely to be more successful. Sometimes such comparisons of μ_m and K_s appear to explain the distribution and abundance of species in relation to nutritional environments (e.g., Paasche, 1975).

In extrapolating the kinetic values obtained in the laboratory to natural environments, however, one must keep in mind that these values can vary with environmental conditions. The ratio of monovalent to divalent ions influences growth characteristics (Miller and Fogg, 1957; Provasoli, 1958); light and temperature alter both μ_m and K_s (see below); and even the rate of agitation affects K_s (Skoglund and Jensen, 1976).

For organisms with diel periodicities in division such as most planktonic algae, equation (3) can describe only nonperiodic and random division under continuous illumination or the relationship between μ and S at a specific time of the cell division cycle. During a division cycle, there is no direct relationship between instantaneous growth rate, $\mu(t)$, and the concentrations of limiting

nutrients, S. Only the integrated growth, $\int_0^t \mu(t) \, dt$, is related to the numerical average of S at the peak and trough of its oscillation. The period average of growth, $1/T \int_0^t \mu(t) \, dt$, where T is the length of period in days, is also related to S at the beginning of the period (Chisholm and Nobbs, 1976; Gotham, 1977; Rhee *et al.*, 1981). If the nutrient uptake and division cycles of competing species are not identical, they may be able to share the supply of limiting nutrients on a different temporal scale unless there are gross differences in the kinetic parameters (Rhee *et al.*, 1981). Without information on the diel periodicities, therefore, μ_m and K_s in equation (3) alone may not be sufficient to explain nutrient-mediated competitive interactions in nature.

Equation (3) predicts that when $S = 0$, there is no growth. However, it is common to observe that S is exhausted long before logarithmic growth ceases in batch cultures (Kuenzler and Ketchum, 1962; Rhee, 1972; Finenko and Kruptkia-Akinina, 1974; Sakshaug and Holm-Hansen, 1977; Nyholm, 1977; Nalewajko and Lean, 1978; Myklestad, 1977). Nyholm (1977) also showed that when the inflow of reservoir medium was stopped for a steady-state culture of *Chlorella pyrenoidosa* in a chemostat, the cells could grow continuously for at least 30 hr at a constant rate. Clearly, such growth is at the expense of intracellular nutrient storage. Indeed, in P-limited batch cultures of *S. obliquus,* the cessation of logarithmic growth coincided with the drop of cellular polyphosphates to zero (Rhee, 1972). This pool can be as large as 70% of the total cell P in *Pavlova (Monochrysis) lutheri* and 50% in *S. obliquus,* respectively (Sakshaug and Holm-Hansen, 1977; Rhee, 1973,1974). It is apparent then that Equation (3) can apply only to steady-state growth, in which there is an equilibrium between S, intracellular pools, and growth.

In natural waters, S cannot be related to the growth rate because an equilibrium between S and the intracellular pool appears to be lacking. In Lake Michigan, where the growth of *Cladophora* is limited by P, no correlation has been observed between the concentrations of molybdate-reactive P and the hot-water-extractable P (surplus P) in the alga (Lin, 1977). In fact, the concentration of molybdate-reactive P was below the limit of detection. (However, there was a correlation between surplus P and total dissolved P, which was mostly unidentified organic fractions.) Seasonal changes in molybdate-reactive P and hot-water-extractable P also showed no correlation during a *Peridinium* bloom in Lake Kinneret (Wynne and Berman, 1980). In this lake, P does not appear to be limiting.

A further problem is that of determining which dissolved organic fractions are biologically available and thus should be included in S. The ambient nutrient concentration also may not provide a clue to the abundance of phytoplankton, because the productivity is regulated by the rate of supply, not by the concentration.

2.1.2. Intracellular Concentrations and Growth

Growth rate can be related to the total cell concentration of the limiting nutrient, or cell quota. Droop (1968) described the vitamin B_{12}-limited growth of *Monochrysis lutheri* by the empirical equation

$$\mu = \mu_m'(1 - q_0/q) \tag{4}$$

where μ_m' is the asymptote of the hyperbola, or μ when $q = \infty$; q is the cell quota; and q_0 is the minimum cell quota, or q when $\mu = 0$. The term μ_m' is related to maximum growth rate μ_m in Eq. (3) by

$$\mu_m = \mu_m'(1 - q_0/q_m). \tag{5}$$

where q_m is q when $\mu = \mu_m$ (Droop, 1974; Goldman and McCarthy, 1978).

Caperon (1968) used an empirical function akin to the Monod equation to describe N-limited growth of *Isochrysis galbana:*

$$\mu = \mu_m'(q - q_0)/[K_q + (q - q_0)] \tag{6}$$

where K_q is the half-saturation constant.

Fuhs (1969) employed an exponential equation to express the P-limited growth of *Thalassiosira fluviatilis* and *Cyclotella nana:*

$$\mu_m' = (1 - 2^{-(q-q_0)/q}) \tag{7}$$

All three equations appear to be equivalent in describing nutrient-limited growth during steady state (Rhee, 1980). The reason why growth rate can be expressed as a function of the total cell nutrient concentration appears to be that during steady state, this concentration increases with growth rate in a manner similar to increases in cellular RNA (Rhee, 1973,1978), which is directly related to growth rate. In *Scenedesmus obliquus,* among various cellular constituents, only the cellular RNA concentration is a function of growth rate irrespective of the type of limiting nutrient (Rhee, 1978).

Of the three empirical equations, Droop's is the simplest and is therefore more frequently used than others. This equation can describe growth limited by various nutrients for various species of phytoplankton (Rhee, 1980; Rhee *et al.,* 1981). Unlike the Monod equation, this function can also apply to batch growth from logarithmic to stationary phase, relating the instantaneous change in cell quota to the growth rate under continuous illumination (Droop, 1975; Lederman, 1979). It was found possible to describe the growth of a nonsteady-state continuous culture of a natural phytoplankton population from Cayuga Lake, New York, when the cell quota was expressed in terms of cell volume (J. P. Barlow, personal communication). Similarly, Jones et al. (1978a,b) found a good fit of equation (5) to data from the large volume continuous flow culture of natural populations from Loch Creran, Scotland, when they plotted $1/q$ against μ, with q expressed as per cell carbon.

However, Droop's equation cannot describe other types of transient growth. When nutrient-depleted cells are supplied with nutrients, there is an appreciable lag between the increases in cell quota and growth rate (Healey, 1979). A similar lag was observed after a large increase in dilution rate in chemostat cultures (Cunningham and Maas, 1978). When the dilution rate for N-limited *Chlamydomonas reinhardtii* was increased from 0.55 to 0.82 day^{-1}, there was an immediate increase in cell N. However, an increase in growth rate lagged behind by about 10 hr (Fig. 1). Before the cells reached a new steady state, there was also a damped oscillation in cell numbers, cell quota, and growth rate.

The Droop equation is not without other drawbacks. μ_m' is the asymptote when q is infinite, but since q has a finite value at q_m, μ_m' is always larger than the maximum growth rate μ_m. The difference between μ_m' and μ_m depends on the ratio of the minimum to the maximum cell quota, q_0/q_m (equation 5), or on the size of the intracellular nutrient pools. For nutrients with relatively low q_0/q_m, such as P, iron, and vitamin B_{12}, the difference between μ_m' and μ_m is less than 3%; but for nutrients such as Si and inorganic carbon the difference can range from 22 to 36% (Goldman and McCarthy, 1978; Goldman and Graham, 1981). Therefore, equation (4) cannot provide information on μ_m or relative growth rate (μ/μ_m) without the use of equation (5), which in turn requires determination of the additional parameter, q_m. Without being related to μ_m, growth rate alone is not sufficient to describe the nutritional status of an organism.

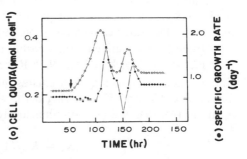

Figure 1. Variation of cell quota and growth rate of *Chlamydomonas reinhardtii* during step experiments in which the dilution rate was increased from 0.55 to 0.82 day^{-1} at the time indicated by the arrow. From Cunningham and Maas, 1978, with permission.

Under a light/dark cycle, there is no correlation between cell quotas and instantaneous growth rate. Only the period average growth is related to the lowest cell quota during the cycle (Rhee *et al.*, 1981; Gotham and Rhee, 1982). Like the Monod equation, therefore, cell quota models are valid only for non-periodic environments or for the comparable time in the division cycle.

Under steady-state conditions, the Monod and cell-quota models are mathematically equivalent in describing growth rate (Burmaster, 1979). Since concentration-dependent nutrient uptake (see below), cell quota, and growth are at equilibrium during steady state, the relationship of growth rate to external substrate concentrations should be compatible with that to cell quota.

2.1.3. Nutrient Uptake and Growth

The rate of change in q can be described as the difference between the uptake rate, v, and the loss rate of q through cell division, μq,

$$dq/dt = v - \mu q \tag{8}$$

Thus, during steady state,

$$v = \mu q \tag{9}$$

The concentration-dependent uptake rate of various nutrients can be described by the Michaelis-Menten equation

$$v = V'_m S/(K_m + S) \tag{10}$$

where V'_m is the apparent maximum rate of uptake and K_m is the half-saturation constant (Droop, 1968; Eppley *et al.*, 1969; MacIsaac and Dugdale, 1969; Fuhs *et al.*, 1972; Paasche, 1973b; Goering *et al.*, 1973; Rhee, 1973, 1978; Nelson *et al.*, 1976; Rigby *et al.*, 1980; Gotham and Rhee, 1981a,b; Healey, 1981).

During steady state, equations (9) and (10) are equivalent, since v in equation (9) is the same as v in equation (10) with S for the steady state. Thus, the steady state satisfies both equations (9) and (10) simultaneously.

Droop (1968) derived a relationship between growth and uptake when μ m $\approx \mu'_m$ (see reviews by Rhee, 1980; Rhee et al., 1981). However, μ_m is quite different from μ'_m for many nutrients as mentioned above. A general relationship between K_s and K_m covering both cases can be obtained by solving the equations (4), (9), and (10) simultaneously (DiToro, 1980):

$$K_s = K_m/(1 + B) \tag{11}$$

where

$$B = V_m/(q_0 \mu'_m) \tag{12}$$

and V_m is the true maximum uptake rate, or V'_m when $q = q_0$.

Thus, when B is large, $K_s = K_m/B$, and K_s is much smaller than K_m. This means that a high rate of uptake can compensate for the inefficiency of a large K_m to supply adequate nutrient for growth (DiToro, 1980).

DiToro (1980) has shown that the values of B for various species are in general largest for P and smallest for Si, with N intermediate between them. As a consequence, the difference between K_m and K_s is largest for P among the three nutrients. In general, the estimated values of K_s agree well with those measured experimentally.

One cannot, then, use K_m as a substitute for K_s. These parameters are identical only when uptake is measured as specific rate per unit of cellular nutrient (v/q) during steady state. This specific rate of uptake is by definition (or by equation 9) the growth rate.

Nutrient uptake is a hyperbolic function of external concentrations (equation 10), but at the same time, it is inversely related to the intracellular nutrient concentrations (see references cited in Rhee, 1980; Gotham and Rhee, 1981a,b). In P-limited chemostat cultures of *S. obliquus,* P uptake formally resembles the kinetics of noncompetitive enzyme inhibition (Rhee, 1973,1974):

$$v = V_m/(1 + K_m/S)(1 + i/K_i) \tag{13}$$

where i is cellular polyphosphate concentration and K_i is the inhibition constant. When S is at a saturation level,

$$V'_m = V_m/(1 + i/K_i) \tag{14}$$

This equation shows that V'_m is higher at lower polyphosphate concentrations or at higher degrees of nutrient limitation. It has been shown that V'_m can vary almost two orders of magnitude depending upon i (Gotham and Rhee, 1981a). The true maximum uptake rate is thus defined as V'_m when $\mu = 0$. However, K_m is not affected by the degree of P limitation. This relationship has recently been demonstrated for *Ankistrodesmus falcatus, Asterionella formosa, Fragilaria crotonensis,* and *Microcystis* sp. (Gotham and Rhee, 1981a). The term i can be replaced by $(q - q_0)$ when a second maximum uptake rate is incorporated into equation (14). This additional term is required probably because the increase of $(q - q_0)$ with growth rate is not proportional to that of the polyphosphates, particularly at high growth rates.

In some species of algae, it has been reported that V'_m remains unchanged with growth rate (Burmaster and Chisholm, 1979) or that V'_m remains constant until q approaches q_m (Healey and Hendzel, 1975; Nyholm, 1977; Brown and Harris, 1978). These observations can also be explained by equation (14): if K_i were so high relative to i that i/K_i approached zero, there would be no change in V'_m. If the change in i with μ were such that i was $<K_i$ at low μ and that i remained relatively constant but approached K_i at high μ, the change in V'_m would be observed only at high μ.

It is now clear that K_m and V_m are insufficient to describe phosphate uptake *in vivo*, because v is also affected by K_i. Consequently, comparisons of uptake between competing species must include all three terms.

Utilizing the inverse relationship between V'_m and the degree of P limitation, Lean and Pick (1981) have determined the nutritional status of natural populations. Since the ^{14}C fixation rate is relatively constant, they have used the ratio of the apparent maximum carbon fixation and P uptake rates as an index to define P limitation into four categories. Ratios below 10 indicate extreme limitation; ratios between 10 and 30, moderate limitation; ratios between 30 and 100, low limitation; and ratios above 100, no deficiency.

Nitrogen uptake is also affected by both external and internal N concentrations. Unlike P uptake, however, both V'_m and K_m change with the degree of N limitation or with growth rate. The likely intracellular feedback components have not been established, although cellular concentrations of free amino acids can be related to changes in V'_m and K_m during steady state (Rhee, 1978). However, by using $(q - q_0)$ as i, the change in V'_m at a saturation level of N can be adequately described by equation (14), as in P uptake (Gotham and Rhee, 1981b).

2.2. Optimum N:P Ratio and Multinutrient Limitation

Use of the Redfield ratio to detect the limiting nutrient implicitly assumes that growth is limited by one nutrient present in minimum relative to its requirement, in agreement with Liebig's law of the minimum. Until recently, however, there was no evidence to show that nutrient limitation occurred in an either-or, or threshold manner. Indeed, Rodhe (1978) claimed that the concept of threshold-type limitation was obsolete.

Experimental results (Rhee, 1974,1978; Droop, 1974), however, showed that growth rate is controlled by the single nutrient in shorter supply in an either-or manner. When chemostat cultures of *Scenedesmus obliquus* (Rhee, 1974,1978) were grown in a medium with various N:P ratios at a constant dilution rate, the steady-state cell number increased linearly with increasing N relative to a fixed P concentration up to an N:P = 30. Above this ratio, the increase stopped abruptly and the cell number remained constant despite further increases in the N concentration in the medium. The linearity of the increase and abruptness of the change in cell number showed that at N:P = 30, N limitation changes to P limitation without a simultaneous overlap of P and N limitations. Droop (1974) also showed that there are no multiple nutrient limitations using P and vitamin B_{12} as limiting nutrients for *Monochrysis lutheri*.

The optimum N:P ratio for *S. obliquus* has turned out to be the same as the ratio of the minimum cell quotas for N and P, q_{0N}/q_{0P} (Rhee, 1978). If $\mu'_{mN} = \mu'_{mP}$, then the ratio of q_N/q_P at the same dilution rate, when either nutrient is limiting, would also be the same as the optimum ratio, because from equation (4), $q_{0N}/q_{0P} = q_N/q_P$. Since $\mu'_{mN} \neq \mu'_{mP}$, however, the steady-state q_N/q_P ratio cannot be the same as the optimum ratio.

The different values of μ'_m for N and P result from differences in the q_0/q_m ratio (equation 5) or from different free pool sizes of the respective nutrients. In such cases, the q_N/q_P ratio deviates significantly from the optimum ratio as μ increases. The steady-state cell quota includes storage or pool fractions (Fuhs, 1969; Rhee, 1973,1978), although their chemical characteristics are not well defined for nutrients other than P, and the sizes of these fractions increase hyperbolically with μ. The increases of pools for N and P are disproportionate, however, being much smaller for N than for P relative to their total cell contents. Thus, cellular N:P ratios decrease with growth rate, becoming progressively smaller than the optimum ratio (Rhee and Gotham, 1980).

As mentioned earlier, a constant growth rate can be maintained at the expense of stored P when external P is depleted. This strongly suggests that all the cell quota in steady-state cells is not needed to maintain a given growth rate. In fact, Droop (1975) and Lederman (1979) have shown that there is a range of values for q_m for the maximum growth rate. As growth rate decreases, q becomes closer to the single minimum value which is necessary to maintain that growth rate, and differences between them disappear when $\mu = 0$ or at q_0. Therefore, the q_{0n}/q_{0P} ratio, but not q_N/q_P, should be the same as the optimum ratio.

The optimum ratio varies widely among species (Table I). An average of these ratios is about 21. If the unusually high value for *Pavlova lutheri* is excluded, the average is about 18, which is not far from the Redfield N:P ratio.

The trend of change in the cellular N:P ratio follows the change in the N:P ratio in the medium (Myklestad, 1977; Rhee and Gotham, 1981a). Thus, the differences in the optimum ratios between species can provide a basis for competitive elimination and coexistence. For example, if the N:P ratio in water results in a cellular ratio between 12 and 30 for both *Asterionella formosa* and *Scenedesmus obliquus,* these two organisms would be limited by different nutrients because of the difference in their optimum N:P ratios. Therefore,

Table I. Optimum N:P ratios

Species	N:P ratio (by atom)	References
Ankistrodesmus falcatus	21	Rhee and Gotham (1980)
Asterionella formosa	12	Rhee and Gotham (1980)
Chaetoceros affinis	24	Myklestad (1977)
Fragilaria crotonensis	25	Rhee and Gotham (1980)
Melosira binderana	7	Rhee and Gotham (1980)
Pavlova lutheri	45	Sakshaug (1978)
Scenedesmus obliquus	30	Rhee (1974,1978)
Selenastrum capricornutum	23	Rhee and Gotham (1980)
Skeletonema costatum	12	Myklestad (1977)
Synedra ulna	10	Rhee and Gotham (1980)

they can coexist. At cellular ratios below 12, both organisms will be N-limited, but the degree of limitation would be greater for *Scenedesmus* because of its higher optimum ratio. Consequently, with a comparable μ_m, *Scenedesmus* will be competitively eliminated. Above a cellular ratio of 30, *Asterionella* will be excluded.

There is also diel periodicity in cellular N:P ratios under a light/dark cycle (Rhee and Gotham, 1980). If the type of limiting nutrient for phased cell division is determined by the ratio of the accumulated nutrients in a cell rela-

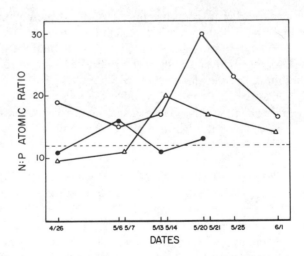

Figure 2. Variations in N:P ratios of natural populations in Trondheim fjord, in which *Skeletonema costatum* constituted more than 99% of the biomass in terms of cell density during spring bloom. The optimal ratio (12) is shown by the horizontal line. Symbols indicate different sampling stations. Drawn from Sakshaug, 1978, with permission.

tive to the optimum ratio, an oscillation of the cellular N:P ratio could occur across the optimum ratio. Such an oscillation could cause a temporal variation in the kind of limiting nutrient. This temporal variation could permit coexistence if the diel change is not identical between competing species, even if their optimum ratios are identical.

Temporal as well as spatial variations in cellular N:P ratios have been found in natural populations (Sakshaug, 1978). The N:P ratios of *Skeletonema costatum* in Trondheim fjord varied across the optimum ratio with time as well as sampling station (12) from April to June (Fig. 2). These changes suggest that the limiting nutrient varies with time and place in the fjord.

3. Light

3.1. Growth and Photosynthesis

Light is the source of energy for photoautotrophic algae, providing electrons to reduce CO_2 to carbohydrates by photosynthesis. These carbohydrates are utilized for biosynthesis of cellular materials. Algal growth is frequently limited by the availability of light in such places as deep or turbid waters. In surface waters, on the other hand, it is not uncommon to observe photoinhibition, i.e., inhibition due to high light intensity (Lund *et al.,* 1963; Talling, 1961; Takahashi *et al.,* 1973; McCarthy and Carpenter, 1979; Li *et al.,* 1980).

3.1.1. Growth Rate

Growth rate is generally related to light intensity (I) by a saturation function below photoinhibition levels (Eppley and Sloan, 1966; Swift and Meunier, 1976; Chan, 1978; Van Liere, 1979; Rhee and Gotham, 1981b). However, when the growth curve is extrapolated to the light axis in a plot of μ versus I, it frequently intercepts I at a value >0. This intercept is the compensation intensity, or the intensity required to maintain the viability of population without net growth. At this point, respiration rate is equal to photosynthetic rate.

Compensation intensity for growth varies over a wide range between species, i.e., from 0.32 to 18 $\mu Ein/m^2/sec$ in *Skeletonema costatum* and *Dunaliella tertiolecta,* respectively (Falkowski and Owens, 1980). In six species of marine phytoplankton, Falkowski and Owens (1978) also observed a variation of the compensation intensity over four orders of magnitude, along with wide differences in dark respiration rates.

In a given species, the compensation intensity for growth varies with the level of light at which it is grown. In the dinoflagellates *Glenodinium* sp. and *Gonyaulax polyedra* and in the green alga *Scenedesmus obliquus,* the respiration rate decreases at low light levels and, thus, the compensation intensity is lower for shade-adapted cells (Prezelin, 1976; Prezelin and Sweeney, 1978; Senger and Fleischhacker, 1978a). *D. tertiolecta* also shows a decrease with light intensity, but in *S. costatum,* the compensation level appears to be unaffected (Falkowski and Owen, 1980).

When growth is expressed as a function of the absorption rate of light energy (q_E), one can obtain the efficiency of an organism to convert light energy to growth. The rate q_E may be expressed as $(dE/dt)(1/X)$, where X is biomass. When X is expressed in the same energy unit as light energy measured by the heat of combustion, q_E has a unit of time^{-1}. Under light-limited conditions, the relationship between μ and q_E is linear:

$$\mu = cq_E + \mu_e \qquad (15)$$

where c is the efficiency factor for growth, and μ_e is the specific maintenance rate, or the negative intercept of the μ axis (Gons and Mur, 1975; Van Liere, 1979). When the $\mu - q_E$ relationship was examined in *Scenedesmus protuberans, Aphanizomenon flos-aquae,* and *Oscillatoria agardhii* (Zevenboom and Mur, 1980), the efficiency factor varied between species, with the highest value for the green alga *S. protuberans* and the lowest for the cyanobacterium *A. flos-aquae.* Thus, to attain the same growth rate, the heterocystous cyanobacterium *A. flos-aquae* and the nonheterocystous *O. agardhii* must absorb more light energy than the green alga. However, the compensation rate of energy absorption, or q_E when $\mu = 0$, was highest in the green alga.

Species differences in the efficiency factor and the compensation rate play an important role in phytoplankton succession. Competition experiments between *S. protuberans* and *O. agardhii* (Mur *et al.,* 1978; Van Liere, 1979; Zevenboom and Mur, 1980) showed that the population dynamics of mixed cultures at various light conditions can be explained by the interplay of the two characteristics of light utilization; when available light becomes low, due either to low incident irradiance or to an increase in population density, *S. protuberans* is not successful because of its high compensation rate of absorption. However, at high irradiance, this alga becomes dominant because of the higher efficiency factor. On the basis of these light characteristics, Mur *et al.* (1978) and Loogman *et al.* (1980) put forth a hypothesis to explain the succession of cyanobacteria after a bloom of green algae in shallow, eutrophic Dutch lakes. In spring, when the water is clear and the underwater light level is high, green algae become dominant because of their high efficiency factor. Their increasing biomass, however, reduces the level of light, and the new condition favors cyanobacteria, which have low compensation rates of absorption. Similarly, the dominance of *Phaeodactylum tricornutum* over *Thallassiosira pseudonana* in nutrient-enriched mass algal culture systems has also been attributed to the former species' higher efficiency of growth at low light conditions (Nelson *et al.*, 1979).

The efficiency of light-limited growth has often been expressed in terms of the level required to support one-half of the maximum growth rate (K_I), in analogy to K_s in nutrient-limited growth. K_I is then a value of $I - I_0$, where I_0 is the compensation intensity. Values for this parameter differ widely between species, ranging from 7.6 μEin/m^2/sec (0.7/W m^2) in *Oscillatoria agardhii, O. redekei,* and *Aphanizomenon flos-aquae* to 54.1 μEin/m^2/sec in *Scenedesmus protuberans* (Van Liere and Mur, 1980). *Fragilaria crotonensis* has an even higher K_I than the green alga at 65 μEin/m^2/sec (Rhee and Gotham, 1981b). However, K_I alone is not sufficient to predict competitive interactions under light limitation, because the maximum growth rates for the organisms with low K_I values are frequently much lower than for those with high K_I values (Van Liere and Mur, 1980). Furthermore, there are also species differences in I_0 values.

Since growth is ultimately the result of photosynthetic carbon fixation, the μ versus I relationship should be predictable from plots of photosynthesis versus I. However, this is not always the case. Tomas (1980) reported that in laboratory cultures of *Olisthodiscus luteus,* the calculated growth from photosynthesis agreed well with actual growth based on cell number, but in outdoor cultures, the calculated value always overestimated the growth rate. In *Gonyaulax polyedra,* the K_I and saturation intensity for photosynthesis and growth were different (Prezelin and Sweeney, 1978). This disagreement is not surprising, since cells excrete organic carbon, which sometimes accounts for as much as 90% of the carbon fixed by photosynthesis, and this excretion is frequently related to photosynthetic rate (Nalewajko and Martin, 1969; Berman and Holm-Hansen, 1974; Hellebust, 1974; Berman, 1976; Mague *et al.*, 1980). Furthermore, dark respiration and the cell carbon content vary with irradiance.

3.1.2. Photosynthesis: Light and Dark Reactions

Photosynthesis can be considered to consist of two phases: light and dark. In the light phase, which is directly dependent upon light energy, photochemical reactions generate ATP and reducing power (NADPH) with the evolution of O_2. In the dark phase, CO_2 is enzymatically reduced to carbohydrates and other products, utilizing the energy and electrons generated in the light reactions. Thus in a plot of I versus photosynthesis, the area in which photosynthetic rate increases linearly with light intensity shows the limitation by light reactions. The area in which the photosynthetic rate is little affected by a further increase of light reflects the limitation by dark reactions. Under light saturation, the dark reactions thus set the upper limit of photosynthesis.

Light reactions occur in the two light-absorbing systems, photosystems (PS) I and II. Each system has its own set of light-absorbing pigments, chlorophyll (chl) a and various accessory pigments, which absorb and transmit light energy like antennas. In addition, PS I contains a specialized chl a-protein complex, P700, which serves as an electron trap. A light-absorbing pigment assembly, or photosynthetic unit, of PS I consists of a single molecule of P700 and a finite number of antenna pigments. Light absorbed anywhere in this unit is transferred to the trap. Electrons in the trap are then expelled, and through a chain of reactions, these electrons reduce NADP. PS II has its own set of pigment units, which include the trap P680 instead of P700. As in PS I, photons absorbed by antenna pigments are transferred to the trap. The electrons in the trap are expelled and passed through a series of redox carriers, generating ATP from ADP by photophosphorylation. They finally come to rest at P700, filling the "hole" left by the expelled electrons. The electron in PS II comes from H_2O, with the release of O_2 (for details see Govindjee and Braun, 1974; Kok, 1976).

The primary pathway for CO_2 fixation in phytoplankton is through the C_3 pathway, the reductive pentosephosphate cycle. In some species, however, a significant proportion of carbon fixation takes place through the C_4 pathway (Morris, 1980). For example, assimilation of CO_2 in *Skeletonema costatum* and *Phaeodactylum tricornutum* is primarily of the C_4 type.

In the pentosephosphate cycle, or Calvin cycle, CO_2 is carboxylated to ribulose-1,5-biphosphate (RuBP) by RuBP carboxylase to yield two molecules of phosphoglycerate. The C_4 pathway incorporates CO_2 primarily into phosphoenolpyruvate (PEP) by PEP carboxylase. In the C_4 pathway, this CO_2 acceptor is generated by the decarboxylation of C_4 compounds, with the release of CO_2. This CO_2 is then incorporated into RuBP by RuBP carboxylase. Therefore, the net CO_2 incorporation in C_4 metabolism takes place through the Calvin cycle. The assimilation products are, however, quite different. The products of C_4 metabolism are primarily proteins and tricarboxylic acid cycle intermediates, whereas those of the C_3 pathway are polysaccharides and sugar phosphates (for details see Raven, 1974; Hatch, 1976; Morris, 1980).

The primary assimilatory pathway appears to vary not only between species but also with the organism's physiological state (Glover *et al.*, 1975; Beardall *et al.*, 1976; Mukerji *et al.*, 1978; Glover and Morris, 1979; Morris, 1980). For example, *Dunaliella tertiolecta* assimilates CO_2 primarily through the C_3 pathway during exponential growth. However, during the stationary phase, when the maximum photosynthetic capacity declines, the reductive pentosephosphate pathway is superimposed by C_4 metabolism.

The assimilatory pathway also appears to vary with environmental conditions. When natural populations of phytoplankton were exposed to reduced light, ^{14}C incorporation into protein increased markedly during photosynthesis (Morris and Skea, 1978). The cyanobacteria *Merismopedia tennissima* and *Oscillatoria rubescens* also showed increased ^{14}C incorporation into protein under reduced light and an inverse relationship between incorporation into protein and carbohydrates (Konopka and Schnur, 1980). The extent of the increase of protein synthesis is also influenced by the population's nutritional state, with those from high-nutrient waters responding more dramatically to reduced light than those from nutrient-poor waters (Morris and Skea, 1978).

The ecological significance of different pathways in phytoplankton photosynthesis is not clear. In higher plants, C_4 photosynthesis is more efficient under high O_2 and low CO_2 concentrations and at high light levels (Hatch, 1967).

3.1.3. Effects of Light Quality

A number of cellular processes of phytoplankton appear to be influenced by the quality of light. The underlying mechanisms of different responses to various light spectra are unclear in many cases. Nonetheless such differences are of ecological importance, since both the intensity and spectral quality change with depth of the water and amount and type of the suspended matter.

It was found that the growth rate of *Chlamydomonas reinhardtii* varies with the wavelength of light. At the same intensity, about 25 $\mu Ein/m^2/sec$ (2.4 $erg/cm^2/sec$), the growth rates in green, white, blue, and red light were 0.28, 0.32, 0.40, and 0.41/day, respectively (Brown and Geen, 1974). In *Dunaliella tertiolecta,* no significant differences were observed at 150 $\mu Ein/m^2/sec$, but at 33 $\mu Ein/m^2/sec$, the growth rate was significantly higher in blue than in white light (Jones and Galloway, 1979). These results with *D. tertiolecta* strongly suggest that the observed growth responses may be related to the action spectra of photosynthesis, since at high intensities, the action spectra become flat and photosynthesis becomes independent of the spectral composition (Smith, 1968). The photosynthetic rates of *C. reinhardtii, Cyclotella nana,* and *D. tertiolecta* also showed maxima in blue and red light (Wallen and Geen, 1971; Brown and Geen, 1974).

Photosynthetic products differ at different light spectra. Blue light enhances the synthesis of amino acids and protein at the expense of carbohydrates, whereas carbohydrate synthesis is stimulated by red light (Soeder and Stengel, 1974; Brown and Geen, 1974). The enhancement of protein synthesis appears to be related to the enhanced PEP carboxylase activity under blue light (Miyachi *et al.,* 1978; Ruyters, 1980). These responses are similar to those of natural populations exposed to low light. Cellular glycerol, which plays a key role in osmoregulation in euryhaline organisms, was lower under blue than under white light (Jones and Galloway, 1979).

Excretion of organic matter is least under blue and red light, and the nature of the products varies with the spectral composition of the light under which cells are grown (Soeder and Stengel, 1974; Brown and Geen, 1974).

Pigmentation appears to change with the spectral quality of light so as to enhance photosynthesis (Haldall, 1970; Krogman, 1973; Govindjee and Braun, 1974). Blue-green algae (Jones and Myers, 1965; Bogorad, 1975; de Marsac, 1977; Van Liere, 1979; Kohl and Nicklisch, 1981), diatoms, dinoflagellates, green algae (Brown *et al.*, 1967; Haldall, 1970; Jeffrey, 1980) all show such changes in pigment composition.

The vertical distribution of the sedentary green, brown, and red algae in benthic habitats was originally explained as an example of chromatic adaptation. However, a recent examination of photosynthesis per unit quantum in these macroalgae indicated no correlationship with their distribution pattern. Rather, the changes in pigment composition in these species were considered to be an adaptation to low light intensity (Dring, 1981).

3.1.4. Photoinhibition

The mechanism of photoinhibition was investigated in depth in *Asterionella formosa* by Belay and Fogg (1978). The inhibition expressed as relative photosynthetic rate measured by ^{14}C incorporation indicates disruptions in both photochemical and dark reactions. Photosynthetic capacity decreases at inhibitory levels of light, but this precedes the destruction of chlorophyll. The inhibition does not appear to involve an accelerated release of photosynthetic products, since the amount of extracellular release was generally constant relative to the total carbon fixation at all intensities. Photoinhibition in *A. formosa* was greater at high temperatures, at high CO_2 concentrations, and when the cells were nutrient-limited (Belay and Fogg, 1978).

Other processes implicated in photoinhibition are the action of ultraviolet light, inhibition of dark respiration, stimulation of chlorophyllase activity (Soeder and Stengel, 1974), and enhancement of photorespiration (Tolbert, 1974; Li *et al.*, 1980).

3.1.5. Cell Composition

In many species cellular contents of chl *a* and accessory pigments such as chl *b, c,* biliproteins, and carotenoids increase with decreasing light levels (e.g., Jørgensen, 1970; Haldall, 1970; Yoder, 1979; Rhee and Gotham, 1981a). However, the responses of other cell constituents to weak light appear to vary between species. Cell carbon increased with decreasing light in *Scenedesmus obliquus, Fragilaria crotonensis, Skeletonema costatum,* and *Dunaliella tertiolecta* (Rhee and Gotham, 1981a; Falkowski and Owen, 1980), but decreased in *S. costatum* in batch cultures (Yoder, 1979). Cell N and P contents were higher at lower light in *S. obliquus* and *F. crotonensis* (Rhee and Gotham, 1981a), but in *S. costatum,* cell N did not vary with light levels (Yoder, 1979). Cell Si content increased with light in *S. costatum,* but in *Rhizosolenia fragilissima,* it was high at a low intensity (Paasche, 1980).

Cell RNA and protein contents increased with decreasing light in *S. obliquus,* but the efficiency of protein synthesis (the rate of protein synthesis per unit of RNA) was lower at low intensities (Rhee and Gotham, 1981a). In *Anacystis nidulans,* on the other hand, cell RNA increased with light intensity but protein concentration was highest at a mid intensity declining on either side (Parrott and Slater, 1980).

3.2. Light Adaptation

In a natural environment, phytoplankton are exposed to various levels of light. The levels vary with season and the depth and turbidity of the water. Vertical transport of planktonic algae in the mixed layer also exposes them to different light regimes on a relatively short time scale. Therefore, the ability to adapt to changing light conditions is an ecological necessity for survival.

Jørgensen (1969,1970) suggested two types of adaptation by planktonic algae: the chlorella and cyclotella types. In the chlorella type, organisms increase their cellular chl content with decreasing light. When they are exposed to saturating light, the photosynthetic rate per cell is lower for the cells adapted to high light levels. At subsaturating levels of light, the slope of a plot of photosynthesis per cell versus *I* is also lower, mainly because of the low cellular chl content. In this type of adaptation, however, there is little difference in photosynthetic rates at adapted light intensities between high- and low-light-adapted cells. This chlorella type of adaptation has been found in *Chlorella pyrenoidosa, Ankistrodesmus falcatus,* and *Synechococcus elongatus* among other species. In light-limited cultures of *Scenedesmus obliquus* and *Fragilaria crotonensis* (Rhee and Gotham, 1981a), net photosynthetic rate per cell also appears to be constant during steady state regardless of the level of light, but the rate per unit chl *a* follows a saturation function.

In the cyclotella type of adaptation, cellular chl content remains unchanged, and only the light-saturated photosynthetic rate varies. For low-light-adapted, or shade-adapted, cells the initial slope of the *I* versus photosynthesis curve remains the same, but the light-saturated photosynthetic rate is much higher when examined at high intensities. The rate at the adapted intensities is also considerably higher for cells grown at higher intensities. This kind of adaptation has been observed in *Cyclotella meneghiniana* and *Skeletonema costatum* among other species. However, there are also various intermediate types of adaptation between the chlorella and cyclotella types.

In recent years, the mechanisms of light adaptation have been studied in greater detail. The adaptive strategies appear to be diverse with wide interspecific differences.

In *Scenedesmus obliquus,* the major mechanism appears to involve the regulation of redox carriers in the photosystems. In this species, cellular chl *a* content did not vary with culture light intensity, although its concentration per unit of cell volume decreased at high intensity (Senger and Fleishhacker, 1978a,b). The initial slope of a plot of photosynthesis per cell versus *I* was little different between cultures grown at low and high intensities, but the light-saturated photosynthetic rate was about three times higher for the cells adapted to high intensities. Yet there was no difference in photosynthetic unit size or quantum yield between the two cultures. Therefore, it has been suggested (Senger and Fleishhacker, 1978a,b) that the difference in photosynthetic capacity is due to a difference in the amount of the redox carriers, i.e., when low-light-adapted cells are exposed to high light, the concentration of redox carrier (plastoquinone) becomes limiting to photosynthesis.

Glenodinium appears to adapt by increasing the efficiency of light absorption. Prezelin (1976) and Prezelin *et al.* (1976) showed that, under low light conditions, *Glenodinium* sp. increased the cellular content of the peridinin-chl *a*-protein complex and of the unidentified chl *a* component of the chloroplast membrane. When these shade-adapted cells were exposed to various levels of light, the maximum photosynthetic rate per cell was little different from the rate in high-light-adapted cells. This maintenance of constant photosynthetic capacity was taken to reflect an increase in photosynthetic unit size, i.e., the number of antenna pigment molecules associated with reaction centers, which would increase the potential for photon capture for photosynthesis. The higher slope for the shade-adapted cells also seemed to support this inference. When photosynthetic rate per unit of chl *a* was measured at various light levels, the slope was identical for cells adapted to various intensities, but the maximum rate was higher for high-light-adapted cells, reflecting their lower chl *a* concentrations.

Phaeodactylum tricornutum also adapts to low light conditions by increasing the efficiency of light absorption. The slope of a plot of photosynthesis versus *I* was higher for shade-adapted cells; in this case, both per cell and per chl *a* (Beardall and Morris, 1976). This suggests an increase in quantum efficiency if one assumes that the quanta absorbed are proportional to the number of chl molecules in a light-limiting range. The saturated rate of photosynthesis per cell for shade-adapted cells was little different from that for light-adapted cells. Since the cell chl *a* content increased at low light, the photosynthetic capacity per unit of chl was less for low-light-adapted cells, as in *Glenodinium* sp.

Falkowski and Owens (1980) suggested two strategies of adaptation on the basis of photosynthetic unit size and number. Measuring the photosynthetic unit size by the chl/P700 ratio, they found that as *Skeletonema costatum* became shade-adapted, the photosynthetic unit size increased, while the reaction center for PS I (P700) per cell decreased. In *Dunaliella tertiolecta,* on the other hand, the photosynthetic unit size decreased, but the number of reaction centers increased. Either change would promote efficient transfer of light energy to reaction centers under low light conditions. In theory, an increase in number would increase photosynthetic capacity, whereas an increase in size would enhance the efficiency of light utilization (Herren and Mauzerall, 1972, quoted from Falkowski and Owens, 1980). However, photosynthetic responses corresponding to these changes in photosynthetic unit size or number were not found. In *S. costatum,* the measured efficiency of light utilization per unit of chl *a* (the slope of photosynthesis versus *I* curve) did not increase, while the chl/P700 ratio increased. In *D. tertiolecta,* the photosynthetic capacity (per cell or unit of chl *a*) decreased despite the increase of PS I reaction centers. The authors attributed these inconsistencies to the probability that chl/P700 did not measure the unit size as defined by the method of O_2 flash yield.

Perry *et al.* (1981) also reported an increase in the photosynthetic unit size in five marine species at low light intensities (*Chaetoceros danicus, C. gracilis, Thalassiosira fluviatilis, T. pseudonana,* and *Isochyrsis galbana*), but no change or a decrease in *Dunaliella euchlora* and *Ditylum brightwellii*. The photosynthetic unit size as well as changes in the size with light varied widely among species. Combined data of various species show that the measured photosynthetic efficiency per chl *a* is an inverse function of the unit size, but the efficiency per P700 is higher with larger unit sizes. It has therefore been suggested (Perry *et al.,* 1981) that organisms with intrinsically larger photosynthetic unit sizes may adapt more readily to the rapid fluctuations of light that occur in the mixed layer.

In *Gonyaulax polyedra,* the strategy of adaptation is rather complex. This species uses different mechanisms in different light levels to optimize photosynthesis at the intensities where they are grown (Prezelin and Sweeney, 1978).

The time course of adaptation to changing light was studied by measuring relative changes in the chl/P700 ratio (Falkowski, 1980). The time course was described by the first-order kinetic equation

$$\ln r = -Kt \tag{16}$$

where *r* is the ratio of chl/P700 ratios at times t_0 and t_1, and *K* is the rate constant, which is inversely related to the half time ($t_{1/2} = \ln 2/K$). The rate constant for the five species examined ranges from 1.7×10^{-2}/hr in *Detonula confervaceae* (at $10°C$) to 5.6×10^{-2}/hr in *S. costatum*.

When light was perturbed from a high to a low level, the growth rate of *Glenodinium* sp. adapted to new light conditions within 2 days. The decrease of growth rate was accompanied by an increase in light-harvesting pigment concentrations to levels far exceeding the steady state level at the new light conditions, and the concentration did not come to a steady state until 40 days after the shift (Prezelin and Matlick, 1980). It appears therefore that the adjustment of growth rate is influenced by, but not coupled to, pigment synthesis.

In euphotic zones the time scale for adaptation should be considered in relation to the rate of mixing, which transports organisms vertically. If the rate of adaptation is slower than the rate of displacement, light adaptation would not be manifested as a function of depth. If the rate of adaptation is faster, cells would exhibit adaptive characteristics. Situations suggesting these two cases have been observed in the New York Bight (Falkowski, 1980). In the summer of 1977, when the water was thermally stratified, clear differences in the photosynthesis versus I profile were observed between populations near pycnocline and those at the surface. In late autumn of the same year, when the water was well mixed, there were no differences in photosynthetic reactions among populations at various depths.

The adaptation of photosynthetic capacity by regulating the concentration or activity of redox components (Senger and Fleischhacker, 1978a,b) occurs in less than one generation time (Vincent, 1980). Chlorophyll fluorescence induced by 3(3,4-dichlorophenyl)-1,1-dimethyl urea (DCMU), which is an indirect measure of the capacity of noncyclic electron flow, shows that the photochemical activity of natural populations is highly responsive to short-term changes in light as well as to nutrient availability (Vincent, 1980, 1981).

There are other types of adaptation, especially at high light levels, which occur on a much shorter time scale. When shade-adapted cells are exposed to strong light, the chl fluorescence yield, measured as the ratio of *in vivo* fluorescence to chl *a,* decreases. This decrease reflects energy spillover (Govindjee *et al.,* 1973), i.e., the energy absorbed by PS II is directed away from its reaction center to the reaction center of PS I. This transfer, which would reduce O_2 production as well as noncyclic electron transport, takes place within a few minutes. Therefore, it has been suggested that energy spillover is a short-term mechanism of adaptation, which may be of particular significance for organisms subject to wide fluctuations of light in a short time period (Govindjee *et al.,* 1973; Papageorgiou, 1975).

This depression of fluorescence yield has been observed in natural populations in near-surface waters of various lakes (Vincent, 1979). The depression was inversely related to light intensity and was reversible at low intensities. This reversibility strongly suggested that the depression was in response to light, rather than to other factors (see below). Similar depressions have been noted in the centric diatoms *Lauderia borealis* and *Cyclotella nana* (Kiefer, 1973). In these organisms, the decrease was attributed to the contraction of chloroplasts. Continued exposure to strong light led to a movement of chloroplasts to the valvar ends, which took 30 to 60 min for completion. This movement was reflected as a second fluorescent depression and appears to be an adaptive mechanism.

It must be noted, however, that fluorescence yield also varies with nutrient limitation if it affects the reductive pentosephosphate cycle (Govindjee *et al.*, 1973, Papageorgiou, 1975). In N-limited *C. nana* (Kiefer, 1973), the light-induced fluorescence decline was greater under nutrient limitation, and similar findings were also reported in *Skeletonema costatum* and *Pavlova lutheri* (Sakshaug and Holm-Hansen, 1977).

3.3. Interactions with Nutrient Limitation and Uptake

Nutrient limitation decreases photosynthetic rate per cell or per dry weight. In *S. costatum* (Jørgensen, 1970), the rate decreased with increasing P or N limitation. This decrease was noted under both saturating and subsaturating light conditions and was greater under N than P limitation. Similar decreases with decreasing growth rate have been observed in N-limited *Chlorella pyrenoidosa, Chaetoceros gracilis,* and *Scenedesmus obliquus* (Thomas and Dodson, 1972; Picket, 1975; Rhee, 1978). In P-limited *Microcystis aeruginosa* and *Anabaena flos-aquae* (Keenan and Auer, 1974), the photosynthetic rate was proportional to cellular concentrations of hot-water-extractable surplus P.

Photosynthetic rate per chl *a* also decreases with increasing nutrient limitation in some species (Jørgensen, 1970; Thomas and Dodson, 1972; Senft, 1978), but seems unaffected in others (see references cited in Rhee, 1980). In the cyanobacterium *Anacystis nidulans* (Öquist, 1974), a decrease in the efficiency of energy transfer in the pigment assemblage appears to be the underlying mechanism for reduction of photosynthesis under iron limitation.

Under saturating light conditions, photosynthetic rate per chl *a* (P'_m) in P-limited *C. pyrenoidosa* and *Anabaena wisconsinense* was related to cell P content by the equation (Senft, 1978):

$$P'_m = P_m(1 - q_0/q) \tag{17}$$

where P_m is the maximum rate of photosynthesis. The value of q may be expressed in terms of cell number, chl *a,* cell volume, or surplus P; the best fit was obtained in terms of chl *a.*

Under light limitation, nutrient requirements increase. This is indicated both by the increasing minimum cell N quota in N-limited *S. obliquus* with decreasing irradiance and by the higher cell N quota for a given growth rate at lower light levels (Rhee and Gotham, 1981a). It has been found in CO_2-limited *A. nidulans* that the maintenance concentration of CO_2, the concentration below which no growth can occur, increases with decreasing irradiance, and the CO_2 requirement to maintain growth rate also increases (Young and King, 1980). If the extent of increased nutrient requirements is species-specific, species differences may influence competitive interactions under the simultaneous limitations of light and nutrient.

The requirements for light and nutrient can compensate for each other in maintaining growth rate (Rhee and Gotham, 1981a). This compensatory relationship suggests that the effects of light and nutrient limitation may be difficult to separate in euphotic zones and that the depth of light saturation and adaptation may be determined in part by the vertical nutrient profile. Conversely, the limiting level of nutrient may increase with depth because of the greater attenuation of light.

The interaction effects of light and nutrient limitations are greater than the sum of their individual effects (Rhee and Gotham, 1981a). For example, under nutrient-sufficient conditions, the growth rate of *S. obliquus* was about 1.34/day at 128 μEin/m^2/sec (11.8 W/m^2). When light was saturating at 185 μEin/m^2/sec but nitrogen was limiting at a cell quota of 0.65 $\times$ 10^{-7} μmol/cell, the growth rate was about 0.4/day. However, when light was 128 μEin/m^2/sec and cell N quota was 0.65 $\times$ 10^{-7}/day, growth rate was zero. The interaction effects are therefore not additive, nor are they multiplicative.

Under severe nutrient depletion, an otherwise optimum light level can be lethal (Talling, 1979). Maddux and Jones (1964) found that at "low" levels of N and P, the optimum intensity was lower than at "high" levels. In their study, however, both the "low" and "high" levels appear to have been in the range of nutrient sufficiency. Therefore, it is not clear what changed the optimum level.

It is well known that light stimulates the assimilation of inorganic nitrogen (Syrett, 1962; Morris, 1974; Anderson and Roel, 1981). MacIsaac and Dugdale (1972) found in natural populations of marine phytoplankton that the uptake of nitrate and ammonium can be described as a saturation function of light intensity. Since the uptake of these nutrients is also a function of their concentrations in water, the uptakes at various depths were explained on the basis of the interactions of light and nutrient at those depths. The maximum uptake rate in oligotrophic waters was found deep in the euphotic zone, but in eutrophic waters, it was found near the surface. These observations were explained on the basis that in oligotrophic waters, N uptake was limited by its concentration, whereas in eutrophic waters, light was the limiting factor.

The stimulation of N uptake by photosynthesis may be explained by the availability of photogenerated electrons and ATP. When N-limited *Chlorella fucusa* (Thomas *et al.*, 1976) was given nitrate or nitrite, the CO_2 fixation rate decreased. In the presence of an inhibitor of photosynthetic electron transport DCMU, on the other hand, nitrite (but not nitrate) reduction decreased by about 50%. Thus, it was concluded that about half of the electrons used for nitrite reduction are generated photochemically (the other half are provided by respiratory electron flow) and that nitrite reduction competes for photoreductant with the reductive pentosephosphate cycle. In *Skeletonema costatum* and natural populations (Falkowski and Stone, 1975), similar inhibition of CO_2 fixation was observed, but it was attributed to the diversion of ATP generated by photophosphorylation to N uptake. In other algae, light stimulation of N uptakc was explained by the indirect role of light in providing carbon skeletons for the assimilation of ammonium into amino acids and proteins (Grant and Turner, 1969).

The effects of light on P uptake are not clear. In some natural communities of phytoplankton, light stimulation of P uptake has been reported (Reshkin and Knauer, 1979), but in other marine (Perry, 1976) and freshwater (Lean and Pick, 1981) communities, no enhancement has been found. If there are stimulatory effects, their underlying mechanisms are not clear.

4. Temperature

The thermal environment of planktonic algae varies from one extreme at polar regions, with temperatures near $0°C$, to the other extreme at hot springs, where temperatures can exceed $70°C$ with little seasonal change. In the Antarctic, the diatom *Navicula glaciei* thrives and develops a dense population (up to 244 mg chl a/m^2) in the coastal tide-crack overflow area above the sea ice during winter, whereas in hot-springs *Synechococcus lividus* and *Chloroflexus aurantiacus* can grow at $73-74°C$ (Tansey and Brock, 1978; Whitaker and Richardson, 1980). However, in most aquatic habitats where phytoplankton production occurs, the temperature changes seasonally within a relatively moderate range. The surface water in the sea at the middle latitudes, for instance, changes from $4°C$ to a moderate $16°C$ (Baross and Morita, 1978). The vertical variation in euphotic zones is less than the seasonal variations.

Nevertheless, even a minor change in temperature can affect competitive abilities of species, since the optimum temperature for growth varies with species (Eppley, 1972; Goldman and Carpenter, 1974). Therefore, temperature changes undoubtedly contribute to the distribution and seasonal succession of species.

4.1. Growth

4.1.1. Growth Rate

In many species, the growth rate decreases exponentially with temperature in suboptimal ranges. This relationship has often been expressed by the Arrhenius equation (e.g., Sorokin, 1960; Goldman, 1979):

$$\mu = Ae^{-E/RT} \tag{18}$$

where R is the gas constant, A is the preexponential factor, T is the absolute temperature, and E is the activation energy. Thus growth responses to temperature have often been characterized by E or Q_{10}, the ratio of reaction rates at temperatures t and $t + 10°C$.

Equation (18) shows that the relationship between $\ln \mu$ and $1/T$ is inversely linear with the value of the slope $-E/R$. When $\ln \mu_m$ of various algal species was plotted against the inverse of optimum temperatures for the respective species, the relationship was linear, following equation (18) (Eppley, 1972; Goldman and Carpenter, 1974). For a single species, however, the relationship between μ and $1/T$ is rarely linear between the minimum and maximum temperatures, and in some species no linear relationship is discernible even within a narrow temperature range (Sorokin, 1960; Fuhs *et al.*, 1972).

In *Skeletonema costatum,* the temperature-growth rate relationship has been described by various exponential equations (Sakshaug, 1978; Yoder, 1979). In *Scenedesmus obliquus, Asterionella formosa* (Rhee and Gotham, 1981b), *Oscillatoria agardhii* (Zevenboom, 1980), and *Anabaena variabilis* (Collins, 1980), it has been described by linear expressions. Thus, the expression of temperature stress on growth cannot be generalized.

An interesting observation has been made with respect to the Arrhenius-type plot for yeasts in which μ versus $1/T$ is linear at moderate temperatures (Shaw, 1967). When the temperature was shifted within the range defined by the linear part of the plot, the organisms attained the normal growth rate at the new temperature without lag. In contrast, when a shift was made to regions outside the linear range, a period of transient growth occurred before reaching the normal growth rate at the new temperature. In phytoplankton ecology, possible interspecific differences in the temperature range within which immediate adaptation can take place may be a contributing factor to temperature-mediated successions, in addition to differences in optimal temperatures (Rhee, 1980).

4.1.2. Cell Composition

At suboptimal temperature ranges, the cell quotas of nutrient-sufficient cultures increase with decreasing temperature, as the cell quotas of limiting nutrients do in nutrient-limited cultures. In *S. obliquus* and *A. formosa,* cell C, N, and P increased in an exponential manner with decreasing temperature (Rhee and Gotham, 1981b). Cell volume did not vary in *A. formosa,* although in *S. obliquus,* it increased linearly with decreasing temperature. Therefore, the higher cell quotas in these species are not associated with changes in cell volume. Increases in cell C and N at suboptimal temperatures have also been reported in other species (Eppley and Sloan, 1966; Jørgensen, 1968; Goldman, 1977). Above optimal temperatures, their concentrations also increased (Goldman, 1977).

When *S. obliquus* and *A. formosa* were grown at a fixed dilution rate (or growth rate) in N- and P-limited chemostats, respectively, cell quotas of the limiting nutrients increased at lower temperatures. The cell quotas at the same relative growth rate (μ/μ_m) also showed increases at lower temperatures, unlike the findings of Goldman (1980). Cell quotas for nonlimiting nutrients in nutrient-limited cultures increased and were identical with those in nutrient-sufficient cultures (Rhee and Gotham, 1981b). Silicification in *Chaetoceros affinis* and *Thalassiosira nordenskioeldii* was also higher at lower temperatures. In *T. pseudonana,* there was an increase, but at a very low temperature, the cell Si content decreased. *Skeletonema costatum,* however, showed little change over a range from 8 to 23°C (Paasche, 1980; Durbin, 1974).

Temperature effects on chl content vary from species to species. In *Nitzschia closterium* (Morris and Glover, 1974), *Oscillatoria luteus* (Tomas, 1980), and *Dunaliella tertiolecta* (Eppley and Sloan, 1966), there was a general increase with temperature. In *Scenedesmus obliquus* (Senger and Fleischhacker, 1978a), the chl content also increased between 20 and 30°C, but it decreased at 35°C. However, in a different strain of *S. obliquus* (Rhee and Gotham, 1981b) it increased with decreasing temperature below 20°C. In *Phaeodactylum tricornutum* (Morris and Glover, 1974), its content was highest at 12°C and lower at both 18 and 7°C.

Cellular RNA and protein concentrations were higher at lower temperatures in *S. obliquus* (Rhee and Gotham 1981b). As temperature was decreased from 30 to 10°C, *Ochromonas danica* (Aaronson, 1973) showed a dramatic increase in carbohydrates, but cell lipid concentrations decreased. Protein was highest at 20°C and RNA at 25°C.

4.2. Temperature Adaptation

The adaptation of algae to changing temperature is poorly understood. Steeman-Nielsen and Jørgensen (1968a,b) and Jørgensen (1968) suggested that when algae are exposed to suboptimum temperatures, they adapt by increasing the concentrations of enzymes associated with carbon fixation. This hypothesis was based on the observation that when *S. costatum* was transferred from 20 to 8°C and fully adapted, its ^{14}C fixation rate was practically unchanged. The authors correlated this finding with the high cell protein contents found at low temperatures.

A similar observation has been made in nutrient-sufficient, steady-state turbidostat cultures of *S. obliquus* (Rhee and Gotham, 1981b). When net carbon fixation rate per cell was calculated, the rate appeared to vary little with temperatures from 10 to 20°C, although the growth rate decreased linearly. Thus, in terms of the production of particulate organic carbon, this alga appears to possess the ability to adapt to suboptimal temperatures. This ability to maintain a constant carbon assimilation rate may be due in part to higher concentrations of chl *a* at lower temperatures. Thus, carbon fixation rate per chl *a* decreases with temperature. Such decreases have been observed in a number of marine species (Li, 1980). However, *P. tricornutum* and *D. tertiolecta* (Morris and Glover, 1974) showed decreasing carbon fixation rate per cell with temperature, but the rate per chl *a* remained unaffected. Therefore, the reduced rates were attributed to decreases in cellular chl *a* concentrations.

In *S. obliquus,* cellular protein concentrations were also higher at lower temperatures, but so were the cellular RNA levels (Rhee and Gotham 1981b). However, the efficiency of protein synthesis, i.e., the rate per unit RNA, decreased. If one requirement to adapt to nonoptimum temperatures is the maintenance of ribosomal activity, this species does not seem to possess adaptive capacity in this respect.

On a biochemical level, the ability to function under extreme temperatures includes the maintenance of (1) the physical stability and biosynthetic activity (protein synthesis) of ribosomes, (2) the conformation of proteins (enzymes) and other cellular structures, and (3) membrane permeability (Inniss and Ingraham, 1978; Ameluxen and Murdock, 1978).

When cells are grown at high temperatures, the ratio of unsaturated to saturated fatty acids in the membrane commonly decreases, whereas at low temperatures, the opposite takes place. This ensures the fluidity and thus the stability of the membrane. When *Synechococcus cedrorum* was grown at 40°C, the ratio of unsaturated to saturated fatty acids was lower than at 30°C, yielding fatty acids with a high melting point, which should be more fluid at the high temperature (Sherman, 1978). The thermophilic cyanobacterium *Mastigocladus* has no polyunsaturated fatty acids, and in a thermophilic strain of *Synechococcus,* they are either lacking or minimally present (Holton *et al.,* 1968; Stanier *et al.,* 1971).

The Antarctic diatom *Navicula glaciei,* on the other hand, has a far higher ratio of unsaturated to saturated fatty acids than other *Navicula* (Whitaker and Richardson, 1980). In this Antarctic diatom, decosahexaenoic acid, important constituents of membrane phospholipids in cold-water organisms, has also been found (Whitaker and Richardson, 1980).

Thus, it appears that in response to temperature changes, the fatty acid composition is altered in the direction which will ensure membrane fluidity.

4.3. Interactions with Nutrient Limitation and Uptake

Temperature has been considered relatively unimportant in natural environments because it was thought that although the maximum growth rate was set by temperature, this maximum could seldom be achieved due to growth limitation by other environmental factors, primarily nutrients (Eppley, 1972). This conclusion assumed that temperature affected only the maximum growth rate and did not interact with other environmental stresses, i.e., that interaction effects of nutrient limitation and temperature stress were multiplicative, with only μ_m, but not K_s (equation 3), being affected by temperature. Recent studies show, however, that temperature interacts with other environmental stresses to modify growth kinetics in a complex manner.

The kinetics of nutrient-limited growth at suboptimal temperatures can also be described by the Monod model. Unlike the above assumption, however, temperature stress alters not only the maximum growth rate, μ_m, but also the half-saturation constant for growth, K_s, a clear demonstration that interaction effects are not multiplicative. When *Thalassiosira nordenskioeldii* (Paasche, 1975) was grown under Si limitation, μ_m was higher at $10°C$ (1.26/day) than at $3°C$ (0.89/day). K_s, however, was higher at the low temperature (0.088 μ M Si) than at the higher temperature (0.022 μM). As Paasche (1975) suggests, there is no physiological or biochemical basis for expecting either an increase or decrease of K_s with temperature. Indeed, various patterns of change in K_s have been observed. In N-limited *Dunaliella* sp. and *Gymnodinium splendens* (Thomas and Dodson, 1974), K_s increased with temperature. In *Dunaliella*, μ_m also increased, but it remained unchanged in *G. splendens*. In ammonium-limited *Selenastrum capricornutum* (Reynolds *et al.*, 1975) and nitrate-limited *Oscillatoria agardhii* (Zevenboom, 1980), K_s does not change significantly with temperature, although μ_m varies. Yet in P-limited *O. agardhii* (Ahlgren, 1978), the half-saturation constant had a high value at $15°C$, the optimum temperature for growth, and decreased at $20°C$ along with μ_m; but at $25°C$, it again increased appreciably, although μ_m was essentially the same as at $20°C$. Therefore, temperature does interact with nutrient limitation and the responses to these double stresses are species-specific.

When nutrient-limited growth is expressed in terms of cell quota (equation 4), temperature stress affects not only μ_m' but also the minimum cell quota, q_0. In P- and N-limited *Scenedesmus obliquus, Asterionella formosa* (Rhee and Gotham, 1981b), *Monochrysis lutheri* (Goldman, 1979), and *Oscillatoria agardhii* (Zevenboom, 1980), the minimum cell quota increased with decreasing temperature. The cell quota at a given growth rate was also higher at lower temperatures. These increases in q_0 and q indicate that suboptimal temperatures increase nutrient requirements. The requirements also seem to increase above the optimum temperature (Goldman, 1977). In nutrient-limited environments, therefore, nonoptimum temperatures would pose a double stress by aggravating the nutrient limitation.

The increase of minimum cell quota per unit decrease of temperature was different for different nutrients and thus, the optimum nutrient ratio seemed to change. In *Scenedesmus obliquus,* the ratio of q_0 for N and P, the optimum N:P ratio, decreased with temperature (Rhee and Gotham, 1981a).

Combined effects of nutrient limitation and temperature stress are greater than the sum of individual effects. In N-limited *Scenedesmus obliquus* (Rhee and Gotham, 1981b), the growth rate at $q = 0.95 \times 10^{-7}$ μmol N/cell at optimum temperature ($20°C$) is 0.75/day. Under nutrient sufficiency, the growth rate at $16°C$ is 1.05/day. However, when $q = 0.95 \times 10^{-7}$ μmol/cell and temperature is 16°C, growth rate is zero. It has been found that the greater the degree of nutrient limitation, the larger is the decrease of temperature required to control growth solely by temperature. In other words, the relative effects of temperature on growth regulation are less with greater degrees of nutrient limitation.

It seems that the optimum temperatures can vary with environmental conditions. In artificial seawater medium enriched with high concentrations of N and P (10 and 0.5 mM, respectively) and at a light intensity of 24.3 $\mu Ein/m^2/$ sec (1883 lux), the optimum growth temperature for turbidostat cultures of *Nitzschia closterium* was about 23°C; with low N and P (8.9 and 0.42 mM) and at 11 $\mu Ein/m^2/sec$ (860 lux), it was 16°C (Maddux and Jones, 1964). Such a dependence of optimum temperature on other environmental factors may in part explain the observation that temperature optima determined in the laboratory are frequently much higher than the temperatures in nature at which species are found most abundant. For example, *Detonula confervaceae* in Narragansett Bay, Rhode Island, reaches maximum abundance below 1°C, yet the optimum temperature for laboratory cultures is 12°C (Smayda, 1969). Similarly, the optimum temperature for *Pseudopendinella pyriformis* (Chrysophyceae) in culture is 15°C, although they were isolated from a habitat at 7°C (Ostroff *et al.*, 1980).

The optimum temperature for nutrient uptake may not necessarily be the same as the optimum temperature for growth. When the maximum nitrate uptake rate was measured in N-limited *Scenedesmus obliquus* growing at a fixed dilution rate in a chemostat at various temperatures, the highest value was observed at 15°C (Rhee and Gotham, 1981b). The optimum temperature for growth was 20–25°C. In N-limited *Oscillatoria agardhii*, the maximum nitrate uptake rate increased with decreasing temperature, whereas growth rate was a linear function of temperature (Zevenboom, 1980). Similar differences have been observed in *Chlorella pyrenoidosa* (Shelef *et al.*, 1970). These differences could be a mechanism to adapt to the increased nutrient requirements for growth at lower temperatures, but the decrease in the uptake rate at temperatures below 15° is then difficult to explain. For P-limited *A. formosa*, the optimum temperature for P uptake coincided with that for growth (Rhee and Gotham, 1981b).

4.4. Temperature–Light Interactions

Photochemical reactions are insensitive to temperature (Govindjee and Rabinowich, 1969). Therefore, light reactions of photosynthesis represented by the slope the photosynthesis (P)-I curve (per chl a) are little affected; i.e., when light is limiting, photosynthetic rate is unaffected by changes in temperature. However, the dark reactions observed as the plateau of the curve of photosynthesis versus I are temperature-sensitive. Therefore, the findings that the saturation irradiance decreases with temperature but not the slope of the *P-I* curve (Smayda, 1969; Yentsch and Lee, 1966; Ostroff *et al.,* 1980; Terlizzi and Karlander, 1980) may be due to these different temperature responses by the light and dark reactions. It appears therefore that at deeper layers of the euphotic zone, the effects of temperature on primary production may be minimal.

Temperature inhibition of dark reactions does not always affect growth rate in the same proportion because excretion of organic carbon and synthesis of storage material also vary with temperature. In *Cryptomonas erosa,* ^{14}C uptake was less adversely affected by low temperature than was cell division; cell division at 4°C was 6–10 times less than at 24°C, but ^{14}C fixation was only 1.6–5 times less (Morgan and Kalff, 1979). This was attributed to the accumulation of photosynthetic products as storage materials and large amounts of organic carbon excreted at the low temperature.

The rate of dark respiration decreases with temperature (Aruga, 1965; Morgan and Kalff, 1979) and, therefore, the compensation intensity for cell division decreases (Morgan and Kalff, 1979). Photorespiration, the light-dependent O_2 uptake, also decreases with temperature (Dökler and Przbylla, 1973). Therefore, although temperature does not affect photosynthetic processes under light limitation, it does influence the net carbon fixation through its effects on dark respiration and photorespiration.

The effects of temperature on growth and photosynthetic rates vary with the length of the photoperiod. At optimum temperature, these rates seem to be controlled by the total light energy received during a 24-hr cycle (Terborgh and Thimann, 1964; Dring, 1970; Hobson, 1974; Loogman *et al.,* 1980). Thus, a plot of growth rate versus the total light energy received during the photoperiod results in a hyperbolic curve (Loogman *et al.,* 1980), and too long a daylength has photoinhibitory effects (Hobson, 1974). In general, temperature has no effect on growth or photosynthetic rates for short photoperiods, which are insufficient to saturate growth or photosynthesis (Tamiya *et al.,* 1955; Hobson, 1974; Durbin, 1974; Yoder, 1979). This is similar in appearance to its effects on light and dark reactions. However, the underlying mechanisms of the effect of photoperiod must be different, since light is saturating during photoperiods. The findings imply that, during winter at high latitudes when light is weak and daylength is short, the role of temperature on primary production may be insignificant.

5. Concluding Remarks

Threshold-type control of growth by a single nutrient lays a foundation for using a "standard" nutrient ratio (e.g., the Redfield ratio) as an index in determining the type of nutrient limitation. The average of the optimum N:P ratios of various species known so far seems to further justify the use of the Redfield ratio. The concept of optimum nutrient ratios also provides a basis for understanding coexistence and competitive exclusion. However, the influence of diel periodicity of biological processes on species interactions (Rhee et al., 1981) has been largely overlooked. Species differences in this periodicity can play a critical role by providing temporal differences in sharing limiting resources.

Basic relationships between the availability of nutrients and phytoplankton growth in general are well understood, mainly through investigations under steady-state conditions. Whether these steady-state relationships apply to natural environments is, however, a subject for debate. The answer may depend largely on the time scale under consideration, but our poor understanding of the limits of environmental fluctuations to which organisms can adapt without a time lag contributes a great deal to this uncertainty. Work such as that by Cunningham and Maas (1978) on growth responses of chemostat cultures to changes in dilution rate provides a useful insight into these limits.

Under constant conditions the effects of light and temperature are generally clear. As with nutrient limitation, however, information on the time course of adaptation to the fluctuations of light and temperature is essential to clearly define their effects on natural populations. There is only scant information on these aspects.

Studies of nutrient-light and nutrient-temperature interactions have demonstrated clearly that the juxtaposition of physical stresses to nutrient limitation can alter growth characteristics in nutrient-limited environments. It is essential to attain a clear understanding of the extent to which such interactions influence primary productivity and succession in natural environments.

ACKNOWLEDGMENT. The author wishes to thank Drs. G. W. Fuhs and I. J. Gotham for their comments on the manuscript. This work was supported in part by U.S. EPA grants R806820 and R807909.

References

Aaronson, S., 1973, Effect of incubation temperature on the macromolecular and lipid content of the phytoflagellate *Ochromonas danica*, *J. Phycol.* **9**:111–113.

Ahlgren, G., 1977, Growth of *Oscillatoria agardhii* Gom. in chemostat culture. 1. Investigations of nitrogen and phosphorus requirements, *Oikos* **29**:209–224.

Ahlgren, G., 1978, Growth of *Oscillatoria agardhii* in chemostat culture. 2. Dependence of growth constants on temperature, *Mitt. Int. Verein. Limnol.* **21**:88–102.

Ameluxen, R. E., and Murdock, A. L., 1978, Microbial life at high temperatures: mechanisms and molecular aspects, in: *Microbial Life in Extreme Environments* (D. J. Kushner, ed.), pp. 217–278, Academic Press, New York.

Anderson, S. M., and Roel, O. A., 1981, Effects of light intensity in nitrate and nitrite uptake and excretion by *Chaetoceros curvisetus*, *Mar. Biol.* **62**:257–261.

Aruga, Y., 1965, Ecological studies of photosynthesis and matter production of phytoplankton. II. Photosynthesis of algae in relation to light intensity and temperature, *Bot. Mag. Tokyo* **78**:360–365.

Baross, J. A., and Morita, R. Y., 1978, Microbial life at low temperatures: ecological aspects, in: *Microbial Life in Extreme Environments* (D. J. Kushner, ed.), pp. 9–72, Academic Press, New York.

Beardall, J., and Morris, I., 1976, The concept of light intensity adaptation in marine phytoplankton: some experiments with *Phaeodactylum tricornutum, Mar. Biol.* **37**:377–387.

Beardall, J., Mukerji, D., Glover, H. E., and Morris, I., 1976, The path of carbon in photosynthesis by marine phytoplankton, *J. Phycol.* **12**:409–417.

Belay, A., and Fogg, G. E., 1978, Photoinhibition of photosynthesis in *Asterionella formosa, J. Phycol.* **14**:341–347.

Berman, T., 1976, Release of dissolved organic matter by photosynthesizing algae in Lake Kinneret, Israel, *Freshwater Biol.* **6**:13–18.

Berman, T., and Holm-Hansen, O., 1974, Release of photoassimilated carbon as dissolved organic matter by marine phytoplankton, *Mar. Biol.* **28**:305–310.

Bogorad, L., 1975, Phycobiliproteins and complementary chromatic adaptation, *Annu. Rev. Plant. Physiol.* **26**:369–401.

Brown, E. J., and Harris, R. F., 1978, Kinetics of algal transient phosphate uptake and the cell quota concept, *Limnol. Oceanogr.* **23**:35–40.

Brown, T. E., Richardson, F. L., and Vaughn, M. L., 1967, Development of red pigmentation in *Chlorococcum winneri* (Chlorophyta chlorococcales). *Phycologia* **6**:167–184.

Brown, T. J., and Geen, G. H., 1974, The effect of light quality on the carbon metabolism and extracellular release of *Chlamydomonas reinhardtii* Dangeard, *J. Phycol.* **10**:213–220.

Burmaster, D. E., 1979, The continuous culture of phytoplankton: mathematical equivalence among three steady-state models. *Am. Nat.* **113**:123–134.

Burmaster, D. E., and Chisholm, S. W., 1979, A comparison of two methods for measuring phosphate uptake by *Monochrysis lutheri* Droop grown in continuous culture, *J. Exp. Mar. Biol. Ecol.* **39**:187–202.

Caperon, J., 1968, Population growth response of *Isochrysis galbana* to variable nitrate environment, *Ecology* **49**:866–872.

Chan, A. T., 1978, Comparative physiological study of marine diatoms and dinoflagellates in relation to irradiance and cell size. I. Growth under continuous light, *J. Phycol.* **14**:396–402.

Chisholm, S. W., and Nobbs, P. A., 1976, Simulation of algal growth and competition in a phosphate-limited cyclostat, in: *Modelling Biochemical Processes in Aquatic Ecosystems* (R. P. Canale, ed.), pp. 337–355, Ann Arbor Press, Ann Arbor, Mich.

Collins, C. D., 1980, Instantaneous and long-term effects of light intensity and temperature on the blue-green alga *Anabaena variabilis,* Ph.D. thesis, Rensselaer Polytechnic Institute, Troy, N.Y.

Cunningham, A., and Maas, P., 1978, Time lag and nutrient storage effects in the transient growth response of *Chlamydomonas reinhardtii* in nitrogen-limited batch and continuous culture, *J. Gen. Microbiol.* **104**:227–231.

Davis, A. G., 1970, Iron, chelation and growth of marine phytoplankton. I. Growth kinetics and chlorophyll production in cultures of euryhaline flagellate *Dunaliella tertiolecta* under iron-limited conditions, *J. Mar. Biol. Assoc. U.K.* **50**:65–86.

D'Elia, C. F., Guillard, R. R. L., and Nelson, D. M., 1979, Growth and competition of the marine diatoms *Phaeodactylum tricornutum* and *Thalassiosira pseudonana*. I. Nutrient effects, *Mar. Biol.* **50**:305–312.

deMarsec, N. T., 1977, Occurrence and nature of chromatic adaptation in cyanobacteria, *J. Bacteriol.* **30**:82–91.

DiToro, D. M., 1980, Applicability of cellular equilibrium and Monod theory to phytoplankton growth kinetics, *Ecol. Model.* **8:**201–218.

Dökler, G., and Przbylla, K.-R., 1973, Einfluss der Termperatur auf die Lichtamung der Blaualge *Anacystis nidulans, Planta* **110:**153–158.

Dring, M. J., 1970, Photoperiodic effects in microorganisms, in: *Photobiology of Microorganisms* (P. Haldall, ed.), pp. 345–368, Wiley-Interscience, New York.

Dring, M. J., 1981, Chromatic adaptation of photosynthesis in benthic marine algae: an examination of its ecological significance using a theoretical model, *Limnol. Oceanogr.* **26:**271–284.

Droop, M. R., 1968, Vit. B_{12} and marine ecology: IV. The kinetics of uptake, growth and inhibition in *Monochrysis lutheri, J. Mar. Biol. Assoc. U.K.* **48:**689–733.

Droop, M. R., 1974, The nutrient status of algal cells in batch culture, *J. Mar. Biol. Assoc. U.K.* **54:**825–855.

Droop, M. R., 1975, The nutrient status of algal cells in batch culture, *J. Mar. Biol. Assoc. U.K.* **55:**541–555.

Durbin, E. G., 1974, Studies on the autoecology of the marine diatom *Thalassiosira nordenskioeldii* Cleve I. The influence of day length, light intensity, and temperature on growth, *J. Phycol.* **10:**220–225.

Eppley, R. W., 1972, Temperature and phytoplankton growth in the sea. *Fish. Bull.* **70:**1063–1085.

Eppley, R. W., and Sloan, P. R., 1966, Growth rates of marine phytoplankton: correlation with light absorption by cell chlorophyll a. *Physiol. Plant.* **19:**47–59.

Eppley, R. W., Rogers, J. N., and McCarthy, J. J., 1969, Half-saturation constants for uptake of nitrate and ammonium by marine phytoplankton, *Limnol. Oceanogr.* **14:**912–920.

Falkowski, P. G., 1980, Light-shade adaptation in marine phytoplankton, in: *Primary Productivity of the Sea* (P. G. Falkowski, ed.), pp. 99–120, Plenum Press, New York.

Falkowski, P. G., and Owen, T. G., 1978, Effects of light intensity on photosynthesis and dark respiration in six species of marine phytoplankton, *Mar. Biol.* **45:**289–295.

Falkowski, P. G., and Owen, T. G., 1980, Light-shade adaptation—two strategies in marine phytoplankton, *Plant Physiol.* **66:**592–595.

Falkowski, P. G., and Stone, D. P., 1975, Nitrate uptake in marine phytoplankton: energy sources and the interaction with carbon fixation, *Mar. Biol.* **32:**77–84.

Finenko, Z. Z., and Kruptkia-Akinina, D. K., 1974, Effects of inorganic phosphorus on growth rate of diatoms, *Mar. Biol.* **26:**193–201.

Fuhs, G. W., 1969, Phosphorus content and rate of growth in the diatom *Cyclotella nana* and *Thalassiosira fluviatilis, J. Phycol.* **5:**312–321.

Fuhs, G. W., Demerle, S. D., Canelli, E., and Chen, M., 1972, Characterization of phosphorus-limited algae, in: *Nutrient and Eutrophication* (G. E. Likens, ed.), Special Symposia, Vol. 1, pp. 113–132, The American Society for Limnology and Oceanography, Allen Press, Lawrence, Kan.

Glover, H. E., and Morris, I., 1979, Photosynthetic carboxylating enzymes in marine phytoplankton, *Limnol. Oceanogr.* **24:**510–519.

Glover, H. E., Beardall, J., and Morris, I., 1975, Effects of environmental factors on photosynthesis patterns in *Phaeodactylum tricornutum* (Bacillariophyceae). I. Effect of nitrogen deficiency and light intensity, *J. Phycol.* **11:**424–429.

Goering, J. J., Nelson, D. M., and Carter, J. A., 1973, Silicic acid uptake by natural populations of marine phytoplankton, *Deep-Sea Res.* **20:**777–789.

Goldman, J. C., 1977, Temperature effects on phytoplankton growth in continuous culture, *Limnol. Oceanogr.* **22:**932–936.

Goldman, J. C., 1979, Temperature effects on steady-state growth, phosphorus uptake, and the chemical composition of a marine phytoplankter, *Microb. Ecol.* **5:**153–166.

Goldman, J. C., 1980, Physiological processes, nutrient availability, and the concept of relative growth rate in marine phytoplankton ecology, in: *Primary Productivity in the Sea* (P. G. Falkowski, ed.), pp. 179–194, Plenum Press, New York.

Goldman, J. C., and Carpenter, E. J., 1974, A kinetic approach to the effect of temperature on algal growth, *Limnol. Oceanogr.* **19**:756–766.

Goldman, J. C., and Graham, S. J., 1981, Inorganic carbon limitation and chemical composition of two freshwater green microalgae, *Appl. Environ. Microbiol.* **41**:60–70.

Goldman, J. C., and McCarthy, J. J., 1978, Steady-state growth and ammonium uptake of a fast-growing marine diatom, *Limnol. Oceanogr.* **23**:695–703.

Gons, H. J., and Mur, L. R., 1975, An energy balance for algal populations in light-limiting conditions, *Verh. Int. Verein. Limnol.* **19**:2729–2733.

Gotham, I. J., 1977, Nutrient-limited cyclostat growth. A theoretical and physiological aspect, Ph.D. thesis, State University of New York, Albany, N.Y.

Gotham, I. J., and Rhee, G. Y., 1981a, Comparative kinetic studies of phosphate-limited growth and phosphate uptake in phytoplankton in continuous culture, *J. Phycol.* **17**:257–265.

Gotham, I. J., and Rhee, G. Y., 1981b, Comparative kinetic studies of nitrate-limited growth and nitrate uptake in phytoplankton in continuous culture, *J. Phyco.* **17**:309–314.

Gotham, I. J., and Rhee, G. Y., 1982, The effects of phosphate and nitrate limitation on cyclostat growth of two freshwater diatoms, *J. Gen. Microbiol.* **128**:199–205.

Govindjee, and Braun, B. Z., 1974, Light absorption, emission, and photosynthesis, in: *Algal Physiology and Biochemistry* (W. D. P. Stewart, ed.), pp. 346–390, University of California Press, Berkeley.

Govindjee, and Rabinowich, R. I., 1969, *Photosynthesis,* John Wiley & Sons, New York.

Govindjee, Papageorgiou, G., and Rabinowich, E., 1973, Chlorophyll fluorescence and photosynthesis, in: *Practical Fluorescence* (G. G. Guilbault, ed.), pp. 543–575, Marcel Dekker, New York.

Grant, B. R., and Turner, I. M., 1969, Light stimulated nitrate and nitrite assimilation in several species of algae, *Comp. Biochem. Physiol.* **29**:995–1004.

Guillard, R. R. L., Kilham, P., and Jackson, T. A., 1973, Kinetics of silicon-limited growth of marine diatom *Thalassiosira pseudonana* Hasle and Heimdal (*Cyclotella nana* Hustedt), *J. Phycol.* **9**:233–237.

Haldall, P., 1970, The photosynthetic apparatus of microalgae and its adaptation to environmental factors, in: *Photobiology of Microorganisms* (P. Haldall, ed.), pp. 1–16, Wiley-Interscience, New York.

Hatch, M. D., 1976, Photosynthesis: the path of carbon, in: *Plant Biochemistry* (J. Bonner and J. E. Varner, eds.), pp. 797–845, Academic Press, New York.

Healey, F. P., 1979, Short-term responses of nutrient-deficient algae to nutrient addition, *J. Phycol.* **15**:289–299.

Healey, F. P., 1980, Slope of the Monod equation as an indicator of advantage in nutrient competition, *Microb. Ecol.* **5**:281–286.

Healey, F. P., 1981, Phosphate, in: *Biology of Cyanobacteria* (N. G. Carr and B. A. Whitton, eds.), in press. Blackwell Publications, Oxford.

Healey, F. P., and Hendzel, L. L., 1975, Effects of phosphorus deficiency on two algae growing in chemostats, *J. Phycol.* **11**:303–309.

Hellebust, J. A., 1974, Extracellular products, in: *Algal Physiology and Biochemistry* (W. D. P. Stewart, ed.), pp. 838–863, University of California Press, Berkeley.

Hobson, L. A., 1974, Effects of interactions of irradiance, day length, and temperature on division rates of three species of marine algae, *J. Fish. Res. Bd. Can.* **31**:391–395.

Holton, R. W., Blecker, H. H., and Stevens, T. S., 1968, Fatty acids in blue-green algae: possible relation to phylogenic position, *Science* **160**:545–547.

Innis, W. E., and Ingraham, J. L., 1978, Microbial life at low temperatures: mechanisms and molecular aspects, in: *Microbial Life in Extreme Environments* (D. J. Kushner, ed.), pp. 73–104, Academic Press, New York.

Jeffrey, S. W., 1980, Algal pigment systems, in: *Primary Productivity in the Sea* (P. G. Falkowski, ed.), pp. 33–58, Plenum Press, New York.

Jones, K. J., Tett, P., Wallis, A. C., and Wood, B. J. N., 1978a, The use of small, continuous multispecies cultures to investigate the ecology of phytoplankton in a Scottish Sea loch, *Mitt. Int. Verein. Limnol.* **18**:71–77.

Jones, K. J., Tett, P., Wallis, A. C., and Wood, B. J. B., 1978b, Investigation of a nutrient-growth model using a continuous culture of natural phytoplankton, *J. Mar. Biol. Assoc. U.K.* **58**:923–941.

Jones, L. W., and Myers, J., 1965, Pigment variations in *Anacystis nidulans* induced by light of selective wavelength, *J. Phycol.* **1**:7–14.

Jones, L. W., and Galloway, R. A., 1979, Effect of light quality and intensity on glycerol content in *Dunaliella tertiolecta* (Chlorophyceae) and the relationship to cell growth/osmoregulation, *J. Phycol.* **15**:101–106.

Jørgensen, E. G., 1968, The adaptation of plankton algae. II. Aspects of the temperature adaptation of *Skeletonema costatum, Physiol. Plant.* **21**:423–427.

Jørgensen, E. G., 1969, The adaptation of plankton algae. IV. Light adaptation in different algal species, *Physiol. Plant.* **22**:1307–1315.

Jørgensen, E. G., 1970, The adaptation of plankton algae. V. Variation in photosynthetic characteristics of *Skeletonema costatum* cells grown at low intensity, *Physiol Plant.* **23**:11–17.

Keenan, J. D., and Auer, M. T., 1974, The influence of phosphorus luxury uptake on algal bioassays, *J. Water Pollut. Cont. Fed.* **46**:532–542.

Kiefer, D. A., 1973, Chlorophyll *a* fluorescence in marine centric diatoms: Responses of chloroplast to light and nutrient stress, Mar. Biol. **23**:39–46.

Kilham, S. S., 1975, Kinetics of silicon-limited growth in the freshwater diatom *Asterionella formosa, J. Phycol.* **11**:396–399.

Kohl, J.-G., and Nicklisch, A., 1981, Chromatic adaptation of the planktonic blue-green alga *Oscillatoria redekei* van Goor and its ecological significance, *Int. Rev. Ges. Hydrobiol.* **66**:83–94.

Kok, B., 1976, Photosynthesis: the path of energy, in: *Plant Biochemistry* (J. Bonner and J. E. Varner, eds.), pp. 846–886, Academic Press, New York.

Konopka, A., and Shnur, M., 1980, Effect of light intensity on macromolecular synthethesis in cyanobacteria, *Microbial Ecol.* **6**:291–301.

Krogman, D. W., 1973, Photosynthetic reactions and components of thylakoids, in: *The Biology of Blue-green Algae* (N. G. Carr and B. A. Whitton, eds.), pp. 80–98, University of California Press, Berkeley.

Kuenzler, E. J., and Ketchum, B. H., 1962, Rate of phosphorus uptake by *Phaeodactylum tricornutum, Biol. Bull.* **123**:134–145.

Lean, D. R. S., and Pick, F. R., 1981, Photosynthetic response of lake plankton to nutrient enrichment: a test for nutrient limitation, *Limnol. Oceanogr.* **26**:1001–1019.

Lederman, T. C., 1979, The effects of irradiance and phosphorus in batch cultures of *Pavlova (Monocrysis) lutheri,* Ph.D. thesis, University of Sterling, Scotland.

Li, W. K. W., 1980, Temperature adaptation in phytoplankton: cellular and photosynthetic characteristics, in: *Primary Productivity in the Sea* (P. G. Falkowski, ed.), pp. 259–280, Plenum Press, New York.

Li, W. K. W., Glover, H. E., and Morris, I., 1980, Physiology of carbon photoassimilation by *Oscillatoria thiebautii* in the Caribbean Sea, *Limnol. Oceanogr.* **25**:447–456.

Lin, C. K., 1977, Accumulation of water soluble phosphorus and hydrolysis of polyphosphate by *Cladophora glomerata* (Chlorophyceae), *J. Phycol.* **13**:46–51.

Lin, C. K., and Schelske, C. L., 1979, Effects of nutrient enrichment, light intensity, and temperature on growth of phytoplankton from Lake Huron, U.S. Environmental Protection Agency Report EPA-600/3-79-049, Environmental Research Laboratory, Duluth, Minn.

Loogman, J. G., Post, A. F., and Mur, L. R., 1980, Influence of periodicity in light conditions, as determined by the trophic state of water, on the growth of the green alga *Scenedesmus protuberans* and the cyanobacterium *Oscillatoria agardhii,* in: *Hypertrophic Ecosystems* (J. Barica and L. R. Mur, eds.), pp. 79–82, Dr. W. Junk bv. Publishers, The Hague.

Lund, J. W. G., Mackereth, F. J. H., and Mortimer, C. H., 1963, Changes in depth and time of certain chemical and physical conditions and of the standing crop of *Asterionella formosa* Hass. in the North Basin of Windermere in 1947, *Phil. Trans. Roy. Soc. London Sci. B* **246:**255–290.

MacIsaac, J. J., and Dugdale, R. C., 1969, The kinetics of nitrate and ammonia uptake by natural populations of marine phytoplankton, *Deep-Sea Res.* **16:**45–57.

MacIsaac, J. J., and Dugdale, R. C., 1972, Interactions of light and inorganic nitrogen in controlling nitrogen uptake in the sea, *Deep-Sea Res.* **19:**209–232.

Maddux, W. S., and Jones, R. F., 1964, Some interactions of temperature, light intensity, and nutrient concentrations during the continuous culture of *Nitzschia closterium* and *Tetraselmis* sp., *Limnol. Oceanogr.* **9:**79–86.

Mague, T. H., Friberg, E., Hughes, D. J., and Morris, I., 1980, Extracellular release of carbon by marine phytoplankton: a physiological approach, *Limnol. Oceanogr.* **25:**262–279.

McCarthy, J. J., and Carpenter, E. J., 1979, *Oscillatoria (Trichodesmium) thiebautii* (Cyanophyta) in the central North Atlantic Ocean, *J. Phycol.* **15:**75–82.

Menzel, D. W., and Ryther, J. H., 1964, The composition of particulate organic matter in the western North Atlantic, *Limnol. Oceanogr.* **9:**179–186.

Miller, J. D. H., and Fogg, G. E., 1957, Studies on the growth of Xanthophyceae in pure culture. I. The mineral nutrition of *Monodus subterraneus* Petersen, *Arch. Mikrobiol.* **28:**1–17.

Miyachi, S., Miyachi, S., and Kamiya, A., 1978, Wavelength effects on photosynthetic carbon metabolism in *Chlorella, Plant Cell Physiol.* **19:**277–288.

Monod, J., 1942, *Recherches sur la Croissance des Cultures Bacteriennes* (2nd ed.), Hermann, Paris.

Morgan, K. C., and Kalff, J., 1979, Effects of light and temperature interactions on growth of *Cryptomonas erosa* (Crytophyceae), *J. Phycol.* **15:**127–134.

Morris, I., 1974, Nitrogen assimilation and protein synthesis, in: *Algal Physiology and Biochemistry* (W. D. P. Stewart, ed.), pp. 583–609. University of California Press, Berkeley.

Morris, I., 1980, Path of carbon assimilation in marine phytoplankton, in: *Primary Productivity of the Sea* (P. G. Falkowski, ed.), pp. 139–160, Plenum Press, New York.

Morris, I., and Glover, H. E., 1974, Questions on the mechanism of temperature adaptation in marine phytoplankton, *Mar. Biol.* **24:**147–154.

Morris, I., and Skea, W., 1978, Product of photosynthesis in natural populations of marine phytoplankton from the Gulf of Maine, *Mar. Biol.* **47:**303–312.

Mukerji, D., Glover, H. E., and Morris, I., 1978, Diversity in the mechanism of carbon dioxide fixation in *Dunaliella tertiolecta* (Chlorophyceae), *J. Phycol.* **14:**137–142.

Müller, H., 1972, Wachstum und Phosphat bedarf von *Nitzschia actinostroides* (Lemm.) v. Goor in statischer und homocontiunierlicher Kulter unter Phosphat-limiting, *Arch. Hydrobiol.* (Suppl.) **38:**399–484.

Mur, L. R., Gons, H. J., and Van Liere, L., 1978, Competition of the green alga *Scenedesmus* and the blue-green alga *Oscillatoria, Mitt. Int. Verein. Limnol.* **21:**473–479.

Myklestad, S., 1977, Production of carbohydrates by marine planktonic diatoms. II. Influence of the N/P ratio on the growth medium, on the assimilation ratio, growth rate, and production of cellular and extracellular carbohydrates by *Chaetoceros affinis* var. *willei* (Gram) Hustedt and *Skeletonema costatum* (Grev.), Cleve. *J. Exp. Mar. Biol. Ecol.* **29:**161–179.

Nalewajko, C., and Lean, D. R. S., 1978, Phosphorus kinetics—algal growth relationships in batch cultures, *Mitt. Int. Verein. Limnol.* **21**:184–192.

Nalewajko, C., and Martin, L., 1969, Extracellular production in relation to growth of four planktonic algae and of phytoplankton populations from Lake Ontario, *Can. J. Bot.* **47**:405–413.

Nelson, D. M., Goering, J. J., Kilham, S. S., and Guillard, R. R. L., 1976, Kinetics of silicic acid uptake and rates of silica dissolution in the marine diatom *Thalassiosira pseudonana,* *J. Phycol.* **12**:246–252.

Nelson, D. M., D'Elia, C. F., and Guillard, R. R. L., 1979, Growth and competition of the marine diatoms *Phaeodactylum tricornutum* and *Thalassiosira pseudonana.* II. Light limitation, *Mar. Biol.* **50**:313–318.

Nyholm, N., 1977, Kinetics of phosphate limited algal growth, *Biotechnol. Bioeng.* **19**:467–492.

Öquist, G., 1974, Iron deficiency in the blue-green alga *Anacystis nidulans:* changes in pigmentation and photosynthesis, *Physiol. Plant.* **30**:30–37.

Ostroff, C. R., Karlander, E. P., and Van Valkenberg, S. D., 1980, Growth rates of *Pseudopendinella pyriformis* (Chrysophyceae) in response to 75 combinations of light, temperature, and salinity, *J. Phycol.* **16**:421–423.

Paasche, E., 1973a, Silicon and the ecology of marine plankton diatoms. I. *Thalassiosira pseudonana (Cyclotella nana)* grown in a chemostat with silicate and limiting nutrient, *Mar. Biol.* **19**:117–126.

Paasche, E., 1973b, Silicon and the ecology of marine plankton diatoms. II. Silicate-uptake kinetics in five diatom species, *Mar. Biol.* **19**:262–269.

Paasche, E., 1975, Growth of the plankton diatom *Thalassiosira nordenskioeldii* Cleve at low silicate concentrations, *J. Exp. Mar. Biol. Ecol.* **18**:173–183.

Paasche, E., 1980, Silicon content of five marine plankton diatom species measured with a rapid filter method, *Limnol. Oceanogr.* **25**:474–480.

Papageorgiou, G., 1975, Chlorophyll fluorescence: an intrinsic probe of photosynthesis, in: *Bioenergetics of Photosynthesis* (Govindjee, ed.), pp. 319–371, Academic Press, New York.

Parrott, L. M., and Slater, J. H., 1980, The DNA, RNA and protein composition of the Cyanobacterium *Anacystis nidulans* growth in light and carbon dioxide-limited chemostats, *Arch. Microbiol.* **127**:53–58.

Parsons, T. R., Takahashi, M., and Hargrave, B., 1978, *Biological Oceanographic Processes* (2nd ed.), Pergamon Press, Elmsford, N.Y.

Perry, M. J., 1976, Phosphate utilization by an oceanic diatom in phosphorus-limited chemostat culture and in the oligotrophic waters of the central North Pacific, *Limnol. Oceanogr.* **21**:88–107.

Perry, M. J., Talbot, M. C., and Albert, R. S., 1981, Photoadaptation in marine phytoplankton: Responses of the photosynthetic unit, *Mar. Biol.* **62**:91–101.

Picket, J. M., 1975, Growth of *Chlorella* in nitrate-limited chemostat, *Plant Physiol.* **55**:223–225.

Plowman, K. M., 1972, *Enzyme Kinetics,* McGraw-Hill, New York.

Prezelin, B. B., 1976, The role of peridinin-chlorophyll *a*-protein in photosynthetic light adaptation of the marine dinoflagellate, *Glenodinium* sp., *Planta* **130**:225–233.

Prezelin, B. B., and Matlick, H. A., 1980, Time-course of photoadaptation in the photosynthesis-irradiance relationship of a dinoflagellate exhibiting photosynthetic periodicity, *Mar. Biol.* **58**:85–96.

Prezelin, B. B., and Sweeney, B. M., 1978, Photoadaptation of photosynthesis in *Gonyaulax polyedra, Mar. Biol.* **48**:27–35.

Prezelin, B. B., Ley, A. C., and Haxo, F. T., 1976, Effects of growth irradiance on photosynthetic action spectra of the marine dinoflagellate, *Glenodinium* sp., *Planta* **130**:251–256.

Provasoli, L., 1958, Nutrition and ecology of protozoa and algae, *Annu. Rev. Microbiol.* **12**:279–308.

Raven, J. A., 1974, Carbon dioxide fixation, in: *Algal Physiology and Biochemistry* (W. D. P. Stewart, ed.), pp. 434–455, University of California Press, Berkeley.

Redfield, A. C., 1958, The biological control of chemical factors in the environment, *Am. Sci.* **46**:205–221.

Reshkin, S. J., and Knauer, G. A., 1979, Light stimulation of phosphate uptake in natural assemblages of phytoplankton, *Limnol. Oceanogr.* **24**:1121–1124.

Reynolds, J. H., Middlebrook, E. J., Porcella, D. B., and Grenney, W. J., 1975, Effects of temperature on growth constants of *Selenastrum capricornutum, J. Water Pollut. Cont. Fed.* **47**:2420–2436.

Rhee, G-Y., 1972, Competition between an alga and an aquatic bacterium for phosphate, *Limnol. Oceanogr.* **17**:505–514.

Rhee, G-Y., 1973, A continuous culture study of phosphate uptake, growth rate and polyphosphate in *Scenedesmus* sp., *J. Phycol.* **9**:495–506.

Rhee, G-Y., 1974, Phosphate uptake under nitrate limitation by *Scenedesmus* sp. and its ecological implications, *J. Phycol.* **10**:470–475.

Rhee, G-Y., 1978, Effects of N:P atomic ratios and nitrate limitation on algal growth, cell composition and nitrate uptake, *Limnol. Oceanogr.* **23**:10–25.

Rhee, G-Y., 1980, Continuous culture in phytoplankton ecology, in: *Advances in Aquatic Microbiology*, Vol. 2 (M. R. Droop and H. W. Jannasch, eds.), pp. 151–203, Academic Press, New York.

Rhee, G-Y., and Gotham, I. J., 1980, Optimum N:P ratios and coexistence in planktonic algae, *J. Phycol.* **16**:486–489.

Rhee, G-Y., and Gotham, I. J., 1981a, The effects of environmental factors on phytoplankton growth: light and the interactions of light with nitrate limitation, *Limnol. Oceanogr.* **26**:649–659.

Rhee, G-Y., and Gotham, I. J., 1981b, The effect of environmental factors on phytoplankton growth: temperature and the interactions of temperature with nutrient limitation, *Limnol. Oceanogr.* **26**:635–648.

Rhee, G-Y., Gotham, I. J., and Chisholm, S. W., 1981, Use of cyclostat cultures to study phytoplankton ecology, in: *Continuous Culture of Cells,* Vol. 2 (P. C. Calcott, ed.), pp. 159–186, CRC Press, Cleveland.

Rigby, L. H., Craig, S. R., and Budd, K., 1980, Phosphate uptake by *Synechococcus leopoliensis* (Cyanophyceae): enhancement by calcium ion, *J. Phycol.* **16**:389–393.

Rodhe, W., 1978, Algae in culture and nature, *Mitt. Int. Verein. Limnol.* **21**:7–20.

Ruyters, G., 1980, Blue light-enhanced phosphoenolpyruvate carboxylase activity in a chlorophyll free *Chlorella* mutant, *Z. Pflanzenphysiol.* **100**:107–112.

Ryther, J., and Dunstan, W. M., 1971, Nitrogen, phosphorus, and eutrophication in the coastal marine environment, *Science* **171**:1008–1013.

Sakshaug, E., 1978, The influence of environmental factors on the chemical composition of cultivated and natural populations of marine phytoplankton, Ph.D. thesis, University of Trondheim, Norway.

Sakshaug, E., and Holm-Hansen, O., 1977, Chemical composition of *Skeletonema costatum* (Grev.) Cleve and *Pavlova (Monochyrsis) lutheri* (Droop) Green as a function of nitrate-, phosphate-, and iron-limited growth, *J. Exp. Mar. Biol. Ecol.* **29**:1–34.

Schindler, D. W., 1977, Evolution of phosphorus limitation in lakes, *Science* **195**:260–262.

Senft, W. H., 1978, Dependence of light-saturated rates of algal photosynthesis on intracellular concentrations of phosphorus, *Limnol. Oceanogr.* **23**:709–718.

Senger, H., and Fleischhacker, Ph., 1978a, Adaptation of the photosynthetic apparatus of *Sce-*

nedesmus obliquus to strong and weak light conditions. I. Differences in pigments, photosynthetic capacity, quantum yield and dark reactions, *Physiol. Plant.* **43**:35–42.

Senger, H., and Fleischhacker, Ph., 1978b, Adaptation of the photosynthetic apparatus of *Scenedesmus obliquus* to strong and weak light conditions. II. Differences in photochemical reactions, the photosynthetic electron transport and photosynthetic units, *Physiol. Plant.* **43**:43–51.

Shaw, M. K., 1967, Effects of abrupt temperature shift on the growth of mesophilic and psychrophilic yeasts, *J. Bacteriol.* **93**:1332–1336.

Shelef, G., Oswald, W. J., and Golenke, C. C., 1970, Assaying algal growth with respect to nitrate concentration by a continuous flow turbidostat, in: *Proceedings of 5th International Conference on Water Pollution Research* (S. H. Jenkins, ed.), pp. III25/1–III25/9, Pergamon Press, Oxford.

Sherman, L. A., 1978, Differences in photosynthesis-associated properties of the blue-green alga *Synechococcus cedrorum* grown at 30 and 40 C, *J. Phycol.* **14**:427–433.

Skoglund, L., and Jensen, A., 1976, Studies of N-limited growth in dialysis culture, *J. Exp. Mar. Biol. Ecol.* **21**:169–178.

Smayda, T. J., 1969, Experimental observations on the influence of temperature, light, and salinity on cell division of the marine diatom, *Detonula confervacea* (Cleve) Gran., *J. Phycol.* **5**:150–157.

Smayda, T. J., 1974, Bioassay of the growth potential of the surface water of lower Narragansett Bay over an annual cycle using the diatom *Thalassiosira pseudonana* (oceanic clone, 13-1) *Limnol. Oceanogr.* **19**:889–901.

Smith, R. C., 1968, The optical characterization of natural waters by means of an extinction coefficient, *Limnol. Oceanogr.* **13**:425–429.

Soeder, C. J., and Stengel, E., 1974, Physico-chemical factors affecting metabolism and growth rate, in: *Algal Physiology and Biochemistry* (W. D. P. Stewart, ed.), pp. 714–740, University of California Press, Berkeley.

Sorokin, C., 1960, Kinetic studies of temperature effects on the cellular level, *Biochim. Biophys. Acta* **38**:197–204.

Stanier, R. Y., Kunisawa, R., Mandel, M., and Cohen-Bazier, G., 1971, Purification and properties of unicellular blue-green algae (Order Chroococcales), *Bacteriol. Rev.* **35**:171–205.

Steeman-Nielsen, E., 1978, Growth of plankton algae as a function of N-concentration, measured by means of a batch technique, *Mar. Biol.* **46**:185–189.

Steeman-Nielsen, E., and Jørgensen, E. G., 1968a, The adaptation of plankton algae. I. General part, *Physiol. Plant.* **21**:401–413.

Steeman-Nielsen, E., and Jørgensen, E. G., 1968b, The adaptation of plankton algae. III. With special consideration of importance in nature, *Physiol. Plant.* **21**:647–654.

Swift, D. G., and Taylor, W. R., 1974, Growth of vitamin B_{12}-limited cultures: *Thalassiosira pseudonana, Monochrysis lutheri* and *Isochrysis galbana, J. Physiol.* **10**:385–391.

Swift, E., and Meunier, V., 1976, Effects of light intensity on division rates, stimulable bioluminescence and cell size of the oceanic dinoflagellates *Dissodinium lunula, Pyrocystis fusiformis* and *P. noctiluca, J. Phycol.* **12**:14–22.

Syrett, P. J., 1962, Nitrogen assimilation, in: *Physiology and Biochemistry of Algae* (R. A. Lewin, ed.), Academic Press, New York.

Takahashi, M., Fujii, K., and Parsons, T. R., 1973, Simulation study of phytoplankton photosynthesis and growth in the Frazer River estuary, *Mar. Biol.* **19**:102–116.

Talling, J. F., 1961, Photosynthesis under natural conditions, *Annu. Rev. Plant Physiol.* **12**:133–154.

Talling, J. F., 1979, Factor interactions for the prediction of lake metabolism, *Arch. Hydrobiol. Beih. Ergebn. Limnol.* **13**:96–109.

Tamiya, H., Sasa, T., Nihei, T., and Ishibashi, S., 1955, Effects of variation in daylength, day and night temperatures, and intensity of daylight upon the growth of *Chlorella, J. Gen. Appl. Microbiol.* **4:**298–307.

Tansey, M. R., and Brock, T. D., 1978, Microbial life at high temperatures: ecological aspects, in: *Microbial Life in Extreme Environments* (D. J. Kushner, ed.), pp. 159–216, Academic Press, New York.

Terlizzi, D. E., and Karlanda, E. P., 1980, Growth of a coccoid nanoplankter (Eustigmatophyceae) from the Chesapeake Bay as influenced by light, temperature, salinity, and nitrogen source in factorial combination, *J. Phycol.* **16:**364–368.

Terborgh, J., and Thiman, K. V., 1964, Interaction between daylength and intensity in growth and chlorophyll content of *Acetabularia crenulata, Planta* **63:**83–98.

Thomas, E. A., 1969, The process of eutrophication in central European lakes, in: *Eutrophication: Causes, Consequences, Correctives,* pp. 29–49, Natl. Acad. Sci./Natl. Res. Council, Publ. 1700.

Thomas, R. J., Hipkin, C. R., and Syrett, P. J., 1976, The interaction of nitrogen assimilation with photosynthesis in nitrogen deficient cells of *Chlorella, Planta.* **133:**9–13.

Thomas, W. H., and Dodson, A. N., 1972, On nitrogen deficiency in tropical Pacific Ocean phytoplankton. II. Photosynthetic and cellular characteristics of a chemostat-grown diatom, *Limnol. Oceanogr.* **17:**515–523.

Thomas, W. H., and Dodson, A. N., 1974, Effects of interactions between temperature and nitrate supply on the cell division rates of two marine phytoflagellates, *Mar. Biol.* **24:**213–217.

Tolbert, N. E., 1974, Photorespiration, in: *Algal Physiology and Biochemistry* (W. D. P. Stewart, ed.), pp. 474–504, University of California Press, Berkeley.

Tomas, C. R., 1980, *Olisthodiscus luteus* (Chrysophyceae). IV. Effects of light intensity and temperature on photosynthesis, and cellular composition, *J. Phycol.* **16:**149–156.

Van Liere, L., 1979, On *Oscillatoria agardhii* Gomont: experimental ecology and physiology of nuisance bloom-forming cyanobacterium, Ph.D. thesis, University of Amsterdam, The Netherlands.

Van Liere, L., and Mur, L. R., 1980, Occurrence of *Oscillatoria agardhii* and some related species, in: *Hypertrophic Ecosystems* (J. Barica and L. R. Mur, eds.), pp. 67–77, Dr. W. Junk bv. Publishers, The Hague.

Vincent, W. F., 1979, Mechanisms of rapid photosynthetic adaptation in natural phytoplankton communities. I. Redistribution of excitation energy between photosystems I and II, *J. Phycol.* **15:**429–434.

Vincent, W. F., 1980, Mechanisms of rapid photosynthetic adaptation in natural phytoplankton communities. II. Changes in photochemical capacity as measured by DCMU-induced chlorophyll fluorescence, *J. Phycol.* **16:**568–577.

Vincent, W. F., 1981, Photosynthetic capacity measured by DCMU-induced chlorophyll fluorescence in an oligotrophic lake, *Freshwater Biol.* **11:**61–78.

Wallen, D. G., and Geen, G. H., 1971, Light quality in relation to growth, photosynthetic rates, and carbon metabolism in two species of marine plankton algae, *Mar. Biol.* **10:**34–43.

Whitaker, T. M., and Richardson, M. G., 1980, Morphology and chemical composition of a natural population of an ice-associated Antarctic diatom *Navicula glaciei, J. Phycol.* **16:**250–257.

Wynne, D., and Berman, T., 1980, Hot water extractable phosphorus—an indicator of nutritional status of *Peridinium cinctum* (Dinophyceae) from Lake Kenneret (Israel), *J. Phycol.* **16:**40–46.

Yontsch, C. S., and Lee, R. W., 1966, A study of photosynthetic light reactions, and a new interpretation of sun and shade phytoplankton, *J. Mar. Sci.* **24:**319–337.

Yoder, J. A., 1979, Effect of temperature on light-limited growth and chemical composition of *Skeletonema costatum* (Bacillariophycea), *J. Phycol.* **15:**362–370.

Young, T. C., and King, D. L., 1980, Interacting limits to algal growth: light, phosphorus, and carbon dioxide availability, *Water Res.* **14**:409–412.

Zevenboom, W., 1980, Growth and nutrient uptake kinetics of *Oscillatoria agardhii,* Ph.D. thesis, University of Amsterdam, The Netherlands.

Zevenboom, W., and Mur, L. R., 1980, N_2-fixing cyanobacteria: why they do not become dominant in Dutch, hypertrophic lakes, in: *Hypertrophic Ecosystems* (J. Barica and L. R. Mur, eds.), pp. 123–130, Dr. W. Junk bv. Publishers, The Hague.

3

Factors Limiting Productivity of Freshwater Ecosystems

HANS W. PAERL

1. A Perspective on Freshwater Ecosystems

In considering environmental factors potentially limiting microbial productivity in fresh water, the spectrum of physical, chemical, and resulting biological properties distinguishing individual ecosystems must be realized and identified. The physical individuality of an aquatic ecosystem is a product of its geological origin and its geophysical (climatic and morphological) conditioning. Specific geochemical events, such as tectonic activity, volcanism, and weathering, have profound impacts on shaping the chemical environment (Hutchinson, 1957; Golterman, 1975a). Biological activity, particularly the redox transformations of growth-limiting elements, additionally modifies the chemical environment (Golterman, 1975a; Wetzel, 1975). Interactions of the general processes outlined above are largely responsible for shaping the biotic environment, including constraints on its productivity and diversity.

The ability of a microorganism to establish itself and proliferate depends on its genetic potential to grow and reproduce under given environmental constraints. Normally, a complex set of potentially limiting environmental factors dictates the success or failure of a microorganism to establish itself. The minimal set of physical, chemical, and biotic factors necessary for growth and reproduction constitutes a niche. Hutchinson (1961) has spatially described a niche as a hypervolume, within which sets of required growth limiting factors must overlap in order to assure the continued existence of a species.

HANS W. PAERL ● Institute of Marine Sciences, University of North Carolina, Morehead City, North Carolina 28557.

The hypervolume is a conceptual means of understanding and identifying environmental factors limiting microbial productivity in fresh water. A great number of microbial species exhibit rather narrow tolerances in terms of physical and chemical parameters which regulate growth. Specific oxygen and nutrient regimes are examples of such parameters. On a macroscale, these parameters often appear homogeneous. For example, in physically mixed water columns, oxygen and nutrient levels often appear uniform from top to bottom. Despite such apparent physical and chemical uniformity, a high diversity of microorganisms, all exhibiting distinctly different growth requirements, can exist. How is this possible?

Two qualities of aquatic environments need to be realized in order to rectify this paradox. One is that, despite the presence of apparent uniformity of environmental parameters in space, such uniformity over a short period of time (hours to days) is often short lived. Variable temperature and light regimes due to climatic changes, fluctuations in chemical regimes due to periods of precipitation and runoff events, and nutrient utilization/mineralization by the biota itself can all contribute to rapid changes among growth-limiting factors in aquatic environments over short (hours to days) time intervals. Hence, even though environmental parameters may appear uniform in space during any given period, they may be rapidly altered over time, still remaining uniform at each time interval within a given space. These dynamic characteristics generate ecologically diverse environments over time, each best suited to specific sets of microorganisms exhibiting optimal growth rates within the spectrum of environmental extremes.

Secondly, within the apparent macrohomogeneity of each environment, for example, a water column, a vast number of microenvironments exist (Hutchinson, 1961; Richerson et al., 1970). Because physical and chemical monitoring techniques are designed to examine ml to liter volumes, the capacity for determining nonhomogeneity or "patchiness" among growth-limiting variables is often limited. Microscopic examinations of natural waters can reveal suspended particles, microbial colonies, and surfaces (inorganic, plant, and animal) exhibiting physical-chemical properties distinct from more homogeneous large scale water samples (Melchiorri-Santolini and Hopton, 1972). The combination of such surface properties leads to a microenvironment, where specific biological redox reactions and the microflora suitable for conducting these reactions are located (Oláh, 1972; Weibe and Pomeroy, 1972; Paerl, 1974, 1975; Bitton and Marshall, 1980). Considering oxygen alone, microenvironments exist which exhibit highly reduced conditions attributable to respiratory activities of concentrated (by attachment) populations of microorganisms (Paerl, 1979). Such microenvironments could promote growth and nutrient transforming processes characteristic of specific groups of microorganisms requiring reduced conditions. Examples are the denitrifying, N_2 fixing, methanogenic, and sulfur reducing bacteria (Alexander, 1971). Deployment of macroenvironmental monitoring techniques would fail to detect such specific microbial habitats and hence overlook the potential for harboring a high diversity of species in apparently "homogeneous" environments.

In terms of the macroenvironment, exclusion of specific groups of microorganisms may occur due to physical, physicochemical, chemical, and biological constraints. Physical constraints include: light availability (quality and quantity), temperature extremes, turbulence, and physical removal of microorganisms due to flushing of water bodies. Physico-chemical constraints often form profound barriers to microbial establishment. Such constraints include; salinity and conductivity gradients, pH and oxygen regimes, and geothermal and geochemical venting. Chemical constraints are commonly encountered; particularly, nutrient limitation and toxic substance inhibition of growth. The establishment of biological constraints is the product of physical-chemical conditions promoting biological interactions, which themselves may limit the growth and reproduction of microorganisms. Animal grazing and predation is one form. Antagonism, either by direct or chemical means, as well as mutualism and symbiosis, are additional means by which growth can be constrained or promoted.

2. The Concept of Limiting Factors

The above constraints can all be considered as growth limiting factors; that is, features of the physical, chemical, and biological environment which limit the establishment and proliferation of specific microorganisms. The first systematic approach to determining which environmental parameters controlled production processes was Justus von Leibig's "Law of the minimum" (Liebig, 1840). This law stated that, given all the factors required for growth of a population or community of organisms, the factor in lowest supply (or exhausted most rapidly) would be the factor limiting the growth of the organisms under consideration. This law has served as the basis for interpreting a wide variety of potentially growth-limiting factors, ranging from growth restricting physical factors (light, temperature, salinity, pH) to nutrient limitation (Blackman, 1905). Batch bioassays and chemostat techniques have relied heavily on the concept of a single most limiting factor regulating microbial growth at any given time. In practice, straightforward interpretations and applications of Leibig's law of the minimum have met with mixed success (Blackman, 1905; Ström, 1933; Ryther and Guillard, 1959; Goldman, 1964; Gerhart and Likens, 1975). In certain freshwater ecosystems having received accelerated nutrient supplies, the absence of a single growth-requiring nutrient or the presence of a deficiency in a specific physical factor (light, temperature, pH) may prove to be obvious and can be readily shown using bioassay or chemostat techniques (Pleisch, 1970; Edmondson, 1972; Schelske and

Stoermer, 1972). In ecosystems deficient in a wide variety of important growth factors, however, the addition or elevation of several factors may be required to stimulate growth (Goldman and Carter, 1965; Goldman, 1966; Thurlow *et al.*, 1975). Oligotrophic freshwater systems often show multiple nutrient deficiencies, presumably because ambient levels of several essential nutrients fall short of the minimum amounts required for adequate growth (Goldman, 1972; Shapiro and Glass, 1975). This is presumably the explanation for the presence of oligotrophic conditions in the first place. Even in these systems, there must exist a single nutrient which is most severely deficient. In other words, if we supplemented multiple nutrient deficient waters with infinitely small increments of each required nutrient (singly), we should find that when the *most* limiting nutrient is added some growth should result. The extent of resultant growth may be quite small since another nutrient would presumably be close to limiting. Once an additional nutrient is supplied (in addition to the most limiting nutrient), another growth increment would result. If other nutrients proved to exist at near growth-limiting concentrations, a similar response would occur following their additions, and so forth. If, by chance, two or more nutrients were equally depleted, no growth response would occur until both nutrients were supplied. Considering the variety in both gradients and levels of nutrients commonly encountered, as well as the diversity of individual nutrient requirements exhibited by microorganisms (Golterman, 1975a; Wetzel, 1975), the chances of equal nutrient limitation occurring among several nutrients would appear to be very small.

Unfortunately, the relative degree of nutrient limitation among several potentially limiting nutrients can be exceedingly difficult to determine in bioassays or chemostat assays. Often differences in nutrient depletion among a variety of nutrients are simply too small to detect. Furthermore, in stepwise nutrient addition experiments, physiological responses of microorganisms under investigation may often be altered once nutrient limitation by the most limiting nutrient has been relieved, making interpretations of growth limitation by a second (or more) nutrient(s) more complicated. For these reasons, as well as others yet to be clarified, combined nutrient additions (added simultaneously) often yield more interpretable results than stepwise nutrient additions in demonstrating multiple nutrient limitation.

In natural systems, several physical, chemical, and biological factors can often interact to control and limit microbial growth. For example, nutrient depleted surface waters often limit the degree of phytoplankton growth, forcing a bulk of the algal cells to deeper, nutrient-rich waters (Kiefer *et al.*, 1972; Paerl *et al.*, 1976). At greater depths, however, access to photosynthetically active radiation (PAR) may become a growth-limiting factor. A depth in the water column that assures nutrient availability as well as adequate radiation for optimal photosynthetic growth would be desirable; it has been repeatedly shown, however, that optimal photosynthetic growth (in terms of light availability) is often sacrificed in order to assure adequate ambient nutrient supplies (Talling, 1961; Tilzer *et al.*, 1975, 1977). Accordingly, phytoplankton are often located in deep nutrient-rich waters where photosynthetically active radiation (PAR) is a growth-limiting factor (Kiefer *et al.*, 1972; Paerl *et al.*, 1976).

In estuarine, as well as hypersaline environments a similar tradeoff, in terms of satisfying interacting limiting factors, can exist. Salinity and nutrient gradients often overlap each other. Situations can arise where suboptimal salinity conditions may be present in regions where nutrient requirements can be met and *vice versa*. Such conditions may persist for days or months or only a few hours per day in the case of salt wedges intruding into low-salinity waters during tidal cycles (Remaine and Schlieper, 1971).

Biological interactions may also complicate otherwise straightforward interpretations of limiting factors. Heavy grazing or predation by protozoans, invertebrates, and fish may occur in pelagic, benthic, epiphytic, or epilithic regions of natural waters having nutrient sufficiency.

3. Geological, Physical, Chemical, and Biological Characteristics of Freshwater Ecosystems in Relation to Nutrient Limitation

On a global scale, the past two decades have seen a flurry of nutrient limitation studies. A wide variety of aquatic ecosystems have been examined, including streams and rivers (Wuhrmann and Eichenberger, 1975; Whitton, 1975; Golterman, 1975b), estuarine and intertidal systems (Williams, 1972; Thayer, 1974), reservoirs (Nusch, 1975; Toerien and Steyn, 1975), small lakes in forested zones (Goldman, 1960; Gerhart and Likens, 1975), alpine and boreal lakes (Goldman, 1960, 1964), agricultural lakes (Jones and Bachmann, 1975), large navigable lakes, including the North American Great Lakes (Beeton, 1969; Shapiro and Glass, 1970; Schelske and Stoermer, 1972; Burns and Ross, 1972), tropical lakes (Melack, 1981), and lakes present in areas of recent geological activity (White *et al.*, 1982). From these studies, it has become evident that nutrient limitation patterns show distinct differences between various geographic regions as well as between diverse geological histories of freshwater basins and catchments. The results of such diverse conditions have fostered disagreement, as brought forth in the literature (Likens, 1972), concerning the order of importance related to specific nutrient input controls and removals in freshwater ecosystems.

In order to ascertain that nutrient deficiencies, and hence limitations, vary among ecosystems and that the spectrum of conclusions on specific limiting nutrients is valid, it would be useful to consider the diverse geological and geographical locations of lakes and rivers examined thus far. There is agreement that geological origins of rivers and lake basins play an important role in determining the relative abundance of mineral nutrients eroded and leached into aquatic ecosystems either through surface runoff or groundwater transport (Golterman, 1975a). An example of such diversity can be found in comparing granitic with metamorphic or sedimentary lake and catchment basins. Very generally speaking, granitic soils yield a relatively greater abundance of phosphates and silicates but often prove to be poor sources of nitrates and nitrites (Holland, 1978). Conversely, metamorphic and sedimentary soils tend to have a higher abundance of nitrogenous nutrients (Holland, 1978). As a result, lake basins located in granitic and particularly mountainous regions may have a tendency to be nitrogen limited. Phosphorus limitation is more common in lakes and rivers present in sedimentary and alluvial catchments.

The relationships between geological origins of basins and nutrient limitations are far from categorical, however. Catchment size, precipitation patterns, evaporation, and land use are integral factors dictating specific trends in nutrient limitation. Small catchments, particularly those supporting large rivers and lakes, tend to receive less nutrient input via precipitation than through other sources such as erosion, sedimentation, leaching action of groundwater, and man-made nutrient sources. As a result, yearly water replenishment by precipitation and runoff is only a small percentage of the volume of the receiving water body. Accordingly, the retention time, which is the average time it takes for the entire water volume to be replenished, is long. It follows that nutrient enrichment by precipitation is small on a yearly scale as well. Precipitation normally carries a much higher nitrogen than phosphorus load and, therefore, is considered to be a more significant source of nitrogen than phosphorus (Hutchinson, 1957). Phosphorus inputs are more often dictated by leaching of rocks and soils (Kolenbrander, 1972), groundwater inputs, and biological cycling (Golterman, 1975a). Considering these various major nutrient sources, one would anticipate that a lake having a small catchment area in proportion to its water volume and a relatively long retention time would tend to be nitrogen limited. Lake Tahoe, California and Nevada (Goldman and Armstrong, 1968) and Crater Lake, Oregon (A. J. Horne, personal communication), as well as a variety of alpine lakes having small catchment area to lake volume ratios, fit this prediction remarkably well. There are numerous exceptions to this trend, however, which at times can be attributed to a variety of alternative nutrient sources. Lakes and rivers in geothermal and volcanic regions can often contain unique proportions of nutrients, vastly different, in terms of ratios, from what might be expected in specific geographical regions. Man's activities can affect patterns of nutrient limitation distinct from the norm in certain localities. Sewage and industrial waste input, as well as agricultural runoff, can lead to a high diversity of nutrient limiting patterns in any geographic locale. Lastly, the microbial communities of lakes can affect the ratios of growth-limiting nutrients. In particular, the presence of nitrogen fixing microorganisms has a profound influence on potential nitrogen limitation since nitrogen fixing cyanobacteria and a variety of heterotrophic, chemolithotrophic, and photolithotrophic (non-O_2 evolving) bacteria can make direct use of atmospheric nitrogen dissolved in water. A tendency towards nitrogen limitation would not apply to such planktonic and benthic microorganisms (Fogg et al., 1973; Newton et al., 1977). Zooplankton and fish also may aid in alleviating both nitrogen and phosphorus limitation through the horizontal and vertical transportation and excretion of these nutrients throughout portions of the water column (Williams, 1974; Lehman, 1980).

The physical features of lakes, including thermal stratification, the formation of density layers, and turbulent events, must be considered as well in assessing potential availability, through circulation, of growth-limiting nutrients.

Phosphorus has been extensively implicated as a key limiting nutrient in a wide variety of geographical and geological regions (Griffith et al., 1973). Abundant rainfall in a large catchment can lead to adequate nitrogen requirements in receiving waters. Furthermore, nitrogen fixing microorganisms can circumvent nitrogen limitation. In agricultural catchments, nitrogen fertilizer applications can prove highly mobile (McColl et al., 1975), either through surface runoff or groundwater inputs. Terrestrial nitrogen fixation could potentially constitute another source of biologically utilizable nitrogen once products of this process are transported into aquatic environments. Singularly or combined, these processes promote phosphorus limitation. A vast body of literature can be utilized to point out the diverse causes and manifestations of phosphorus limitation (see Golterman, 1975a). Striking relationships between average chlorophyll a concentrations and phosphorus loading throughout a large range of lakes of differing trophic states (Vollenweider, 1968; Patalas, 1972; Dillon and Rigler, 1974) provide strong evidence that phosphorus can limit freshwater primary productivity, albeit not exclusively.

Examples of phosphorus sensitivities of freshwater ecosystems are present in a variety of localities. Foremost is Edmondson's (1977) documentation of excessive phosphorus inputs (sewerage) leading to the accelerated eutrophication of Lake Washington, near Seattle, Washington. A long list of studies yielding similar conclusions is still growing in support of phosphorus limitation. Among these is the implication of phosphorus as the most important nutrient stimulating primary productivity of Lake Erie (Beeton, 1969; Burns and Ross, 1972) and the subsequent reversal of this lake's eutrophicating trends through phosphorus removal from industrial effluent and improved sewage treatment procedures. Similarly, phosphorus has been implicated as the nutrient controlling the eutrophication of Lake Michigan (Schelske and Stoermer, 1972). The studies of Schindler et al (1973) on the role of phosphorus in the eutrophication of experimental Canadian Shield lakes have been instrumental in pointing out the dangers associated with excessive phosphorus loading of our inland waters. Vollenweider (1968), Shapiro and Glass (1970), Rigler (1973), Lean (1973), and Dillon and Rigler (1974) have all demonstrated the strong link between accelerated phosphorus input and eutrophicating trends in freshwater. In many lakes showing advanced eutrophication in the form of N_2 fixing cyanobacterial blooms, decreased phosphorus inputs or phosphorus removal are the only remedies remaining in efforts to reverse such deterioration (Cronberg et al., 1975).

Research in the areas of nutrient sensitivities of natural waters in specific geographical locales has paid off, both in terms of avoiding environmental degradation and formulating environmentally compatible products. Detergents, which were initially identified as major constituents of municipal and industrial phosphorus loading (Griffith et al., 1973), have through federal, state, and local pressures been reformulated to substitute phosphorus with nonphosphorus cleaning agents and surfactants. The results of eliminating phosphorus from detergents, as well as related cleansers, have been beneficial. Recent studies have revealed dramatic decreases and in some cases reversals of nuisance blooms and other indicators of eutrophication (high biological oxygen demand, for example) since such phosphorus sources have decreased (Thomas, 1973; Edmondson, 1977; Anderson et al., 1973).

One nutrient which bears geological and geographical relevance, but whose biostimulatory effects remain the subject of disagreement, is inorganic carbon (Likens, 1972). It is well known that dissolved inorganic carbon (DIC) levels can vary dramatically between geographic regions or even locally. Furthermore, depending on the ambient pH, the proportions of CO_2, HCO_3^-, and CO_3^{2-} making up the DIC pool vary (Hutchinson, 1957). In the late 1960s and early 1970s the possibility was raised that DIC might potentially limit productivity, especially in waters having excess phosphorus and nitrogen levels (Kuentzel, 1969; Lange, 1970; Kerr et al., 1972). The possibility of specific forms of DIC being preferred for photosynthetic growth was also raised (Morton et al., 1972). The question of DIC, and specifically CO_2, limitation of primary productivity will be discussed elsewhere.

Of the mineral nutrients, silicon is seldom considered as a growth-limiting nutrient. In most aquatic ecosystems, levels of the biologically utilizable form of silicon, silicate (SiO_2), exceed the 2 ppm level considered adequate for phytoplankton nutrition and growth (Lund, 1950). However, care must be taken in assuming adequate SiO_2 levels in all lakes and rivers, for a substantial and growing list of SiO_2 limited conditions have been documented. Silicon limitation is an important factor dictating algal community composition and primary productivity in waters supporting periodic diatom blooms. Silicon frustules, containing the cellular protoplasm in diatoms, form a substantial bulk of diatom biomass; accordingly, it is crucial for adequate SiO_2 supplies to be present in order to assure proper frustule development and concomitant diatom growth. A classic example of silicon controlled productivity was revealed in Lake Windermere, England, by Lund (1950) and Lund et al. (1963). In these studies, a strong inverse relationship was found between SiO_2 concentrations and blooms of the dominant pennate diatom Asterionella formosa Hass.; as diatom blooms commenced during spring months, SiO_2 levels dropped to 0.5–1 ppm levels. When SiO_2 supplies become exhausted, the *Asterionella* bloom showed a dramatic decline and was replaced by low silicon-requiring green and cyanobacterial populations. Following this initial study, silicon depletion was shown to affect phytoplankton community composition elsewhere. Notable are studies by Kilham (1971), Schelske and Stoermer (1971), Tilman and Kilham (1976) and Tilman *et al.* (1981), all demonstrating that silicon depletion in lakes leads to periodic declines in diatom populations having high silicon requirements. In their studies on eutrophicating trends in Lake Michigan, Schelske and Stoermer (1971, 1972) demonstrated that when adequate amounts of phosphorus and nitrogen were present in lake water, SiO_2 proved to be a limiting nutrient. Although phosphorus can generally be considered to be the most important limiting nutrient in Lake Michigan, late summer depletion of SiO_2 (when phosphorus supplies were still adequate) had a profound effect on shaping algal community structure, bringing about replacement of diatom with green algal and cyanobacterial species.

In subsequent work, Tilman *et al.* (1981) have shown that among diatom species competing at growth-limiting SiO_2 concentrations, the species having the lowest requirements will dominate. The SiO_2 requirements exhibited by competing species appear temperature independent for temperatures within or below optimal growth temperatures for the species, but become temperature reliant at temperatures above the optimal ranges. These results point out that care must be taken to interpret silicon, as well as other nutrient, competition conditions in natural waters, taking into consideration optimal growth temperatures of the species involved.

The abundance of minor or trace elements is directly related to parent rock types in lake and river basins (Golterman, 1975a). It has been established that aquatic primary producers have requirements for a wide variety of minor elements, including iron (Arnon, 1958; Menzel and Ryther, 1961; Schelske, 1962; Goldman, 1964), manganese (Hopkins, 1930; Arnon, 1958), cobalt (Holm-Hansen *et al.*, 1954; Benoit, 1957), copper (Walker, 1953; A. J. Horne and N. J. Williams, personal communications; H. W. Paerl *et al.*, unpublished data), zinc (Goldman, 1964); molybdenum (Bortels, 1940; Goldman, 1960), boron (Eyster, 1952, 1968; Wetzel, 1964), and vanadium (Arnon and Wessel, 1953). The relative abundance of such trace elements in catchment and basin rocks and soils varies dramatically on a geographical scale (Holland, 1978). In many instances, even though ratios of these elements vary, adequate amounts of trace elements necessary to complement optimal growth exist (Golterman, 1975a, b; Wetzel, 1975). Hence, nutrients or environmental factors other than trace elements will limit growth under such circumstances.

There are cases, however, where a chronic lack of specific trace elements does limit the growth of primary producers. Of all trace elements, iron deficiencies appear most common (Menzel and Ryther, 1961; Schelske, 1962;

Goldman, 1964; Goldman and Carter, 1965; Thurlow *et al.*, 1975). This is due
to the diverse metabolic demands for iron, including production and functioning of cytochromes, ferredoxin, flavoproteins, nitrogenase, nitrate reductase,
and chlorophyll synthesis. Because iron, as well as the related trace element
manganese, readily precipitate under oxidized conditions, these elements are
often not present in soluble forms, making direct biological utilization difficult
(Golterman, 1975a). Coupled with the fact that iron can be chelated by a variety of organic ligands, particularly colored humic substances (Shapiro, 1964,
1969), as well as coprecipitated with phosphates and carbonates (Tsusue and
Holland, 1966), and that total iron content of natural waters is commonly less
than 10 ppb, biological availability is often severely limited. Certain groups of
microorganisms, most notably cyanobacteria, are known to excrete powerful
hydroxymate chelators capable of sequestering iron when ambient levels may
become limiting to growth (Neilands, 1974; Murphy *et al.*, 1976).

A variety of other trace elements have exhibited growth-limiting characteristics, in the use of either large enclosure (limnocorrals, polyethylene columns) or bioassay experiments. Goldman (1960) observed that a molybdenum
deficiency in Castle Lake, California, limited primary production. Following
this conclusion, Goldman (1964) tested various North American and New
Zealand lakes, to discover that a wide array of trace elements controlled the
magnitude of primary productivity. Copper (Cu^{2+}), although toxic at levels
exceeding a few ppb (Steemann Nielsen and Wium-Andersen, 1971), has
recently been found to stimulate both photosynthetic CO_2 fixation and nitrogen
fixation at levels not exceeding 3 ppb among natural cyanobacterial populations (A. J. Horne and N. J. Williams, personal communication; H. W. Paerl
et al., unpublished data). It is known that copper is essential for cytochrome *a*
and plastocyanin (photosystem I) activity (Golterman, 1975a). Since both
nitrogen and CO_2 fixation are dependent on photosystem I activity in cyanobacteria, the combined copper demands of these processes may, at times,
exceed the natural availability of Cu^{2+}, especially during bloom periods. It
remains unclear how Cu^{2+} availability is regulated in natural waters, since
numerous natural chelators of this element are normally present in lakes and
rivers (Sunda and Hanson, 1978). In particular, humic substances derived
from soil leachate and from decomposition of plant materials are capable of
chelating copper and other trace elements lowering their biological availability
in the free ionic form (Sunda and Hanson, 1978). In the Chowan River, North
Carolina, virtually all the measurable copper is chelated by humic substances
leaching from the Piedmont forest and as pulp mill effluent (W. A. Sunda and
H. W. Paerl, unpublished data). In this same river, 1–3 ppb Cu^{2+} additions
stimulated CO_2 and nitrogen fixation, an indication that the free ionic form
was unavailable due to chelation. Clearly, more work is required to clarify and
distinguish between Cu^{2+} depletion or a lack of availability due to chelation.

Recent work has emphasized the potential for productivity limitation by a wide variety of trace elements, depending on geographic locale. Lindstrom (1980) has obtained evidence for selenium playing a role in algal growth limitation in a Swedish lake. Manganese is suspected of limiting primary productivity in the epilimnia of a variety of New Zealand lakes (Goldman, 1964; W. F. Vincent, unpublished data) and Lake Superior (Shapiro and Glass, 1975). Deficiencies of such micronutrients are often difficult to substantiate in natural environments because (1) natural levels of micronutrients of interest often approach analytical limits of detection, (2) chemical forms of micronutrients being added in bioassay experiments often contain contaminants in the form of other potentially limiting micronutrients, and (3) chelation or precipitation may occur in nutrient limitation assays, thereby masking a biological response.

Synergistic effects of macro- and micronutrients in natural ecosystems are an important factor in the assessment of the nutrient limitation. As mentioned in the discussion of Liebig's Law of the Minimum and the evaluation of the limiting nutrient concept, growth limitation is often difficult to attribute to a single nutrient. Either several nutrients are potentially limiting at their natural concentrations, or separate metabolic processes, each necessary for increased growth yields, are simultaneously limited by different nutrients.

More often than not, multiple nutrient additions are more effective than single nutrient additions in stimulating primary productivity. Commonly, simultaneous nitrogen and phosphorus additions lead to a synergistic growth response. That is, growth response is greatly enchanced when a single nutrient is supplemented with another. Such results have been obtained when a trace element accompanies nitrogen or phosphorus additions. The supplementation of phosphorus addition with nitrogen has also elicited large growth responses (Goldman and Carter, 1965; Schelske and Stoermer, 1972; Schelske et al., 1978). The most common explanation is that several nutrients instrumental in major metabolic growth-related processes are chronically deficient in natural waters. In considering limitation of photosynthetic CO_2 fixation alone, numerous macro- and micronutrients must function in a coordinated manner to assure proper synthesis of enzymes and constitutents of the photosynthetic apparatus. During nutrient-deficient periods, as indicated by suboptimal CO_2 fixation rates and depressed chlorophyll a production, nitrogen often proves stimulatory in the short run. However, given the fact that iron and manganese are also often found in short supply, these micronutrients can quickly replace nitrogen as the growth-limiting nutrient in the event of nitrogen enrichment. It follows that the most profound and sustained stimulation of photosynthesis is often in response to a combined nitrogen-iron or nitrogen-manganese addition (Goldman and Carter, 1965). Shapiro and Glass (1975) have found a similar synergistic effect on photosynthetic growth when phosphorus and manganese are simultaneously supplied to Lake Superior phytoplankton.

In hardwater lakes, coprecipitation of limiting nutrients is a common occurrence (Tsusue and Holland, 1966; Otsuki and Wetzel, 1972). Under such conditions, multiple nutrient additions often yield increases in primary productivity, whereas single nutrient additions fail to elicit a response.

4. The CO_2 Limitation Question

The most plentiful constituent in photosynthetically produced organic matter is carbon, being derived from the dissolved inorganic carbon (DIC) pool. Pool constituents are CO_2, HCO_3^-, and CO_3^{2-}, the relative proportions of each being pH dependent (Hutchinson, 1957). The total DIC pool, as well as the proportionality of pool constituents, is in dynamic equilibrium with geochemical, meteorological, physicochemical, and biological carbon transformation processes. In discussing DIC with reference to the regulation of primary productivity, the rates of photosynthetic utilization are compared to various nonbiological and biological means of replenishment.

Total DIC levels can vary substantially between as well as within aquatic ecosystems. Variability occurs both spatially and temporally. Large order differences are most often attributable to compositional differences in regional rock types (ranging from highly carbonaceous limestone and sedimentary rocks to igneous rocks) forming a lake or river basin (Golterman, 1975b; Holland, 1978). The DIC load in rainfall is another key factor regulating DIC concentrations (Hutchinson, 1957; Holland, 1978). Within the water column, numerous physical-chemical processes dictate the DIC quantities and availability. Variations in pH, which to a large extent are determined by specific DIC inputs, control the proportionality of DIC constituents (Stumm and Morgan, 1970). The limited buffering capacities of other chemical constituents such as humic and fulvic acids, silicates, sulfates, and a wide variety of charged organic molecules also can affect the pH regime (Stumm and Morgan, 1970). Temperature, altitude, and water density regulate the solubility of DIC. Lastly, the respiratory activities of aquatic biota, as well as photo- and chemolithotrophic fixation of CO_2, can alter both total DIC levels and relative proportions of pool constituents.

The most readily utilized form of DIC is CO_2, since it is easily transported across cell membranes and is directly fixed or reduced in the Calvin cycle of photosynthesis (Moss, 1973). Bicarbonate (HCO_3^-) can be assimilated as a photosynthetic substrate as well (Ingle and Colman, 1975), but its transport and utilization by phytoplankton are not as rapid as for CO_2 (Ingle and Colman, 1975; Paerl and Ustach, 1982). Carbonates (CO_3^{2-}) do not appear to be directly utilized by a wide variety of phytoplankton (Moss, 1973); in fact, uncertainty exists whether any photosynthetic organisms can utilize CO_3^{2-} directly. Furthermore, many CO_3^{2-} salts are highly insoluble at neutral to basic pH regimes.

The sufficiency of the most useful forms of DIC has been the subject of numerous studies as well as controversies regarding nutrient limitation of freshwater productivity (Likens, 1972). Because nutrient limitation of productivity is dictated by the nutrient present in lowest quantities, attention has been focused on nitrogen and phosphorus as well as a variety of trace elements, because their natural concentrations often range from the sub-ppb to the 10 ppb range. Natural DIC levels are virtually always in excess of the 1000 ppb level; hence, the assumption for inorganic carbon sufficiency. Normal cellular growth, however, demands a C:N:P ratio of approximately 100:15:1 (Stanier et al., 1963). Coupled with the fact that in a medium to high pH situation, from 5 to nearly 95% of the DIC can be present in a largely unavailable form (CO_3^{2-}) (Hutchinson, 1957), it is not unreasonable to suspect that the availability of utilizable forms of DIC (CO_2 and HCO_3^-) may control aquatic primary productivity.

Several investigators have proposed that under certain environmental circumstances, CO_2 may be acting as a factor limiting primary productivity (Kuentzel, 1969; Kerr et al., 1972). Circumstances promoting CO_2 limitation included: favorable light and temperature conditions for growth, high pH levels, low total DIC content, excessively high nitrogen and phosphorus concentrations, and trace metal sufficiency. Several additional factors enhanced CO_2 limitation. They included strong thermal stratification of surface waters, countering wind-induced mixing of atmospheric CO_2 into the euphotic zone, and low respiration rates or a scarcity of heterotrophic biota, which could reintroduce CO_2. Unfortunately, many of the initial studies designed to muster evidence for CO_2 limitation were done with the specific intention of disproving the convincing evidence that either phosphorus or nitrogen or both were largely responsible for limiting aquatic primary production (see Likens, 1972). Experiments proving CO_2 limitation were not carefully designed and, as pointed out by Golterman (1975a), numerous contradictions were embodied in the arguments brought forth by Lange (1970, 1971) and Kerr et al. (1972) in support of CO_2 limitation. This work, largely supported by the Soap and Detergent Association of the U.S.A., was conducted on highly specialized waters not necessarily indicative or representative of natural aquatic ecosystems. These included sewage oxidation ponds and test ponds artificially enriched with excess phosphorus and nitrogen in some cases. Furthermore, much of the critical work was published in a report not readily available to the public (Kerr et al., 1970). Examination of this report revealed contradictions between experimental results and conclusions. Because of this chain of events, the question of CO_2 limitation was largely left unappreciated for its potential research value; aquatic ecologists lined up either pro or con on the issue without crucial and objective evidence in an emotional deadlock over whether or not CO_2 limitation could occur in nature (Likens, 1972).

An overwhelming battery of investigators, having accumulated voluminous data on phosphorus and or nitrogen limitation of freshwater, effectively

conveyed the opinion that CO_2 limitation could only occur under truly excep-
tional circumstances. Little doubt exists that in most aquatic ecosystems an
adequate pool of DIC exists to offset growth limitation by CO_2 (Wetzel, 1975).
However, the potential availability of DIC cannot be solely judged from static
measurements of either DIC, pH, or a combination of the two on a macroscale.
During stratified periods (stagnation of the water column), exceedingly high
demands for DIC can occur on a horizontal and vertical small-scale basis.
Examples are: near-surface or metalimnetic blooms of phytoplankton, particularly cyanobacteria, as well as patchy distributions of either colonial algae or
concentrations of unicellular algae in distinct layers throughout the water column. Small-scale (within 1–10 cm layers) pH and DIC measurements indicate
that steep DIC and CO_2 gradients can occur in such microenvironments. Furthermore, dense populations of the bloom forming cyanobacteria *Anabaena,
Aphanizomenon,* and *Microcystis* located in near-surface layers are CO_2 limited following periods of heavy photosynthetic CO_2 demand (Paerl and Ustach,
1982). These demands may be met by migration to the surface (in the form of
surface blooms), where CO_2 requirements are satisfied through the exchange
of atmospheric CO_2 with surface waters (see Section 6). Any CO_2 enrichment
of subsurface nonturbulent waters, however, must occur by way of molecular
diffusion. As a CO_2 transport process, molecular diffusion is extremely slow,
allowing for no more than a few cm of surface water to be enriched by atmospheric CO_2 on a daily basis (assuming turbulence-free conditions in surface
waters) (Hutchinson, 1957).

Schindler and Fee (1973) and Weiler (1975) have examined CO_2 invasion, or flux from the atmosphere, into surface waters of lakes with the goal of
determining whether invasion rates could meet photosynthetic CO_2 demands
during diurnal phytoplankton growth periods. In experimental lakes which had
been fertilized with nitrogen and phosphorus, net primary production rates
during stratified periods were equal to the invasion rates of CO_2 from the atmosphere (Schindler and Fee, 1973). Lehman *et al.* (1975), in a simulation study
relating phytoplankton standing crop to nutrient availability, essentially came
to the same conclusion; namely, that CO_2 invasion was closely matched by CO_2
demands. In specific cases, as illustrated by the study of Schindler *et al.* (1973)
on nutrient-enriched experimental lakes, the invasion of atmospheric CO_2 into
surface waters controlled daily productivity. Conversely, in lakes clearly limited by either nitrogen or phosphorus availability, CO_2 invasion was high
enough to exceed photosynthetic demands by phytoplankton. Schindler *et al.*
(1973) indicated that CO_2 limitation of phytoplankton growth may have been
in effect during specific daily photosynthetic periods, particularly during afternoon hours following morning and midday optima. Such limitation would only
be expected if nitrogen, phosphorus, or any potential noncarbonaceous limiting
factor was not operative. The overall conclusion to be drawn is that CO_2 availability can potentially restrict primary productivity if enrichment of surface

waters, to the point of alleviating noncarbonaceous nutrient limitation, has occurred.

Lake 227, which received extensive nitrogen and phosphorus enrichment (Schindler *et al.*, 1973), represents such a situation, periodically forcing it into CO_2 limitation. Fertilization led to massive standing crops of phytoplankton, ranging from 50 to near 200 μg chlorophyll *a* per liter. Such high standing crops would lead to extremely high CO_2 demands during a daily photosynthetic period. This form of fertilization may be considered excessive when compared to conditions in natural waters, where either nitrogen, phosphorus, or trace element availability might normally limit the magnitude of photosynthetic production. On the other hand, severely polluted systems, receiving sewage, industrial, or agricultural wastes, would be likely candidates for full nutrient sufficiency or nutrient overloads, forcing these systems' productivities to be constrained by other environmental factors. If the polluted system reveals strong surface stratification, dense phytoplankton blooms, periods of elevated pH, and depressed DIC levels, the possibility exists that CO_2 limitation may be operative. A lack of critical and objective work in determining potential CO_2 limitation under the diverse conditions present in lakes and rivers has hampered a clear-cut conclusion on the potentials for CO_2 limitation in natural waters.

5. Vitamins and Growth Promoting Substances as Limiting Factors

Provasoli (1960) and Carlucci and Bowes (1972) have provided useful information on the ubiquity of vitamin deficiencies among primary producers. It appears that many freshwater algae do not possess the enzymatic capabilities for synthesizing essential vitamins, making it necessary to rely on ambient sources or associations with vitamin producing microorganisms. In particular, external requirements for vitamin B_{12}, thiamin, and biotin are commonplace among the Chlorophyta, Euglenophyta, Cryptophyta, Pyrrophyta, Chrysophyta, and Bacillariophyta (Provasoli, 1960). In this regard, vitamins and related growth promoting substances, such as choline and inositol, may be regarded as growth limiting substances in natural aquatic ecosystems. As mentioned in the section dealing with CO_2 limitation, numerous algal species have been shown to be incapable of growth free of bacterial associations. Potential vitamin exchange should receive consideration as a factor enhancing the growth of the host algae in such associations (Burkholder, 1963; Provasoli, 1963). Of the primary producers, the cyanobacteria do not appear to experience vitamin deficiences; studies have indicated that these genera generally do not require exogeneous vitamins and presumably are capable of producing their own supplies of vitamins, or are deriving vitamins from associated bacteria (Alexander, 1971).

6. Microbial Strategies Counteracting Growth Limitations

Deficiencies in environmental factors governing microbial growth may be met in a variety of ways, through the use of physical and biochemical-physiological strategies. Genetic flexibility has led to sets of alternative strategies which can be deployed in the event of growth limitation brought on by environmental factors.

Primary producers have developed several means of minimizing nutrient limitation. Alteration of morphology and size to suit nutrient sufficiency or depletion is commonplace among aquatic microorganisms. Numerous studies, involving a wide array of photosynthetic and heterotrophic microorganisms, have illustrated the ability of microorganisms to alter their surface-to-volume ratios in relation to ambient nutrient conditions (Alexander, 1971; Banse, 1976). The basic conclusion is that low ambient nutrient conditions bring about high surface-to-volume ratios and vice versa. Various taxa have developed cellular projections, including spines (Round, 1965), mucoid webs (Paerl, 1974), and flagella, as well as gas vacuoles (Walsby, 1975), in order to resist sinking and to promote orientation of microorganisms towards regions of the water column supporting optimal growth. Required nutrients which can, at times, be present in the ambient environment in growth limiting concentrations can be stored intracellularly. By having storage capabilities, microorganisms can survive and grow during periods of severe nutrient deficiency. Phosphorus can be accumulated and stored in long-chain phosphate polymers which appear intracellularly as polyphosphate bodies (Kuhl, 1968). Nitrogen can be stored in the form of cyanophycin granules in certain cyanobacteria (Fogg et al., 1973; Stewart, 1977) as well as in proteinaceous matter in eucaryotic phytoplankton. Additional nutrients which can be maintained in a stored state include sulfur (granules and droplets) and organic carbon compounds (as lipids, polyhydroxybutyrate bodies, starches). Iron can be chelated extracellularly by a wide variety of microorganisms (Nielands, 1974), thereby maintaining this required nutrient in the water column in a biologically available form.

During ambient phosphorus (phosphate) depletion, both algae and bacteria are capable of cleaving phosphate groups off organic molecules. This is achieved through the excretion of alkaline and acid phosphatases which enzymatically liberate phosphate from a wide variety of organic compounds. The ecological significance of phosphatase activity in supplementing phosphorus requirements in aquatic ecosystems has been discussed by Berman (1970) and Wetzel (1981).

Selective nutrient transforming processes can be instrumental in assuring adequate supplies of potential growth-limiting nutrients. A good example of this is nitrogen fixation by certain heterotrophic, chemolithotrophic, and photolithotrophic bacteria (Bjälfve, 1962; Newton et al., 1977), as well as by cyanobacteria (Stewart, 1977). This process assures adequate supplies of biologically available nitrogen (as ammonia) during periods of nitrogen deficiency and is an important factor promoting cyanobacterial dominance and blooms during periods of nitrogen limitation in stratified eutrophic surface waters (Fogg *et al.*, 1973).

The transformation of sulfur compounds via oxidative and reductive reactions serves the dual purpose of providing oxidizable and reducible substrates, as well as readily utilizable sulfur sources, for photosynthetic and heterotrophic microorganisms. Because sulfur is readily transformed through a variety of microbial reactions, it is normally available in several forms, both organic and inorganic (Alexander, 1971). In addition, abundant terrigenous sulfur sources exist. Hence, the combination of efficient microbial cycling plus adequate ambient concentrations often circumvents sulfur limitation in freshwater.

Numerous photosynthetic algae and bacteria are able to utilize organic compounds as energy and nutrient sources in the event of (1) a cutoff of photosynthetically active radiation, and (2) an abundance of organic matter in the ambient environment (Stewart, 1974). The ability to utilize organic matter during cessation of photosynthesis gives microorganisms access to (1) a carbon pool available for energy use, and (2) a vast array of major and minor potentially growth-limiting elements often associated with organic carbon compounds. These elements include nitrogen, phosphorus, sulfur, iron, and trace elements.

Mobility is instrumental in assuring spatial access to nutrients and favorable physical regimes (light, temperature, pH, turbulence, and salinity). For example, flagellated green and dinoflagellate algal species, as well as gas-vacuolated cyanobacteria, are often observed vertically migrating through the water column (Tilzer, 1973; Reynolds and Walsby, 1975; Kellar and Paerl, 1980). Such migration promotes exposure to a variety of favorable environmental conditions within a daily photosynthetic period. When light intensity is low, as in early morning or late afternoon hours, upward migration dominates, whereas downward migration prevails during high light intensity midday hours (Tilzer, 1973; Reynolds and Walsby, 1975). Downward migration often exposes algal cells to nutrient rich metalimnetic or hypolimnetic waters. While present in deeper waters, phytoplankton consume phosphorus and nitrogen in excess of immediate cellular requirements (luxury consumption), assuring adequate (stored) supplies of such nutrients during periods of ambient nutrient depletion. Vertical migration often appears to satisfy several basic sets of ecological growth requirements simultaneously, allowing for optimal exposure to and utilization of physical factors, such as light, salinity, turbulence, and temperature, while assuring adequate nutrient supplies (Paerl and Kellar, 1979; Kellar and Paerl, 1980).

Often a tradeoff exists between a planktonic and epiphytic or epilithic existence. A planktonic existence is attractive in the sense that microorganisms are free to move or be moved along a variety of environmental gradients, making optimal use of conditions favoring high growth rates in such a heterogenous environment. Alternatively, attachment to a surface has its attractive features as well. In streams and rivers, attached microorganisms are exposed to a high degree of turbulence, enhancing contact with soluble nutrients and promoting gaseous exchange. Such factors have been shown to promote growth (Madsen, 1972; Paerl and Goldman, 1972). Microorganisms are often attached to sources of nutrition such as plants, animals, or minerals. As a result, epiphytic, epilithic, and epibenthic microbial communities are often exposed to higher concentrations of nutrients and organic matter than their planktonic counterparts (Jannasch and Pritchard, 1972; Paerl and Goldman, 1972). Surface microenvironment production rates can exceed plankton production rates (per unit volume) by several orders of magnitude (Paerl, 1974). High localized production of organic matter also promotes high standing stocks of herbivores and detritivores (Madsen, 1972). The activities of such consumers promotes nutrient cycling, leading to a highly dynamic situation where both production and mineralization rates are enhanced (Alexander, 1971). In terms of growth strategies, surface environments are often preferred over planktonic environments, particularly in oligotrophic waters, where low dissolved nutrient concentrations in the water column may severely limit autotrophic and heterotrophic microbial production (Paerl and Goldman, 1972).

Microbial mutualistic and symbiotic associations with other organisms enhance both partners (Alexander, 1971). It appears that the exchange of metabolites is a key set of processes promoting enhancement of partners. Often energy and nutrient sources limiting the growth of one partner can be supplied by the other and *vice versa*. Removal of potentially toxic substances, such as H_2S and O_2, can be achieved in associations because one of the partners in the association exhibits a metabolic use or need for such substances, thereby removing the toxic effect from the more susceptible partner (Paerl and Kellar, 1978).

It has been proposed that nutrient limitation can be relieved through symbiotic associations. Examples are phosphorus exchange between mineralizing epiphytic bacteria and host algae (which are phosphorus limited) and CO_2 exchange between mineralizing (heterotrophic) bacteria and algal hosts whose photosynthetic production may be limited by CO_2 availability (Kuentzel, 1969; Lange, 1973).

With regard to the question of potential CO_2 limitation, primary producers appear to have developed individual strategies aimed at dealing with this problem. Recent work has shown that certain phytoplankton groups are more effective than others in utilizing bicarbonate, and possibly carbonate, under high pH (low CO_2) conditions. In particular, the cyanobacteria and certain green algal genera show good growth yields in high pH environments (Shapiro, 1973). In contrast, diatoms and flagellates often show a preference for more neutral pH conditions. The ability of cyanobacteria and green algae to dominate in high pH environments has been attributed to their ability to utilize carbonic anhydrase (CA), an enzyme which can catalyze the reversible hydration of carbon dioxide, making CO_2 available for photosynthesis under CO_2 limited conditions. The enzyme reaction ($CO_2 + H_2O \lessgtr HCO_3^- + H^+$) can be utilized to produce CO_2 from HCO_3^- under high pH conditions. In algae, CA levels are related to ambient CO_2 concentrations. For example, Nelson *et al.* (1969) found that *Chlamydomonas reinhardtii* Dang. exhibited a 10-fold increase in CA when supplied with atmospheric CO_2 (0.03%) as opposed to air enriched with 1% CO_2. Confirmation of this relationship was supplied by Graham *et al.* (1971) examining green algae and cyanobacteria and Ingle and Colman (1975) examining cyanobacteria.

The ability to utilize HCO_3^- during either high pH periods, low total DIC periods, or both, is of distinct benefit in circumventing CO_2 limitation. Recent work in this laboratory on *Anabaena oscillarioides* and *Anabaena flos aquae* indicates that these cyanobacteria adapt to CO_2 limited conditions by increasing the synthesis of CA (H. W. Paerl and P. T. Bland, unpublished data). Rapidly growing, low density populations of both species grown under adequate CO_2 supplies showed virtually no ability to utilize DIC when switched to high pH (pH 10.00) conditions. However, dense cultures previously grown on low DIC and high pH conditions revealed an ability to utilize DIC at high pH levels. The conclusion reached from these experiments is that under low density conditions, when adequate DIC supplies are obtainable and pH levels have not been elevated due to high photosynthetic demands, production of CA is not essential for optimal photosynthetic growth. Conversely, high density conditions, where pH rapidly rises in response to a high photosynthetic demand for DIC, promote the synthesis of CA. This induced synthesis plays an important role in maintaining optimal photosynthetic growth.

Other adaptive features come into play when considering potential CO_2 limitation. Diurnal migration patterns among flagellate (Tilzer, 1973) and cyanobacterial populations (Reynolds and Walsby, 1975) have been observed. During high photosynthetic demand periods, pH levels increase dramatically in stratified water layers supporting high densities of algal biomass. Migration out of such layers would be advantageous as long as favorable light and nutrient conditions exist at alternative depths. Among the cyanobacteria an extreme migratory response is often observed; this is the formation of a surface scum. Buoyancy alteration due to the formation and loss of gas vacuoles is responsible for these migratory events (Walsby, 1975). Gas vacuolation is related to photosynthate buildups in *Anabaena* (Reynolds and Walsby, 1975). An active photosynthetic period leads to a relatively large pool of intracellular photosynthate. This in turn leads to increased cellular turgor pressure. High turgor pressure causes a decrease in gas vacuolation. As a result, cells that have undergone this chain of events will have a tendency to sink. Conversely, a lack of photosynthesis leads to low intracellular photosynthate levels and increased vacuolation. The result is an increase in buoyancy, often manifested as a surface scum.

The rationale for surface scum formation is not clearly understood (Reynolds and Walsby, 1975). However, ambient CO_2 deficiencies are known to lower photosynthetic potentials which in turn lower intracellular photosynthate production and enhance gas vacuolation. This response was demonstrated in the laboratory using natural populations of *Anabaena spiroides* and *Aphanizomenon flos aquae* (Paerl and Ustach, 1982). When populations were grown under high pH and low total DIC levels, gas vacuolation increased and scum formation was promoted in both genera. We have examined the possible ecological rationale for this response. During blooms the high CO_2 demand which exists during daylight hours drives the ambient pH up, often in excess of 10 (Kellar and Paerl, 1980). At such pH levels, free CO_2 is no longer available (Hutchinson, 1957). When this situation occurs, photosynthesis can be shown to be CO_2 limited, at least during the remaining high light intensity period of that day. This limitation can be alleviated by CO_2 additions (Paerl and Ustach, 1982).

A tendency for cyanobacteria to form scums during such CO_2 limited events is evident, and is brought on by increased vacuolation in response to decreasing levels of photosynthate (due to depressed photosynthesis). Scum formation exposes cyanobacteria to the air-water interface, where atmospheric supplies can readily invade surface waters (Schindler and Fee, 1973). On calm days when epilimnetic turbulent mixing is minimal, atmospheric CO_2 exchange is confined to the upper few centimeters of the water column. CO_2 is mainly dispersed throughout the water column by molecular diffusion during such periods. It can be shown that scum formation is often most profound during such periods. It appears that scum formation is a response to CO_2 limitation of photosynthesis, bringing cyanobacteria in contact with the air-water interface, where CO_2 supplies remain plentiful.

Several N_2 fixing genera known to form profound scums during calm periods also possess rich supplies of carotenoids (Clayton, 1966; Asato, 1972). In the laboratory, Paerl and Kellar (1979) found a diurnal increase in cellular carotenoid content of *Anabaena* in response to high light and high oxygen levels, common environmental features of surface waters supporting cyanobacterial blooms. In the field, steady increases in carotenoid to chlorophyll *a* and carotenoid to biomass ratios were common features of cyanobacterial blooms as biomass increased and surface scums become more intense (Kellar and Paerl, 1980). Carotenoids have been implicated as photoprotective agents in plants, including cyanobacteria, quenching certain excited states of chlorophyll *a* which promote photooxidation (Clayton, 1966). Asato (1972) discovered a resistance to ultraviolet radiation and photooxidation among several cyanobacterial genera. Carotenoids are also capable of transferring photoreducing power to chlorophyll *a* (Clayton, 1966), thereby conserving radiant energy used to power photosynthesis. Hence, the appearance of cyanobacterial scums, instead of being a prelude to bloom senescence and death, appears to be an ecological strategy poised at making optimal use of photoreducing power and potential CO_2 supplies. During wind-mixed periods, atmospheric CO_2 invasion in surface waters is greatly enhanced, alleviating periodic CO_2 limitation and eliminating the necessity for scum formation.

Kuentzel (1969) and Lange (1971) raised the possibility that bacteria associated with phytoplankton provided CO_2 through decomposition of organic matter, some of which possibly consisted of extracellular products of phytoplankton photosynthesis. Lange (1976) further speculated that mucilagenous sheaths produced by numerous cyanobacterial bloom species were providing nutrient-rich microenvironments in which bacteria could thrive. Release of CO_2 and perhaps other nutrients (PO_4^{3-}, NH_3, and trace metals) by bacteria growing in such microenvironments could promote host growth as well. Clearly, more work and finer detail are required to substantiate these largely unproven hypotheses.

Despite the difficulties in proving such nutrient cycling mechanisms, there is little doubt that growth of algal and cyanobacterial hosts can be dramatically enhanced in the presence of bacteria. In particular, cyanobacteria are virtually never found free of bacteria in nature, and heavy bacterial colonization during peak bloom periods has been reported on numerous occasions (Caldwell and Caldwell, 1978; Paerl and Kellar, 1978). The conclusion from these observations is that the presence of bacteria on cyanobacterial hosts does not necessarily have an adverse effect on host metabolism; on the contrary, growth enhancement of hosts may be due to symbiotic associations with bacteria. In the laboratory, it can be shown that the growth of *Anabaena* and *Microcystis*, as well as the marine cyanobacterium, *Trichodesmium,* is greatly enhanced by the presence of bacteria. Vance (1966) noted that *Microcystis aeruginosa* failed to grow in culture in the absence of bacteria. Recent work by Paerl and Kellar (1978), Gallucci (1981), and F. S. Lupton and K. C. Marshall (unpublished data) has indicated that nitrogen fixation (acetylene reduction) in various species of *Anabaena* is stimulated in the presence of bacteria known to form close associations with this genus. The mechanisms mediating such stimulation are not clear. However, the fact that bacteria-free *Anabaena* often reveal profound growth limitation strongly suggests that chemical exchange may be a functional aspect of such relationships.

7. Determining Factors Limiting Primary Productivity in Freshwater: Bioassay Techniques

The most common means of determining which environmental factors limit freshwater primary productivity is through the use of bioassay techniques. Bioassays using either test organisms or natural communities are routinely employed in investigations of aquatic nutrient limitation or growth restrictions due to suboptimal environmental conditions (light, pH, salinity, oxygen content). In 1968, the National Eutrophication Research Program prepared a preliminary procedure for a standardized algal assay based on a single test species. The assay was designed to: identify nutrients limiting algal growth, establish the biological availability of specific limiting nutrients, yield quantitative results on the degree of nutrient limitation, and be relevant and applicable to solving water quality problems (Maloney *et al.*, 1972). Three basic tests have been established: (1) a bottle test, employing cultures of a single test alga, such as *Scenedesmus, Chlorella,* or *Selenastrum* [see the *Selenastrum capricornutum* Printz Algal Assay Bottle Test; (Shiroyama *et al.*, 1976; Miller *et al.*, 1978)], (2) a continuous flow chemostat test, also employing a single test alga, and (3) an *in situ* test designed to measure the response of a natural primary producer community to nutrient enrichment. Procedures for conducting these three basic sets of tests have been published in the "Provisional Algal Assay Procedure" (Joint Industry/Government Task Force on Eutrophication, 1969). In-depth evaluations, in terms of applicability and reliability of respective tests, have been numerous (Wetzel, 1975). They will be summarized in general form here.

The deployment of single species of test algae has distinct benefits and drawbacks. Weiss and Helms (1971) found that comparative algal assays performed in a single laboratory were highly repetitive and precise. Tests demonstrated a good linear relationship between nutrient concentrations and growth. However, when the same test organisms were employed in different laboratories, a lack of uniformity in terms of growth response to identical nutrient addition was reported. Interlaboratory precision was highly variable as well. Although test organisms can be maintained in pure cultures with few problems and are relatively easy to enumerate, they often bear little resemblance to the natural flora of an aquatic ecosystem under investigation. Single species tests are effective in detecting general nutrient deficiencies. But since predictability, in terms of natural primary productivity or potential species changes, is central to assessments of eutrophication processes, natural microflora must also be examined in parallel. Hence, it is at times inapplicable to extrapolate single species nutrient response data to natural situations, where up to 100 algal species can often coexist.

Flow chemostats are excellent tools in describing and understanding physiological responses (including growth) of test algae to specific physical, chemical, and biological stimuli promoting growth limiting conditions. However, as with the bottle test, it is difficult in many cases to substantiate a case for growth limitation in natural systems due to specific environmental variables based solely on experimental evidence obtained from single species cultures. Furthermore, flow chemostats operate on the principle of nutrient replenishment, the replenishment being varied by medium turnover rates and flushing/dilution

rates. In natural systems, especially large lakes that reveal strong vertical stratification and long water retention periods, natural flora are more likely to be
exposed to batch culture conditions, where nutrient inputs occur during a distinct period (i.e., runoff, turnover, or mixing of the water column) followed by
gradual depletion without significant and constant replenishment.

Bioassays using natural flora are often desirable for examining physical
and chemical limitation of growth and are preferred by most limnological laboratories. Ideally, techniques should place priority on duplicating natural conditions, including incubating algae where they have been sampled, with minimal disturbance and alteration of light, temperature, and dissolved gaseous
regimes. Criteria by which to judge biostimulation must be readily detectable,
meaningful, related to degrees of algal growth, and specific for microbial biomass. Although bioassays using *in situ* microbial assemblages have a high
degree of relevance in determining limiting factors in nature, they suffer from
potential sources of artifacts and yield interpretive problems as well. Often,
natural microbial communities exposed to growth limitation tests are severely
altered (in terms of species diversity and total biomass) by being confined to
enclosures. This is particularly evident with small enclosures, which appear to
promote a "bottle effect" (Vollenweider and Nauwerck, 1961), whereby the
confinement to a bottle for a period as brief as several hours can alter community structure and, hence, dominance.

The danger in altering microbial community structure lies in the fact that
specific metabolic reactions, each controlled by specific nutrient requirements,
may change during the course of an enclosed bioassay. An example of this
would be a decrease or rise in dominance by nitrogen fixing cyanobacteria. A
decrease in dominance by these microorganisms would eliminate nitrogen fixation and increase exogenous nitrogen requirements. An increase in dominance
would increase nitrogen fixation potentials, leading to lower exogenous nitrogen
requirements and increased phosphorus and trace element requirements. It is
also known that the bottle effect often leads to increased dominance by nano-
and ultraplankton (over micro- and netplankton) through time, distinctly different from *in situ* assemblages. The increase in smaller planktonic forms may
be due to their ability to maintain growth under decreasing concentrations of
specific nutrients often encountered in small enclosures, whereas micro- and
netplankton are less effective in competing for the same nutrients below a certain threshhold level. Alternatively, if zooplankton grazing plays a factor in cell
turnover, it may be concentrated on the larger phytoplankton species, favoring
the survival and proliferation of smaller species. Other factors, such as surface
areas offering growth substrates to specific phytoplankton species and the alter
ation of turbulent regimes by enclosures, are likely to affect species composition
and related metabolic functions as well.

One obvious solution to the bottle effect problem is the use of larger enclosures placed in the systems under investigation. Large enclosures have been

successfully tested in a variety of aquatic ecosystems, with the advantages that such enclosures have a minimal effect on altering microbial community structure during the enrichment bioassay period. Large enclosures have included: polyethylene bags holding from several to hundreds of liters of water (Schelske and Stoermer, 1972), polyethylene columns having a 1 m diameter placed vertically in the water column (Goldman, 1964), "limnocorrals," of from 3–10 m in diameter also placed vertically in the water column (D. R. S. Lean, personal communication), and "sea-curtains," vertically bisecting small lakes (Schindler *et al.*, 1973). Lastly, whole lake experiments have been attempted in areas where a large number of lakes having relatively similar physical/chemical and biological characteristics have been available for testing (Schindler *et al.*, 1973).

Criteria by which biostimulation is to be judged are a critical and often a problematic part of enrichment bioassay techniques. Commonly used parameters employed as criteria include: (1) cell and particle counts, (2) plating on growth medium, (3) chlorophyll *a* and other pigment analyses, (4) particulate organic carbon, nitrogen, or phosphorus analyses, (5) $^{14}CO_2$ incorporation through photosynthetic growth, and (6) assaying concentrations of specific cellular compounds, including adenosine triphosphate (ATP) and other adenylates, deoxyribonucleic acid (DNA), and muramic acids. These parameters have one thing in common; they can be accurately measured at the low concentration in which they are found in natural waters. The rationale for choosing these parameters is that they all reflect cellular growth. However, because a majority of the above parameters predominate in specific cellular pools, differential enhancement of such pools will reflect as differences in agreement between parameters on the degrees of biostimulation.

Recently, we (H. W. Paerl, G. W. Payne, and L. Mackenzie, unpublished data) compared various parameters as indicators of biostimulation, using Lake Taupo, New Zealand, microbial communities placed *in situ* in 4-liter polyethylene bags. Varying agreement among the above parameters resulted. Noticeable enhancement compared to no additions of four parameters was observed in response to nitrogen (as NO_3^-) plus phosphorus (as PO_4^{3-}) enrichment. Biomass increases were consistently observed among phytoplankton cell counts, ATP assays, chlorophyll *a* content, and ^{14}C uptake (Fig. 1). These results are not surprising since both nitrogen and phosphorus are present in the diverse cellular pools being assayed. For example, ATP contains both nitrogen and phosphorus, whereas chlorophyll *a* contains nitrogen and no phosphorus. When nitrogen was added in some experiments, the degree of chlorophyll *a* stimulation relative to ATP stimulation was often inflated (Fig. 2). Likewise, when phosphorus was added alone, cellular ATP levels revealed a more profound increase than chlorophyll *a* on several occasions (Fig. 2). Caution must be practiced, therefore, in interpreting nutrient addition by the use of a single biochemical parameter containing that nutrient.

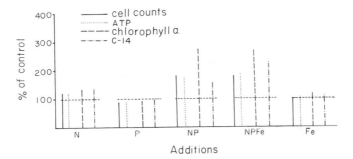

Figure 1. Nutrient stimulation of a natural phytoplankton community from Lake Taupo, New Zealand. Water samples were taken at 2 m depth, dispensed in acid and deionized water cleaned 4-liter clear polyethylene bags, and incubated *in situ* at 2 m for 5 days. Four growth parameters were examined. They included: cell counts, determined before and after nutrient additions; ATP content of particulate matter; chlorophyll *a* content of particulate matter; and $^{14}CO_2$ assimilation. Cell counts were performed on acetone cleared HA Millipore (0.45 μm porosity) filters, examined microscopically at 1000X (Paerl, 1974). In order to monitor $^{14}CO_2$ assimilation, 15 μ Ci ^{14}C-bicarbonate solution was added to each polyethylene bag at the start of *in situ* incubation. All results are given as percent of control (no additions). Results represent the average of triplicate samples. Nutrient concentrations added were: N = 100 μg nitrogen as NO_3^- per liter, P = 30 μg phosphorus as PO_4^{3-} per liter, Fe = 20 μg iron as Fe citrate per liter.

To some extent, ^{14}C incorporation also reflects these problems. For example, if the synthesis of a particular cellular constituent is enhanced by a nutrient addition, ^{14}C incorporation into that constituent takes place. If that cellular component alone is stimulated (i.e., only enrichment of a specific cellular constituent representing a small fraction of the cellular biomass), but no overall

Figure 2. Stimulation of Lake Taupo phytoplankton by nitrogen and phosphorus alone, combined, and together with iron. Nutrient concentrations and growth parameters are identical to those given in Fig. 1. Nitrogen additions caused highly significant ($p < 0.005$) stimulation of chlorophyll *a* and ^{14}C assimilation (compared to control),

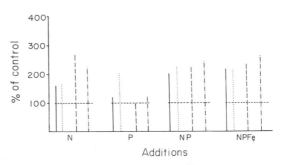

whereas phosphorus additions significantly ($p < 0.01$) stimulated ATP synthesis without parallel stimulation of other parameters. In combination, nitrogen and phosphorus significantly ($p < 0.01$) stimulated all four parameters, with iron having no additional stimulatory effect.

growth occurs, ^{14}C incorporation results (over controls) will show false enhancement. This can be seen in the case of chlorophyll a enhancement by iron additions at a time when no parallel biomass enhancement was recorded. Two distinct periods of enhancement were noticed. During fall and winter over-turn of the water column, iron, as Fe^{3-} citrate, combined with nitrogen and phosphorus caused significant stimulation of all four parameters (Fig. 3). In addition, a trace element solution, which included iron, also showed stimulation and good agreement between all parameters. These results appeared convinc-ing since trace elements, and iron in particular, are not directly assimilated into ATP and chlorophyll a (although trace elements are components of enzymes and cofactors involved in ATP and chlorophyll a synthesis). Clearly, increases in cellular ATP and chlorophyll a content reflected real growth in these experiments.

A contrasting set of results was obtained during the summer months. Iron, individually added with nitrogen and phosphorus, yielded dramatic increases in cellular chlorophyll a and ^{14}C assimilation, whereas both cell counts and particulate ATP failed to show significant increases over nitrogen and phos-phorus without iron (Fig. 4). The trace element solution behaved identically to iron. The discrepancies in stimulation proved significant ($p < 0.01$). It appeared that the presence of minor elements enhanced pigmentation without concommitant increases in real growth. Why should such a discrepancy occur in summer when during winter and spring all parameters were in agreement? During winter and spring, it appeared that biologically available trace ele-ments, and particularly iron, were present in growth-limiting concentrations. When examined by autoradiography, significant increases in algal biomass

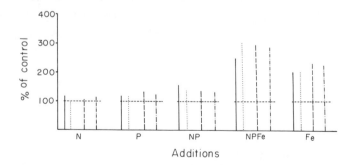

Figure 3. Stimulation of Lake Taupo phytoplankton by iron additions. This bioassay was con-ducted shortly after the fall turnover of the water column. Due to vertical mixing following turn-over, adequate supplies of nitrogen and phosphorus were introduced into the near-surface waters. As a result, nitrogen and phosphorus additions had no significant stimulatory effect. Alone and combined with nitrogen and phosphorus, iron had a profound stimulatory effect, which was reflected by all four parameters. Hence, it appeared that Lake Taupo waters were iron-deficient during this period.

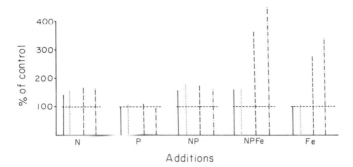

Figure 4. Results from a bioassay conducted during a thermally stratified summer period in Lake Taupo. Nitrogen as well as nitrogen and phosphorus significantly ($p < 0.01$) stimulated all four growth parameters. When added alone or in combination with nitrogen and phosphorus, iron showed profound stimulation of chlorophyll a and ^{14}C assimilation ($p < 0.001$), whereas cell counts and ATP content revealed no stimulation. Results from this bioassay could be interpreted in a contrasting manner. If chlorophyll a or ^{14}C assimilation are the sole parameters used to assess nutrient stimulation, iron would be judged to be a growth-limiting factor. If, on the other hand, cell counts and ATP are included as growth criteria, it would be concluded that no real growth occurred, but rather that the synthesis of certain cellular components (namely chlorophyll a) was enhanced by iron additions.

were recorded for a wide range of species present at that time, including both nannoplankton and netplankton. In contrast, iron and trace element additions in midsummer clearly favored the growth of nannoplankton with relatively little enhancement of netplankton biomass (N + P additions alone stimulated both nanno- and netplankton biomass). Nannoplankton characteristically exhibit high chlorophyll a to cell carbon (biomass) ratios (Paerl *et al.*, 1976). It appeared that the addition of trace elements enhanced phytoplankton chlorophyll a levels without necessarily enhancing total algal biomass. Cells appeared more heavily pigmented. Autoradiographs revealed that over 75% of the newly fixed ^{14}C went into algal chloroplasts, presumably representing newly synthesized chlorophyll a. It appeared that trace element additions during this period stimulated pigment synthesis but had little effect on overall biomass increases.

Such results are troublesome if we are to adapt pigment analyses as accurate indicators of biomass. It is easy to see that during this particular period, differing conclusions on growth limiting nutrients might arise from the use of contrasting parameters. An investigator employing chlorophyll measurements could conclude that trace elements are highly stimulatory while the alternatives, choosing cell counts or ATP analyses, would indicate no stimulation by such nutrients. It appears that, at least in the case of trace elements, cell counts and ATP analyses would be a more reliable indicator of biostimulation than chlorophyll analyses.

Ideally, ATP, chlorophyll *a*, and ^{14}C results should always be compared to biomass as determined by cell counts, since algal volumes determined by these counts represent direct measures of cellular matter. However, species enumeration is very time-consuming and often subjective among different investigators, making it prohibitive and at times impractical on a routine basis.

Bioassay parameters must be chosen which are (1) sensitive, (2) easily executed, and (3) representative of true growth increases of test organisms. Parameters reflecting the buildup of specialized biochemical pools, such as chlorophyll *a*, must be carefully interpreted, for they often do not clearly parallel phytoplankton growth patterns in natural lake waters. In this regard, the use of ATP as a biostimulation indicator is especially attractive. ATP assays are sensitive, easily executed, and appear indicative of true biomass increases. The extent to which ATP:biomass ratios vary among a vast range of phytoplankton is not completely known, although cumulative evidence on diverse cultured and natural phytoplankton populations thus far indicates that this ratio remains remarkably constant. Earlier biochemical evidence likewise shows that a vast range of microorganisms is able to maintain reasonably constant ATP:biomass ratios during periods of environmental stress, including nutrient depletion (Chapman *et al.*, 1971). On the other hand, there is evidence that the introduction of high ambient phosphorus levels tends to increase the ATP content of cells (Karl, 1980). This was not observed in bioassays reported here, perhaps because relatively small additions of P were made.

A major problem with ATP analyses is the inability of the method to separate algal from bacterial biomass. Parallel algal and bacterial counts normally reveal that bacterial biomass is less than 5% of the total microbial biomass. The inclusive measurement of bacterial biomass, therefore, leads to only a small over-estimate of phytoplankton biomass and in a great majority of cases means an insignificant source of error.

The fact that ATP methods measure both algal and bacterial biomass may be beneficial, however, in bioassay studies designed to monitor eutrophicating trends in lakes. Algae and bacteria are often spatially and functionally closely related and together constitute the basic food available to the grazing fauna. Also, both groups of microorganisms utilize phosphate, nitrate, ammonia, iron, and other nutrients suspected of accelerating lake eutrophication. Thus, on a trophic basis, it would seem important to monitor all forms of microbial life responsible for enhancing the overall production of aquatic ecosystems receiving nutrient enrichment.

8. Concluding Remarks

Primary productivity can be restricted through a variety of ways, including physical, chemical, and biological limitation or through a combination of factors. The types and magnitudes of limitation vary substantially in time and space, depending on the combined influences of geological, physicochemical, climatological, and biotic characteristics of individual freshwater basins and catchments.

Despite the presence of multiple potential and active limiting factors, phytoplankton, including bacteria, are flexible in dealing with unfavorable (for growth) environmental conditions. Orientation in microenvironments and mobility along gradients are characteristics poised at making optimal use of resources in the aquatic environment. In addition, the biochemical abilities to sequester, store, produce (fix), and recycle essential nutrients and energy sources are additional attributes, which necessarily evolved as naturally abundant nutrient sources dwindled with the development and diversification of freshwater microflora.

Opportunism is widespread among primary producers, particularly those forming extensive blooms under favorable chemical and physical environmental conditions. Man has had a profound effect on promoting opportunism in nuisance species, especially among the cyanobacteria. These species show strong positive responses to eutrophicating conditions brought on by human activities, whether they be industrial, agricultural, or municipal. Because many cyanobacterial species have the ability to actively migrate throughout the water column, efficiently assimilate and store nutrients (particularly phosphorus and nitrogen), fix atmospheric nitrogen, utilize diverse forms of DIC, function through either autotrophic or heterotrophic metabolism, and sequester trace elements by chelation, they can positively respond to a wide variety of environmental alterations resulting from human as well as geochemical and climatological activities (Fogg *et al.*, 1973).

The development and application of enrichment bioassays, although not without current interpretive problems, has yielded urgently needed answers. This comes at a time when assessment of the relationships between man's increasing water use and aquatic primary productivity as a food and recreational resource is critical. Major strides have been made in resolving the relationships between nutrient inputs and eutrophication and in correcting imbalances. Although the implication of single nutrients as agents causing accelerated eutrophication has proven invaluable, the effects of multiple parameter or multifactor alterations of primary productivity need to be better understood, clarified, and dealt with in preserving our freshwater resources. Much work lies ahead for microbial ecologists, limnologists, and environmental management, for the equilibria between man and the earth's aquatic environments will continue to be fragile and unstable in the future.

ACKNOWLEDGMENTS. The valuable input of numerous coworkers is appreciated. In particular, I thank P. E. Kellar, D. B. Shindler, E. White, and D. R. S. Lean for their insight and advice. Work in the area of primary productivity limitation was funded by the Department of Scientific and Industrial Research, New Zealand, The Canada Centre for Inland Waters, Burlington, Ontario, Canada, and the Water Resources Research Institute, University of North Carolina, Raleigh, N.C. Many thanks are due to P. Bland and J. Garner for aiding with the preparation of this manuscript, and to the Staff of the Institute of Marine Sciences, University of North Carolina, for enthusiastic support.

References

Alexander, M., 1971, *Microbial Ecology,* John Wiley & Sons, New York.

Anderson, G., Cronberg, G., and Gelin, C., 1973, Planktonic changes following the restoration of Lake Trummen, Sweden, *Ambio* **2**:44–47.

Arnon, D. I., 1958, Some functional aspects of inorganic micronutrients in the metabolism of green plants, in: *Perspectives in Marine Biology* (A. A. Buzzati-Traverso, ed.), pp. 351–383, Univ. of California Press, Berkeley.

Arnon, D. I., and Wessel, G., 1953, Vanadium as an essential element for green plants, *Nature* **172**:1039–1041.

Asato, Y., 1972, Isolation and characterization of ultraviolet light-sensitive mutants of the blue-green alga *Anacystis nidulans, J. Bacteriol.* **110**:1058–1064.

Banse, K., 1976, Rates of growth, respiration and photosynthesis of unicellular algae as related to cell size—a review. *J. Phycol.* **12**:135–140.

Beeton, A. M., 1969, Changes in the environment and biota of the Great Lakes, in: *Eutrophication: Causes, Consequences, Correctives,* pp. 150–187, Natl. Acad. Sci., Washington, D.C.

Benoit, J. R., 1957, Preliminary observations on cobalt and vitamin B_{12} in freshwater, *Limnol. Oceanogr.* **2**:233–240.

Berman, T., 1970, Alkaline phosphatases and phosphorus availability in Lake Kinneret, *Limnol. Oceanogr.* **15**:663–674.

Bitton, G., and Marshall, K. C. (eds.), 1980, *Adsorption of Microorganisms to Surfaces,* Wiley-Interscience, New York.

Bjälfve, G., 1962, Nitrogen fixation in cultures of algae and other microorganisms, *Physiol. Plant.* **15**:122–129.

Blackman, F. F. 1905, Optima and limiting factors, *Ann. Bot.* **19**:281–295.

Bortels, H., 1940, Uber dei bedeutung des molybdäns fur stickstoffbindende Nostocaceen, *Arch. Microbiol.* **11**:155–186.

Burns, N. M., and Ross, C., 1972, Oxygen-nutrient relationships within the central basin of Lake Erie, in: *Nutrients in Natural Waters* (H. E. Allen and J. R. Kramer, eds.), pp. 193–250, John Wiley & Sons, New York.

Burkholder, P. R., 1963, Some nutritional relationships among microbes of sea sediments and water, in: *Symposium on Marine Microbiology* (C. H. Oppenheimer, ed.), pp. 133–150, Charles C Thomas, Springfield, Ill.

Caldwell, D. E., and Caldwell, S. J., 1978, A *Zoogloea* sp. associated with blooms of *Anabaena flos aquae, Can. J. Microbiol.* **24**:922–931.

Carlucci, A. F., and Bowes, P. M., 1972, Determination of vitamin B_{12}, thiamine, and biotin in Lake Tahoe waters using modified marine bioassay techniques, *Limnol. Oceanogr.* **17**:774–776.

Chapman, A. G., Fall, L., and Atkinson, D. E., 1971, Adenylate energy charge in *Escherichia coli* during growth and starvation, *J. Bacteriol.* **108**:1072–1086.

Clayton, R. K., 1966, Physical processes involving chlorophyll *in vivo,* in: *Chlorophylls: Physi-*

cal, Chemical, and Biological Properties (I. P. Vernon and G. R. Seely, eds.), pp. 523–568, Academic Press, New York.

Cronberg, G., Gelin, C., and Larsson, K., 1975, Lake Trummen restoration project. II. Bacteria, phytoplankton and phytoplankton productivity, *Verh. Int. Verein. Limnol.* **19**:1088–1096.

Dillon, P. J., and Rigler, F. H., 1974, The phosphorus-chlorophyll relationship in lakes, *Limnol. Oceanogr.* **19**:767–773.

Edmondson, W. T., 1972, Nutrients and phytoplankton in Lake Washington, in: *Nutrients and Eutrophication: The Limiting Nutrient Controversy* (G. E. Liken, Ed.), pp. 172–193, Am. Soc. Limnol. Oceanogr. Spec. Symp. 1., Allen Press, Lawrence, Kans.

Edmondson, W. T.., 1977, Trophic equilibrium of Lake Washington, EPA Report 600/3-77-087.

Eyster, C., 1952, Necessity of boron for *Nostoc muscorum, Nature* **170**:755–756.

Eyster, C., 1968, Microorganic and microinorganic requirements for algae, in: *Algae, Man and the Environment* (D. F. Jackson, ed.), pp. 27–36, Syracuse Univ. Press, Syracuse, N.Y.

Fogg, G. E., Stewart, W. D. P., Fay, P., and Walsby, A. E., 1973, *The Blue-Green Algae,* Academic Press, New York.

Gallucci, K. K., 1981, Algal-bacterial symbiosis in the Chowan River, N.C., MSc. thesis, Univ. of North Carolina, Chapel Hill, N.C.

Gerhart, D. W., and Likens, G. E., 1975, Enrichment experiments for determining nutrient limitation: four methods compared, *Limnol. Oceanogr.* **20**:649–653.

Goldman, C. R., 1960, Molybdenum as a factor limiting primary productivity in Castle Lake, California, *Science* **132**:1016–1017.

Goldman, C. R., 1964, Primary productivity and micronutrient limiting factors in some North American and New Zealand Lakes. *Verh. Int. Verein. Limnol.* **15**:365–374.

Goldman, C. R., 1966, Micronutrient limiting factors and their detection in natural phytoplankton populations, in: *Primary Productivity in Aquatic Environments* (C. R. Goldman, ed.), pp. 123–125, Univ. of California Press, Berkeley,

Goldman, C. R., 1972, The role of minor nutrients in limiting the productivity of aquatic ecosystems, in: *Nutrients and Eutrophication* (G. E. Likens, ed.), pp. 21–38, Am. Soc. Limnol. Oceanogr. Spec. Symp. 1. Allen Press, Lawrence, Kans.

Goldman, C. R., and Carter, R., 1965, An investigation by rapid C-14 bioassay of factors affecting the cultural eutrophication of Lake Tahoe, California–Nevada, *J. Water Pollut. Contr. Fed.* **37**:1044–1059.

Goldman, C. R., and Armstrong, R., 1968, Primary productivity studies in Lake Tahoe, California, *Verh. Int. Verein. Theor. Angew. Limnol.* **17**:49–71.

Golterman, H. L., 1975a, *Physiological Limnology,* Elsevier, New York.

Golterman, H. L., 1975b, Chemistry of running water, in: *River Ecology* (B. Whitton, Ed.), pp. 234–271, Blackwell, Oxford.

Graham, D., Atkins, C. A., Reed, M. L., Patterson, B. D., and Smillie, R. M., 1971, Carbonic anhydrase, photosynthesis and light-induced pH changes, in: *Photosynthesis and Photorespiration* (M. D. Hatch, C. B. Osmond, and R. A. Slayter, Eds.), pp. 267–274, Wiley-Interscience, New York.

Griffith, E. J., Beeton, A. M., Spencer, J. M., and Mitchell, D. T. (eds.), 1973, *Environmental Phosphorus Handbook,* John Wiley & Sons, New York.

Holland, H. D., 1978, *The Chemistry of the Atmosphere and Oceans,* Wiley-Interscience, New York.

Holland, R. E., 1969, Seasonal fluctuations of Lake Michigan diatoms, *Limnol. Oceanogr.* **14**:423–436.

Holm-Hansen, O., Gerloff, G. C., and Skoog, F., 1954, Cobalt as an essential element for blue-green algae, *Physiol. Planta.* **7**:665–675.

Hopkins, E. F., 1930, The necessity and function of manganese in the growth of *Chlorella* sp., *Science* **72**:609–610.

Hutchinson, G. E., 1957, *A Treatise on Limnology,* Vol. 1, John Wiley & Sons, New York.

Hutchinson, G. E., 1961, The paradox of the plankton, *Am. Nat.* **95:**137–145.
Ingle, R. K., and Colman, B., 1975, Carbonic anhydrase levels in blue-green algae, *Can. J. Bot.* **53:**2385–2387.
Jannasch, H. W., and Pritchard, P. H., 1972, The role of inert particulate matter in the activity of aquatic microorganisms, in: *Detritus and its Role in Aquatic Ecosystems* (U. Santollini and J. W. Hopton, eds.), pp. 289–308, Ist. Ital. Idrobiol. 29 Suppl., Pallanza, Italy.
Joint Industry/Government Task Force on Eutrophication, 1969, Provisional algal assay procedure, Washington, D.C.
Jones, J. R., and Bachmann, R. W., 1975, Algal response to nutrient inputs in some Iowa lakes, *Verh. Int. Verein. Limnol.* **19:**904–910.
Karl, D. M., 1980, Cellular nucleotide measurements and applications in microbial ecology, *Microbiol. Rev.,* **44:**739–796.
Kellar, P. E., and Paerl, H. W., 1980, Physiological adaptations in response to environmental stress during an N$_2$-fixing *Anabaena* bloom, *Appl. Environ. Microbiol.* **40:**587–595.
Kerr, P. C., Paris, D. F., and Brockway, D. L., 1970, *The Interrelation of Carbon and Phosphorus in Regulating Heterotrophic and Autotrophic Populations in Aquatic Ecosystems,* U.S. Govt. Printing Office, Washington, D.C.
Kerr, P. C., Brockway, D. C., Paris, D. G., and Barnett, J. T., 1972, The interrelation of carbon and phosphorus in regulating heterotrophic and autotrophic populations in an aquatic ecosystem, Shriners Pond, in: *Nutrients and Eutrophication: The Limiting Nutrient Controversy* (G. E. Likens, ed.), pp. 41–62, Am. Soc. Limnol. Oceanogr. Spec. Symp. 1., Allen Press, Lawrence, Kans.
Kiefer, D. A., Holm-Hansen, O., Goldman, C. R., Richards, R., and Berman, T., 1972, Phytoplankton in Lake Tahoe: deep living populations, *Limnol. Oceanogr.* **17:**418–422.
Kilham, P., 1971, A hypothesis concerning silica and the freshwater planktonic diatoms, *Limnol. Oceanogr.* **16:**10–18.
Kolenbrander, G. J., 1972, The eutrophication of surface water by agriculture and the urban population, *Stikstof* **15:**56–67.
Kuentzel, L. E., 1969, Bacteria, carbon dioxide, and algal blooms, *J. Water Poll. Contr. Fed.* **41:**1737–1747.
Kuhl, A., 1968, Phosphate metabolism of green algae, in: *Algae, Man and the Environment* (D. F. Jackson, ed.), pp. 37–52, Syracuse Univ. Press, Syracuse, N.Y.
Lange, W., 1970, Cyanophyta-bacteria systems: effects of added carbon compounds or phosphate on algal growth at low nutrient concentrations, *J. Phycol.* **6:**230–234.
Lange, W., 1971, Enhancement of algal growth in cyanophyta-bacteria systems by carbonaceous compounds, *Can. J. Microbiol.* **17:**303–314.
Lange, W., 1973, Bacteria assimilable organic compounds, phosphate and enhanced growth of bacteria associated blue-green algae, *J. Phycol.* **9:**507–509.
Lange, W., 1976, Speculations on a possible essential function of the gelatinous sheath of blue-green algae, *Can. J. Microbiol.* **22:**1181–1185.
Lean, D. R. S., 1973, Movements of phosphorus between its biologically important forms in lake water, *J. Fish. Res. Bd. Can.* **30:**1525–1536.
Lehman, J. T., 1980, Release and cycling of nutrients between planktonic algae and herbivores, *Limnol. Oceanogr.* **25:** 620–632.
Lehman, J. T., Botkin, D. B., and Likens, G. E., 1975, Lake eutrophication and the limiting CO$_2$ concept: a simulation study, *Verh. Int. Verein. Limnol.* **19:**300–307.
Liebig, J., 1840, *Organic Chemistry in its Applications to Agriculture and Physiology* [English translation by L. Playfair], Taylor and Walton, London.
Likens, G. E., 1972, Eutrophication and aquatic ecosystems, in: *Nutrients and Eutrophication: The Limiting Nutrient Controversy* (G. E. Likens, ed.), pp. 3–13 Am. Soc. of Limnol. Oceanogr. Spec. Symp. 1., Allen Press, Lawrence, Kansas.

Lindstrom, K., 1980, Bioassays of selenium in Lake Erken, Sweden, *Arch. Hydrobiol.* **89:**110–117.

Lund, J. W. G., 1950, Studies on *Asterionella formosa* Hass. II. Nutrient depletion and the spring maximum, *J. Ecol.* **38:**1–14.

Lund, J. W. G., Mackereth, F. J. H., and Mortimer, C. H., 1963, Changes in depth and time of certain chemical and physical conditions and of the standing crop of *Asterionella formosa* Hass in the North Basin of Windermere in 1947, *Phil. Trans. Roy. Soc., B.* **246:**255–290.

Madsen, B. L., 1972, Detritus on stones in small streams, in: *Detritus and its Role in Aquatic Ecosystems* (U. Santolline and J. W. Hopton, Eds.), pp. 385–403, Mem. Ist. Ital. Idrobiol. 29 Suppl., Pallanza, Italy.

Maloney, T. E., Miller, W. E., and Shiroyama, T., 1972, Algal responses to nutrient additions in natural waters. I. Laboratory assays, in: *Nutrients and Eutrophication: The Limiting Nutrient Controversy* pp. 134–140, (G. E. Likens, ed.), Am. Soc. Limnol. Oceanogr. Spec. Symp. 1., Allen Press, Lawrence, Kans.

McColl, R. H. S., White, E., and Waugh, J. R., 1975, Chemical run-off in catchments converted to agricultural use, *New Zealand J. Sci.* **18:**67–84.

Melack. J., Kilham, P., and Fisher, T. R., 1982, Responses of phytoplankton to experimental fertilization with ammonium and phosphate in an African soda lake, *Oecologia* (in press).

Melchiorri-Santolini, U., and Hopton, J. W. (eds.), 1972, *Detritus and its Role in Aquatic Ecosystems,* Mem. Ist. Ital. Idrobiol. 29 Suppl., Pallanza, Italy.

Menzel, D. W., and Ryther, J. H., 1961, Nutrients limiting production of phytoplankton in the Sargasso Sea with special reference to iron, *Deep-Sea Res.* **7:**276–281.

Miller, W. E., Green, J. C., and Shiroyama, T., 1978, The *Selenastrum capricornutum* Printz algal assay bottle test, U.S. Environmental Protection Agency, Corvallis, Oreg.

Morton, S. D., Sernau, R., and Derse, P. H., 1972, Natural carbon sources, rates of replenishment, and algal growth, in: *Nutrients and Eutrophication: The Limiting Nutrient Controversy* (G. E. Likens, ed.), pp. 197–204, Soc. Limnol. Oceanogr. Spec. Symp. 1., Allen Press, Lawrence, Kans.

Moss, B., 1973, The influence of environmental factors on the distribution of freshwater algae: an experimental study. II. The role of pH and the carbon dioxide-biocarbonate system, *J. Ecol.* **61:**157–177.

Murphy, T. P., Lean, D. R. S., and Nalewajko, C., 1965, Blue-green algae: their excretion of iron-selective chelators enables them to dominate other algae, *Science* **192:**900–901.

Neilands, J. B. (ed.), 1974, *Microbial Iron Metabolism: A Comprehensive Treatise,* Academic Press, New York.

Nelson, E. B., Cenedella, A., and Tolbert, N. E., 1969, Carbonic anhydrase levels in *Chlamydomonas, Phytochemistry* **8:**2305–2306.

Newton, W., Postgate, J. R., and Rodriquez-Barrucco, C., (eds.), 1977, *Recent Developments in Nitrogen Fixation,* Academic Press, London.

Nusch, E. A., 1975, Comparative investigations of extent, causes and effects of eutrophication in Western German reservoirs, *Verh. Int. Verein. Limnol.* **19:**1871–1879.

Oláh, J., 1972, Leaching, colonization and stabilization during detritus formation, in: *Detritus and its Role in Aquatic Ecosystems* (U. Santollini and J. W. Hopton, eds.), pp. 105–127, Mem. Ist. Ital. Idrobiol. 29 Suppl., Pallanza, Italy.

Otsuki, A., and Wetzel, R. G., 1972, Coprecipitation of phosphate with carbonates in a marl lake, *Limnol. Oceanogr.* **17:**763–767.

Paerl, H. W., 1974, Bacterial uptake of dissolved organic matter in relation to detrital aggregation in marine and freshwater systems, *Limnol. Oceanogr.* **19:**966–972.

Paerl, H. W., 1975, Microbial attachment to particles in marine and freshwater ecosystems, *Microbiol. Ecol.* **2:**73–83.

Paerl, H. W., 1979, Optimization of carbon dioxide and nitrogen fixation by the blue-green algae *Anabaena* in freshwater blooms, *Oecologia* **32:**135–139.

Paerl, H. W., and Goldman, C. R., 1972, Stimulation of heterotrophic and autotrophic activities of a planktonic microbial community by siltation of Lake Tahoe, California, in: *Detritus and its Role in Aquatic Ecosystems* (U. Santollini and J. W. Hopton, eds.), pp. 129–147, Mem. Ist. Ital. Idrobiol. 29 Suppl., Pallanza, Italy.

Paerl, H. W., and Kellar, P. E., 1978, Significance of bacterial-*Anabaena* (Cyanophyceae) associations with respect to N_2 fixation in freshwater, *J. Phycol.* **14:**254–260.

Paerl, H. W., and Kellar, P. E., 1979, N_2-fixing *Anabaena:* Physiological adaptations instrumental in maintaining surface blooms, *Science* **204:**620–622.

Paerl, H. W., and Ustach, J. F., 1981, Blue-green algal scums: an explanation for their occurrence during freshwater blooms, *Limnol. Oceanogr.* **27:**212–217.

Paerl, H. W., Richards, R. C., Leonard, R. L., and Goldman, C. R., 1975, Seasonal nitrate cycling as evidence for complete vertical mixing in Lake Tahoe, California–Nevada, *Limnol. Oceanogr.* **20:**1–8.

Paerl, H. W., Tilzer, M. M., and Goldman, C. R., 1976, Chlorophyll *a* versus adenosine triphosphate as algal biomass indicators in lakes, *J. Phycol.* **12:**242–246.

Patalas, K., 1972, Crustacean plankton and the eutrophication of St. Lawrence Great Lakes, *J. Fish. Res . Bd. Can.* **29:**1451–1462.

Pleisch, P., 1970, Die herkunft eutrophierender stoffe beim Pfäffiker- und Greifensee, *Vierteljahresschr. Naturforsch. Ges. Zurich* **115:**127–129.

Provasoli, L., 1960, Micronutrients and heterotrophy as possible factors in bloom production in natural waters, in: Trans. Semin. Algae and Metropolitan Wastes, U.S. Public Health Service, R. A. Taft Sanit. Eng. Center, Cincinnati, Ohio.

Provasoli, L., 1963, Growing marine seaweeds, *Proc. Int. Seaweed Symp.* **4:**9–17. Pergamon Press, Oxford.

Remane, A., and Schlieper, C., 1971, *Biology of Brackish Water,* Wiley-Interscience, New York.

Reynolds, C. S., and Walsby, A. E., 1975, Water blooms, *Biol. Rev.* **50:**437–481.

Richerson, P. J., Armstrong, R., and Goldman, C. R., 1970, Contemporaneous disequilibrium, a new hypothesis to explain the "paradox of the plankton," *Proc. Natl. Acad. Sci. U.S.A.* **67:**1710–1714.

Rigler, F. H., 1973, A dynamic view of the phosphorus cycle in lakes. in: *Environmental Phosphorus Handbook* (E. J. Griffith, A. Beeton, J. M. Spencer, and D. T. Mitchell, eds.) pp. 539–572, John Wiley & Sons, New York.

Round, F. E., 1965, *The Biology of the Algae,* St. Martin's Press, New York.

Ryther, J. H., and Guillard, R. R. L., 1959, Enrichment experiments as a means of studying nutrients limiting to phytoplankton production, *Deep-Sea Res.* **6:**65–69.

Schelske, C. L., 1962, Iron, organic matter, and other factors limiting primary productivity in a marl lake, *Science,* **136:**45–46.

Schelske, C. L., and Stoermer, E. F., 1971, Eutrophication, silica depletion and predicted changes in algal quality in lake Michigan, *Science* **173:**423–424.

Schelske, C. L., and Stoermer, E. F., 1972, Phosphorus, silica, and eutrophication of Lake Michigan, in: *Nutrients and Eutrophication* (G. E. Likens, ed.), pp. 157–171, Amer. Soc. Limnol. Oceanogr. Spec. Symp. 1., Allen Press, Lawrence, Kans.

Schelske, C. L., Rothman, E. D., and Simmons, M. S., 1978, Comparison of bioassay procedures for growth-limiting nutrients in the Laurentian Great Lakes, *Mitt. Intern. Limnol.* **21:**65.-80.

Schindler, D. W., and Fee, E. J., 1973, Diurnal variations of dissolved inorganic carbon and its use in estimating primary production and CO_2 invasion in Lake 227, *J. Fish. Res. Bd. Can.* **30:**1501–1510.

Schindler, D. W., Kling, H., Schmidt, R. V., Prokopowich, J., Frost, V. E., Reid, R. A., and Capel, M., 1973, Eutrophication of Lake 227 by addition of phosphate and nitrate: the second, third and fourth years of enrichmant 1970, 1971, and 1972, *J. Fish. Res. Bd. Can.* **30:**1415–1440.

Shapiro, J., 1964, Effect of yellow acids on iron and other metals in water, *J. Am. Water Works Assoc.* **56:**1062–1082.

Shapiro, J., 1969, Iron in natural waters—its characteristics and biological availability as determined with the ferrigram. *Verh. Int. Verein. Theor. Angew. Limnol.* **17:**456–466.

Shapiro, J., 1973, Blue-green algae: why they become dominant, *Science* **179:**382–384.

Shapiro, J., and Glass, G. E., 1970, Chemical factors stimulating growth of Lake Superior algae, in: *Proceedings of the Conference on the Biology of Lake Superior* (G. E. Glass, ed.), p. 17, Environmental Protection Agency, Duluth, Minn.

Shapiro, J., and Glass, G. E., 1975, Synergistic effects of phosphate and manganese on growth of Lake Superior algae, *Verh. Int. Verein. Limnol.* **19:**395–404.

Shiroyama, T., Miller, W. E., and Green, J. C., 1976, Comparison of the algal growth responses of *Selenastrum capricornutum* Printz and *Anabaena flos-aquae* (Lyngb.) DeBrebisson in waters collected from Shagawa Lake, Minnesota, in: *Biostimulation and Nutrient Assessment* (E. J. Middlebrooks, D. H. Falkenborg, and T. E. Maloney, eds), pp. 127–148, Ann Arbor Science Publishers, Ann Arbor, Mich.

Stanier, R. Y., Douderoff, M., and Adelberg, E. A., 1963, *The Microbial World* (2nd ed.), Prentice-Hall, Englewood Cliffs, N.J.

Steeman Nielsen, E., and Wium-Anderson, S., 1971, The influence of Cu on photosynthesis and growth in diatoms, *Physiol. Plant.* **24:**480–484.

Stewart, W. D. P., 1974, *Algal Physiology and Biochemistry, Botanical Monographs,* Blackwell, Oxford.

Stewart, W. D. P., 1977, A botanical ramble among the blue-green algae, *Br. Phycol. J.* **12:**89–115.

Ström, K. M., 1933, Nutrition of algae. Experiments upon: the feasibility of the Schreiber method in fresh waters; the relative importance of iron and manganese in the nutritive medium; the nutritive substance given off by lake bottom muds, *Arch. Hydrobiol.* **25:**38–47.

Stumm, W., and Morgan, J. J., 1970, *Aquatic Chemistry,* Wiley-Interscience, New York.

Sunda, W. G., and Hanson, P. J., 1978, Chemical speciation of copper in river waters, in: *Chemical Modelling in Aqueous Systems* (Everett A. Jenne, ed.), pp. 147–180, ACS Symposium No. 93., Miami, Fla.

Talling, J. F., 1961, Photosynthesis under natural conditions, *Ann. Rev. Plant Physiol.* **12:**133–154.

Thayer, G. W., 1974, Identity and regulation of nutrients limiting phytoplankton production in the shallow estuaries near Beaufort, N.C., *Oecologia* **14:**75–92.

Thomas, E. A., 1973, Phosphorus and eutrophication, in: *Environmental Phosphorus Handbook* (E. J. Griffith, A. Beeton, J. M. Spencer, and D. T. Mitchell, eds.), pp. 585–611, John Wiley & Sons, New York.

Thurlow, D. L., Davis, R. B., and Sassevilk, D. R., 1975, Primary productivity, phytoplankton populations and nutrient bioassays in China Lake, Maine, U.S.A., *Verh. Int. Verein. Limnol.* **19:**1029–1036.

Tilman, D., and Kilham, S. S., 1976, Phosphate and silicate growth and uptake kinetics of the diatoms *Asterionella formosa* and *Cyclotella meneghiniana* in batch and semicontinuous culture, *J. Phycol.* **12:**375–383.

Tilman, D., Mattson, M., and Langer, S., 1981, Competition and nutrient kinetics along a temperature gradient: an experimental test of the mechanistic approach to niche theory, *Limnol. Oceanogr.* (in press).

Tilzer, M. M., 1973, Diurnal periodicity in the phytoplankton assemblage of a high mountain lake, *Limnol. Oceanogr.* **18:**15–30.

Tilzer, M. M., Goldman, C. R., and De Amezaga, E., 1975, The efficiency of photosynthetic light energy utilization by lake phytoplankton, *Verh. Int. Verein. Limnol.* **19:**800–807.

Tilzer, M. M., Paerl, H. W., and Goldman, C. R., 1977, Sustained viability of aphotic phytoplankton in Lake Tahoe (California–Navada), *Limnol. Oceanogr.* **27:**84–91.

Toerien, D. F., and Steyn, D. J., 1975, The eutrophication levels of four South African impoundments, *Verh. Int. Verein. Limnol.* **19:**1947–1956.

Tsusue, A., and Holland, H. D., 1966, The coprecipitation of cations with $CaCO_3$. 3, *Geochim. Cosmochim. Acta.* **30:***439–453.*

Vance, B. D., 1966, Sensitivity of *Microcystis aeruginosa* and other blue-green algae and associated bacteria to selected antibiotics, *J. Phycol.* **2:**125–128.

Vollenweider, R. A., 1968, *Water Management Research: Scientific Fundamentals of the Eutrophication of Lakes and Flowing Waters, with Particular Reference to Nitrogen and Phosphorus as Factors in Eutrophication,* Technical Report of the Organization for Economic Cooperation and Development, Paris, DAS/CSI/68.27.

Vollenweider, R. A., and Nauwerck, A., 1961, Some observations on the C-14 method for measuring primary production, *Verh. Int. Verein. Limnol.* **14:**134–139.

Walker, J. B., 1953, Inorganic micronutrient requirements of *Chlorella.* I. Requirements for calcium (or strontium), copper and molybdenum, *Arch. Biochem. Biophys.* **46:**1–11.

Walsby, A. E., 1975, Gas vescicles, *Ann. Rev. Plant Physiol.* **36:**427–439.

Weibe, W. J., and Pomeroy, L. R., 1972, Microorganisms and their association with aggregates and detritus in the sea: a microscopic study, in: *Detritus and its Role in Aquatic Ecosystems* (U. Santollini and J. W. Hopton, Eds.), pp. 325–352. Mem. Ist. Ital. Idrobiol. 29 Suppl., Pallanza, Italy.

Weiler, R. R., 1975, Carbon dioxide exchange and productivity in Lake Erie and Lake Ontario, *Verh. Int. Verein. Limnol.* **19:**694–704.

Weiss, C. M., and Helms, R. W., 1971, The Interlaboratory Precision Test. An Eight Laboratory Evaluation of the Provisional Algal Assay Procedure Bottle Test, E. P. A. Report.

Wetzel, R. G., 1964, A comparative study of the primary productivity of higher aquatic plants, periphyton, and phytoplankton in a large, shallow lake, *Int. Rev. Ges. Hydrobiol.* **49:**1–61.

Wetzel, R. G., 1975, *Limnology,* W. B. Saunders, Philadelphia.

Wetzel, R. G., 1981, Longterm dissolved and particulate alkaline phosphatase activity in a hardwater lake in relation to lake stability and phosphorus enrichments, *Verh. Int. Verein. Limnol.* **21:**337–349.

White, E., Payne, G., Pickmere, S., and Pick, F., 1982, The relative importance of nitrogen and phosphorus in eutrophication of Canadian and New Zealand lakes, *Can. J. Fish. Aquat. Sci. (in press).*

Whitton, B. A. (ed.), 1975, *River Ecology,* Blackwell, Oxford.

Williams, N. J. 1974, Zooplankton transport of nutrients into the epilimnion of Castle Lake, California, Ph.D. thesis, Univ. of California, Davis.

Williams, R. B., 1972, Nutrient levels and phytoplankton productivity in the estuary, in: *Proceedings of the Coastal Marsh and Estuary Management Symposium* (R. H. Chabreck, ed.), Louisiana State Univ. Div. of Contin. Education, Baton Rouge, La.

Wuhrmann, K., and Eichenberger, E., 1975, Experiments on the effects of inorganic enrichment of rivers on periphyton primary productivity, *Verh. Int. Verein Limnol.* **19:**2028–2034.

4

Biomass and Metabolic Activity of Heterotrophic Marine Bacteria

F. B. VAN ES AND L.-A. MEYER-REIL

1. Introduction

In the years since 1946, when ZoBell published his seminal book on marine microbiology, it has become evident that heterotrophic bacteria play an important role in nutrient cycling in the sea and that they are also basic members of the food web (Petipa *et al.*, 1970; Sorokin, 1971, 1978; Pomeroy, 1970, 1974; Sieburth, 1976; Williams, 1981). Much evidence for this was obtained in the late 1960s and early 1970s, using the ^{14}C-tracer technique introduced by Parsons and Strickland (1962). This technique enables the uptake of selected organic substrates by natural communities to be studied under *in situ* conditions. It could be shown that organisms smaller than 1–3 μm did indeed assimilate the largest fraction of these dissolved organic substrates when added to seawater samples in concentrations of a few micrograms per liter. As the information provided by the Parsons and Strickland technique is restricted to the fate of the selected substrates, several other methods for the determination of bacterial activity have been developed, giving more direct information on the dynamics of bacterial communities.

Our understanding of the role of heterotrophic bacteria in (marine) ecosystems has also been improved by the development of methods to determine total bacterial numbers and biomass. Although microscopic counting tech-

F. B. VAN ES ● Department of Microbiology, Kerklaan 30, 9751 NN Haren, The Netherlands.   L.-A. MEYER-REIL ● Institut für Meereskunde, Abteilung Marine Mikrobiologie, 2300 Kiel 1, Federal Republic of Germany.

niques have been used for a long time, the problem of recognizing very small bacterial cells in the presence of detrital and inorganic particles of comparable size rendered these methods rather inaccurate. Total microbial biomass parameters, such as ATP, reflect the biomass of at least three different trophic levels; algae, bacteria, and protozoa, and hence are of limited value in studies of the relations between them. However, once specific staining of bacterial cells with fluorescent dyes was possible and the necessary optical instrumentation available, a highly specific counting technique using epifluorescence microscopy could be developed (Francisco *et al.*, 1973; Zimmermann and Meyer-Reil, 1974; Hobbie *et al.*, 1977). Though very time-consuming, this technique also enables a reliable estimate to be made of the biovolume and, hence, biomass of the bacterial population in the presence of other organisms.

We propose to discuss current methods of studying aerobic heterotrophic bacteria in the sea and their role in carbon flux. In Fig. 1, the main relations of bacteria with other components of the ecosystem are summarized. The thick arrows represent the processes of main interest in this context. The abiotic parameters DOM (dissolved organic matter) and its "modified" form (which results from bacterial action) DOM* can be replaced by DIN (dissolved inorganic nutrients) and POM (particulate organic matter) and, in fact, by every substance that takes part in the flow of material through the food web. There are mutual relations between bacteria that are free or attached to particles and other components of the ecosystem. Phytoplankton may provide bacteria directly with organic matter through excretion and through decay of dead phytoplankton cells. On the other hand, bacteria, in their well-known role of mineralizers, provide the phytoplankton with nutrients that may be available in limiting amounts. In such cases, phytoplankton growth can even by directly dependent on the bacterial activity.

The relations between bacteria and zooplankton are less well documented. Losses of dissolved organic carbon derived from phytoplankton during grazing are considerable (Lampert, 1978; Copping and Lorenzen, 1980; see Fenchel and Jørgensen, 1977, for a review of earlier data) and fecal pellets are rapidly colonized by bacteria (Newell, 1965). Moreover, the stimulation of microbial

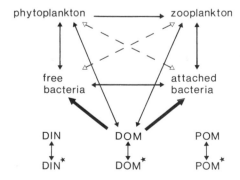

Figure 1. Simplified diagram of relations between bacteria and other components of the aquatic ecosystem. Thick arrows indicate processes of main interest. DIN (dissolved inorganic nutrients), DOM (dissolved organic matter), and POM (particulate organic matter) may, as a result of bacterial action, undergo chemical modifications (indicated by DIN*, etc.).

activity by grazing bacteriovores has been described for many environments (Javornitsky and Prokesova, 1963; Hamilton, 1973; Harrison and Mann, 1975; Fallen and Pfaender, 1976; Fenchel and Harrison, 1976; Abrams and Mitchell, 1980). Conversely, bacteria can serve as the sole or an additional source of food for many protozoa as well as metazoans, as was already pointed out more than 45 years ago (ZoBell and Feltham, 1937; and several papers cited therein). In this respect, the relation between free-living bacteria and those attached to surfaces is of importance. It may well be that filter-feeding zooplankters are capable of filtering bacteria attached to small particles, and incapable of catching free-living ones (Fenchel, 1980a,b,c).

In his book *The Structure of Marine Ecosystems,* Steele (1974) ignored heterotrophic bacteria and other microheterotrophs: he judged their role to be insignificant. It is hoped that through the following sections the reader will obtain a more detailed insight into the role of heterotrophic bacteria in marine ecosystems. In this respect we may also recommend the paper of Williams (1981) which recently came to our attention. In Sections 2 and 3, bacterial number and biomass determinations, and activity measurements will be discussed respectively, with the aim of facilitating a proper ecological interpretation of their applications and results. In Section 4, some results are brought together in order to discuss some long-standing questions and to enable general conclusions to be drawn.

2. Determination of Bacterial Number and Biomass

Most marine pelagic bacteria are small (< 1 μm), and their total numbers are large (see Table III below). Two difficulties arise from these facts. First, it is difficult to estimate the total number accurately, because the bacteria can easily escape attention. Second, large errors may arise when bacterial numbers are, via biovolume, converted into biomass, because small errors in estimating the dimensions of each cell may result in large errors in biomass calculations. For that reason, much effort has been put into developing alternative methods, based on the chemical determination of cell constituents. These indirect approaches are based on the assumption that the constituent analyzed is by weight a constant fraction of the cell mass. Table I lists the most commonly used direct and indirect methods.

2.1. Bright Field and Phase Contrast Microscopy

Since 1676, when Antonie van Leeuwenhoek first saw and described bacteria in water samples, the microscope has been an indispensable tool for microbiologists. Van Leeuwenhoek even tried to quantify those small animals, as he called them, and occasionally found millions of them in a drop of water. More quantitative procedures were described much later (Rasumov, 1932).

Table I. Bacterial Standing Stock Measurements in Natural Waters

Parameter	Method	Units	Derived units	Remarks/ limitations	References
Total number	Light microscopy	N/ml		Microscopy	Sorokin and Kadota, 1972
Total number	Epifluorescence microscopy (EFM)	N/ml		Microscopy	Zimmermann and Meyer-Reil, 1974; Hobbie et al., 1977
Total number	Scanning electron microscopy	N/ml		Microscopy	Zimmerman, 1977; Bowden, 1977; Larsson et al., 197
Total biomass	EFM	μm^3/ml	μg/ml	Microscopy/ conversion factor	Meyer-Reil, 1977
Total biomass	ATP	μg/liter	μgC/liter	Conversion factor	Holm-Hansen and Booth, 1966; Holm-Hansen and Paerl, 1972; Paerl and Williams, 1976
	Muramic acid	μg/liter	μgC/liter	Conversion factor	Moriarty, 1979
	Lipopolysaccharides	μg/liter	μgC/liter	Conversion factor	Watson et al., 1977
Active part of the community	Viable counts + MPN[a]	N/ml			Sorokin and Kadota, 1972
	MPN with ^{14}C substrates	N/ml			Ishida and Kadota, 1979; Lehmicke et al., 1979
	Autoradiography	N/ml		Microscopy	Brock and Brock, 1968; Hoppe, 1976
	Autoradiography + EFM	N/ml	% of total	Microscopy	Meyer-Reil, 1978a
	ETS[b]-active cells + EFM	N/ml	% of total	Microscopy	Zimmermann et al., 1978
	Naladixic acid incubation + EFM	N/ml	% of total	Microscopy	Kogure et al., 1979

[a]MPN = most probable number technique.
[b]ETS = electron transport system.

For a discussion on technical aspects and counting errors see Sorokin and Kadota (1972). Microscopy has the major disadvantage that even after the samples have been stained or viewed under the phase-contrast microscope, many bacteria are unrecognizable. A further disadvantage that all microscopic methods have in common is a certain degree of subjectivity.

2.2. Epifluorescence Microscopy

Bacteria can be stained with fluorescent dyes; for example, with acridine orange (Strügger, 1949), which greatly facilitates the counting of bacteria in natural samples, even in the presence of particles (Francisco et al., 1973). An improved procedure (Zimmermann and Meyer-Reil, 1974), using polycarbonate filters, allows small (< 0.5 μm) cells to be counted and cells to be differentiated according to size classes for biovolume estimation. Polycarbonate filters are superior to filters based on cellulose esters (such as Millipore and Sartorius) because of their plain surface, definite pore size, and low retention of fluorescent dyes (Zimmermann and Meyer-Reil, 1974; Hobbie et al., 1977).

To reduce errors resulting from subjectivity, counting has to meet certain requirements. Only those forms that have a clear outline, a recognizable bacterial shape, and bright orange or green fluorescence should be counted as bacterial cells (Meyer-Reil, 1977). Earlier suggestions that the cells that fluoresce green are alive, and those that fluoresce orange are dead (Pugsley and Evison, 1974) have proved to be wrong (Zimmermann et al., 1978). The difference in color is a result of the relative contents of RNA and double and single strand DNA within the cells (Dalay and Hobbie, 1975; Hobbie, 1976). Fast-growing cells contain relatively more RNA and single-strand DNA (which absorb more dye) and, therefore, fluoresce orange. Moreover, the color depends greatly on how the samples are treated and prepared. A change from orange to green fluorescence can be obtained by thoroughly washing the stained cells with ethanol or even with seawater. Several variations on the staining and destaining procedure have been described (Hobbie et al., 1977; Zimmermann et al., 1978). However, polycarbonate filters are preferable to filters of cellulose derivates, as many of the numerous small cocci may penetrate into the rather spongy structure of the latter and escape observation.

Epifluorescence microscopy allows the spectrum of microbial cells occurring in natural samples (e.g., rods, cocci, filamentous cells, cf. Fig. 2) to be studied. The spectrum and abundance of forms vary locally and seasonally even on a short time scale (Meyer-Reil, 1977; Meyer-Reil et al., 1979; C. Krambeck, personal communication). Moreover, the ratio of free-floating and attached bacteria can be determined relatively easily. Although some years ago, it was generally thought that most bacteria in seawater are attached to surfaces (ZoBell, 1946; Darnell, 1967; Odum and De la Cruz, 1967; Seki, 1972), with the microscopic counting techniques now available, it appears that in earlier investigations, many of the very small free cells may have been overlooked. Many recent observations lead to the conclusion that most cells are not attached to particles (Wiehe and Pomeroy, 1972; Hobbie et al., 1972; Zimmermann, 1977; Ferguson and Palumbo, 1979; Hollibaugh et al., 1980), though in estuarine and coastal waters, the relative abundance of cells attached to particles has been reported to be considerably higher (Wilson and Stevenson,

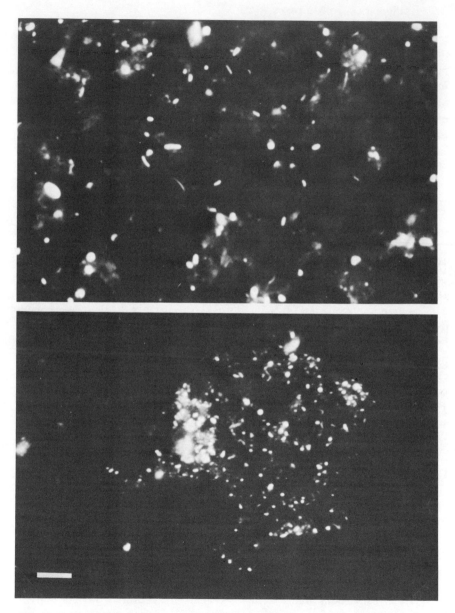

Figure 2. Epifluorescence photograph of bacteria in natural samples. The photograph above shows mainly free floating bacteria from the water overlying beach sediments (Kiel Fjord, Baltic Sea, FRG). The photograph below shows bacteria colonizing a detritus particle (Pacific Ocean; water depth, 500 m). Note the differences in spectrum and size of bacteria. Bar represents 3 μm.

1980; J. McN. Sieburth *et al.*, unpublished data). Epifluorescence microscopy has also been used to determine bacterial growth rates under natural conditions: for this method, the percentage of dividing cells must be ascertained microscopically (see Section 3.7).

Epifluorescence microscopy has been used successfully in combination with other techniques. The determination of the number of respiring cells (Iturriaga and Rheinheimer, 1975; Zimmermann *et al.*, 1978) is based on the observation that aerobic cells can reduce artificial electron acceptors, such as tetrazolium salts,to insoluble forms (formazans) which are deposited as granulae within the cell. These deposits can be seen by light microscopy. As the presence or absence of such an essential requirement as an active respiratory system is tested, this is a good estimate of actively metabolizing cells, provided that all active cells do take up the tetrazolium salt added and the formazan granulae are big enough to be seen. This may be difficult in samples with very many small cells, as in those from offshore waters. With this technique, it is possible within the same preparation to count the total numbers of cells by epifluorescence, and the number of these cells respiring, without having to culture them.

Kogure *et al.* (1979) supplied samples with yeast extract (0.025%) and naladixic acid (0.002%) and incubated them for 6 hr. Naladixic acid is a specific inhibitor of DNA synthesis and prevents the cell division of gramnegative bacteria. As a result, active cells grow out to elongated forms, easily recognizable under the microscope. The total number of cells and the elongated cells were counted using epifluorescence microscopy. This direct count of viable cells, probably a minimum estimate of viable cell numbers, was 5–10% of total counts in open Pacific Ocean water, and roughly 1000 times higher than viable counts on agar plates (Kogure *et al.*, 1979). Most of the cells counted as viable with this direct method were free-living and about 70% were rod-shaped.

Fluorescence microscopy has been combined with the autoradiographic determination of the uptake of ^{3}H- or ^{14}C-labeled organic substrates by the active fraction of the bacterial population. Although autoradiographic methods of studying bacteria (Brock, 1967; Brock and Brock, 1968; Munro and Brock, 1968; Hoppe, 1974, 1976, 1977) have been used successfully, problems arise in relating spots caused by radioactivity on the film of indivdual cells. Single silver grains or small groups of silver grains, often regarded as background, seem to be an additional problem. As an alternative, autoradiographs of samples have been stained with traditional dyes (Paerl, 1974; Ramsey, 1974; Faust and Corell, 1977), fluorescent antibodies (Fliermans and Schmidt, 1975) or fluorescent dyes (Meyer-Reil, 1978a). As antibody reactions are very specific, the combination of this staining and autoradiography makes it possible to follow the activity and population dynamics of one or a few species in the natural habitat. However, when a property of a total bacterial community has to be determined, for example, the fraction of active cells or the fraction of cells capable of assimilating specific substances, the total number of cells present needs to be counted concomitantly. As pointed out before, this can be done in

preparations combining microautoradiography and epifluorescence microscopy. A double filter technique may be used to differentiate between tritium
and other β-emitters (Field *et al.*, 1965). The potential applications of the elegant microautoradiographic technique have certainly not been fully explored
in ecological studies.

The microscopic determinations of the biomass require conversion factors
to convert from cell number to total biovolume and, subsequently, to biomass.
The correct measurement of cell dimensions has been verified with Scanning
Electron Microscopic (SEM) studies (Zimmerman, 1975, 1977; Bowden,
1977; Larsson *et al.*, 1978; Fuhrman, 1981). Fuhrman (1981) has concluded
that epifluorescence microscopy is more accurate than SEM for the determination of bacterial size and volume, as during preparation for SEM, cells
shrink a variable amount. Reported average biovolume values for estuarine and
coastal waters correspond fairly well: from 0.05 to 0.09 μm^3 (Zimmermann,
1975, 1977; Ferguson and Rublee, 1976; Bowden, 1977; Fuhrman *et al.*, 1980;
J. McN. Sieburth *et al.*, unpublished data), though Watson *et al.* (1977)
observed a decrease from 0.28 to 0.09 μm^3 within a week in Woods Hole
(U.S.A.) coastal water. The average biovolume spectrum within single communities ranges from 0.01 μm^3 or less for the smallest forms (Zimmermann,
1977; Meyer-Reil *et al.*, 1979; Wilson and Stevenson, 1980) to over 0.5 μm^3
for long rods and *Hyphomicrobium* cells (Watson *et al.*, 1977; Zimmermann,
1977). Bacteria from natural environments generally are considerably smaller
than cultured cells, used as a reference (Watson *et al.*, 1977; Wilson and Stevenson, 1980; see also Fig. 2). Proposed values for the conversion from biovolume to biomass range from 82×10^{-15} to 121×10^{-15} g C/μm^3 (see Bowden, 1977; Watson *et al.*, 1977; Zimmermann, 1977; Sorokin, 1978).
Romanenko and Dobrynin (1978) and Van Veen and Paul (1979) discussed
the specific weight values of bacterial cells, used in the conversion factor from
biovolume to biomass, and arrived at averaged values of 1.6×10^{-12} and $0.8
\times 10^{-12}$ g/μm^3, respectively.

2.3. Viable Count Techniques

The classic method of counting bacteria is to determine colony forming
units (CFU) of viable cells on agar plates. Although this has the serious disadvantage that only a small fraction (0.0001–10%) of a total population is
counted (ZoBell, 1946; Jannasch and Jones, 1959; Taga and Matsuda, 1974;
Meyer-Reil, 1977), the method is still widely used. The number of colony-
forming units may serve as an additional parameter in the study of various
aspects of natural communities; it can be taken to represent those bacteria that
need high nutrient concentrations for their growth (saprophytic bacteria).
Moreover, indicator organisms such as *Escherichia coli* and *Streptococcus faecalis* are counted as CFU on selective media, in order to evaluate the sanitary
status of water bodies (APHA, 1971). CFU are enumerated on media solidi-
fied with agar or silica, enabling one to count separate colonies of an appro-

priate dilution of a sample. With the most probable number (MPN) technique, the concentration of cells capable of growth in an enriched medium is determined. The technique is based on a serial dilution of the sample down to theoretically one cell per volume. After incubation, the number of positive and negative scores (growth in replicate tubes, inoculated from three or more succeeding dilutions) is recorded. From this, the statistically most probable number is calculated or derived via a table (Meynell and Meynell, 1965; APHA, 1971).

There may be many reasons why a difference of several orders of magnitude is frequently observed between total and viable counts:

1. Some of the cells in the natural habitat are inactive.
2. Bacteria may not be able to grow on the substrate or combination of substrates available.
3. In principle, the substrate may be suitable, but the concentration may either be too high, resulting in substrate-accelerated death (Postgate and Hunter, 1964) or too low (which will not often be the case). Bacteria may not be able to grow at the boundary between a solid and a gaseous (air) phase, because too much oxygen is available, or because of surface tension.
4. The cells grow too slowly to produce visible colonies (even when viewed with a binocular microscope) within the chosen incubation period (Torrella and Morita, 1981).
5. Cells are inactivated by other (possibly daughter) cells in their immediate vicinity, resulting in colonies of only a few cells, invisible under a binocular microscope (Kuznetsov et al., 1979).
6. Cells are inactivated, e.g., because of changes in temperature or pressure during sampling or mixing with melted agar.
7. Bacteria may tend to clump, or already be clumped in the original sample (e.g., microcolonies), resulting in only one visible colony originating from more than one cell.
8. Cells adhere to the glass walls of dilution tubes and pipettes and are thus removed from the sample.

See Jensen (1967), Lewin (1974), Litchfield et al. (1975), and Väätänen (1977) for a review of some of these points and optimization procedures. Some of the reasons listed here may be true without being very important per se. But if each led to the total being underestimated by half, together they could account for a difference of more than two orders of magnitude.

Ishida and Kadota (1979) and Lehmicke et al. (1979) have overcome some of the difficulties listed above by using liquid media and testing the production of $^{14}CO_2$ from traces of labeled organic substrates added to the incubation tubes. As Lehmicke et al. (1979) have pointed out, this counting technique would be excellent to count bacteria that actively degrade xenobiotic substances at μg/liter concentrations in natural waters.

Sieburth and coworkers (Sieburth, 1979; Sieburth et al., unpublished data) hypothesized that there are two physiologically distinct groups of heterotrophic bacteria in the sea. The first group (comprising the vast majority of bacteria) are free-living cells, capable of growing at environmental levels of substrate concentrations. They are assumed to be unable to grow at the high substrate concentrations used on agar plates. The cells that do grow on agar plates are thought to be recruited from the small fraction that grows attached to particles and requires higher substrate concentrations.

The fact that duplicate viable counts from the same sample mostly give comparable results indicates that viable counts can give a reproducible reflection of a population characteristic. In that respect, a comparison of data from different areas or from different times of the year may make sense. We will return to the discrepancy between total and viable counts in Section 4. However, for the reasons mentioned above, CFU cannot serve as an estimate of bacterial biomass in samples from natural environments.

2.4. Chemical Determinations

The determination of bacterial standing stock by microscopy has several limitations: it requires much time, suffers from a certain degree of subjectivity on the part of the microscopist, and conversion from numbers to biomass requires the determination of biovolume and a factor to convert the data from μm^3 to μg. For these reasons, much effort has been put into the development of alternative methods, based on the measurement of the concentration of a cell component and a previously determined relation between this cell component and biomass. The prerequisites for such a cell component were formulated by Holm-Hansen (1973, p. 215):

(a) The measured compound must be present in all living cells, and must be absent from all dead cells. (b) It must not be found associated with nonliving, detrital material. (c) It must exist in fairly uniform concentrations in all cells, regardless of environmental stresses. (d) The analytical technique for the measurement of the biomass indicator must be capable of measuring submicrogram quantities and must be relatively quick and easy to use for large scale programs involving numerous samples.

The requirement that the component to be measured should be absent from all dead cells is necessary in studies on population dynamics. But in food-web studies, cells that are dead but otherwise intact may be just as valuable as food as are living cells. Requirement (d) is not a basic one, but even its partial fulfilment would serve to make the method more convenient.

2.4.1. ATP

Holm-Hansen (1973) and many others have found that ATP fulfills all these requirements, and much literature on ATP measurements in marine and other environments has appeared since the method was developed (Holm-Hansen and Booth, 1966). The subject was recently reviewed by Karl (1980). The conversion factor from ATP to cell carbon has been studied in many organisms under various conditions and has been found to be rather variable, because of the dependence of the physiological conditions and the differences between organisms (Karl, 1980; Tables III–VI). For the overall community C/ATP ratio, however, a factor of 250 may be a reasonable approximation. The total content of adenine nucleotides (AMP + ADP + ATP) may be a more stable measure of biomass (Karl, 1980; Davis and White, 1980), though AMP is sometimes present outside the cells in significant quantities (Davis and White, 1980).

As already pointed out by Holm-Hansen and Booth (1966), it is difficult, if not impossible, to determine the ATP content of a bacterial population in the presence of phytoplnakton (Jassby, 1975; Paerl and Williams, 1976). Consequently, though the ATP concentration (or total adenine nucleotides) is a useful measure of total microbial biomass (algal + bacterial + microzooplankton), it gives little information on the bacterial biomass.

2.4.2. Bacterial Cell Wall Constituents

Two other chemical parameters for the determination of bacterial biomass have received increasing attention because they are, in contrast to ATP, specific for bacteria: muramic acid and lipopolysaccharide (LPS). Muramic acid is a major component of the cell wall in Gram-positive bacteria, but is also present in the peptidoglycan layer of Gram-negative bacteria; LPS is associated exclusively with the outer cell membrane in Gram-negative bacteria (Raetz, 1978). Consequently, both are a measure of the amount of cell envelope present in a sample, and average cell dimensions have to be assumed or determined microscopically before the conversion to biomass can be done. For example, cells are generally larger in sediment than in the overlying water, and larger in coastal water than in offshore water (Meyer-Reil et al., 1978, 1980). Several authors have reported that the surface-to-volume ratio can change with the physiological state of the bacteria (Matin and Veldkamp, 1978; Novitsky and Morita, 1976). Seasonal changes in mean cell sizes in natural samples have been reported for freshwater (Coveney et al., 1977; Krambeck, 1978), as well as for coastal seawater (Meyer-Reil, 1977). This means that the conversion factor from LPS or muramic acid to biomass will change.

The majority of the prokaryotes in marine waters are Gram-negative (Sieburth, 1979); so the concentration of LPS may reflect almost the total bacterial population. Several methods have been proposed for its determination (see Sullivan et al., 1976; Ulitzur et al., 1979; Maeda and Taga, 1979). The very simple and sensitive procedure described by Sullivan et al. (1976) and Watson et al. (1977) is based on the activating effect of LPS on a pro-clotting enzyme from horseshoe crab amebocytes. It cannot easily be calibrated with accuracy because the catalytic function of LPS in the gelation reaction that forms the basis of the determination is rather variable, depending on the bacterium it is isolated from (Sullivan *et al.*, 1976). Nevertheless, Watson *et al.* (1977) found a fairly good correlation between LPS content and epifluorescent bacterial counts in 188 seawater samples. They determined on LPS to cell carbon ratio of 6.35 in bacterial cultures *(E. coli)* and of 5.8–7.5 in seawater samples [carbon estimates based on aeridine orange direct count (AODC) determinations].

Moriarty (1977), taking into account the relative proportion of Gram-positive and Gram-negative bacteria, proposed a conversion factor of 100 from muramic acid content to biomass carbon. Apparently, the LPS content of marine bacteria is approximately 15 times higher than that of muramic acid. Because the LPS determination is easier, faster, and more sensitive, it appears more suitable for rapid biomass measurements, especially when only comparative values are required. But for conversion to absolute biomass data, the conversion factor and its constancy have to be determined by independent measurements (e.g., AODC; see Watson *et al.*, 1977).

3. Determination of Bacterial Activity

Though bacterial numbers or biomass determinations yield valuable information on bacterial communities, they do not necessarily reflect the dynamics of these communities and their interdependence with other organisms. Many ecological studies are concerned with the flow of energy or carbon through the food web, and in that respect it is essential to know the metabolic activity of each separate group of organisms of interest.

One of the basic problems in the study of *in situ* bacterial activity is that bacteria respond very quickly to relatively slight changes in their environment. For laboratory cultures it is well-known that, even during slow growth in chemostats, many organisms can increase their growth rate considerably upon suddenly increased substrate concentrations, without measurable delay (Mateles *et al.*, 1965; Harvey, 1970; Koch, 1979; F. B. Van Es and H. Veldkamp, unpublished data).

As has been shown on numerous occasions, natural populations exhibit a saturation type uptake and utilization curve (Hobbie and Crawford, 1969). Williams and Gray (1970) and Vaccaro (1969) showed that the addition of micromolar concentrations of glucose or amino acids to seawater samples could greatly influence the physiological status of natural communities (actual and maximum uptake rates of the added substrates, as well as growth rate) within 24 hr. This demonstrates that the activity is governed by the availability of substrate. A "bottle effect," as first noted by ZoBell and Anderson (1936, cited in ZoBell, 1946), may be caused by the adsorption of bacteria and dissolved organic substances to the wall of the incubation vessel, resulting in locally raised concentrations of both, and hence higher activity. Though several authors have reported a significant bottle effect for bacterial numbers (for a review, see Kunicki-Goldfinger, 1974) and also for oceanic phytoplankton (e.g., Gieskes *et al.*, 1979), others have failed to measure it during the first 24 hr of incubation (Williams and Gray, 1970; Fuhrman and Azam, 1980).

Because of the pronounced response of bacteria to changing environmental conditions, the prerequisite of a measurement for *in situ* bacterial activity should be that during the measurement, the natural situation is not significantly altered. Another fairly obvious requirement is that, in calculating actual activity or biomass production, possible conversion factors are correct. As will be shown below, these prerequisites are hard to fulfil. These problems are not at all unique to the study of microbial ecology. Although the measurement of primary production in aerobic environments should be comparatively simple, as it only involves algae, and one, well-quantifiable energy source (solar radiation), there are still many discussions on how to quantify it (for reviews see Hall and Moll, 1975; Harris, 1978; Peterson, 1980). A fundamental objection to incubation in closed containers is that this changes an open system into a closed system (Jannasch, 1967); e.g., the equilibrium between the influx of substrates that limit the growth rate and the response in activity of organisms may be interrupted, resulting in an immediate change in the metabolic rates.

Apart from the problems arising from the sensitivity of many bacteria to physical and chemical manipulations (sampling, filtering, incubation in a container, disruption of dynamic equilibria), a further basic problem in studying the heterotrophic processes is that they are carried out by almost all organisms present (including the primary producers). Of the bulk of organic matter, only a small fraction is mineralized within an incubation time of up to one day. Increases in bacterial biomass at the expense of dead organic matter are often small because a considerable fraction of the community does not react at all and part of the community will be removed during the incubation by grazers. Eliminating bacteriovores through filtration may significantly affect the activity of the bacterial community (Javornitsky and Prokesova, 1963; Fenchel and Harrison, 1976; Morrison White, 1980; Abrams and Mitchell, 1980). Moreover, particles that could influence the activity of microorganisms are removed (Jannasch and Pritchard, 1972). The other reaction components in the mineralization process (concentrations of electron acceptor, CO_2, minerals, and heat production) are common to the mineralization process carried out by all heterotrophs and cannot solely be used for measuring bacterial processes.

The most relevant attempts to circumvent these problems are summarized below. In Table II, some commonly used methods for assessing pelagic microbial activity are listed. A subdivision is made into methods that trace the fate of "model" substrates, and methods that follow changes within the community. Although there are considerable differences between these two approaches, this does not mean that one of them is superior. Much, if not all, depends on the

Table II. Activity, Growth Rate, and Biomass Production Measurements

A. Methods that trace the fate of specific substrates
  a. Biological oxygen consumption rate (BOC) and diurnal oxygen curve analysis
  b. Heterotrophic $^{14}CO_2$ fixation
  c. Uptake kinetics of organic substrates
  d. Turnover time of substrates added at trace concentrations
  e. Rate of nucleotide incorporation into RNA and DNA

B. Methods that follow changes within the bacterial community
  f. Increase in cell number during incubation of a sample without predators present
  g. Increase in cell length after incubation with naladixic acid
  h. Increase in cell numbers on membrane filters in contact with natural water
  j. Increase of biomass in dialysis containers in contact with natural water
  k. Determination of washout rate in a continuous culture
  l. Increase in cell numbers or cell length on submerged glass slides
  m. Determination of frequency of dividing cells

kinds of questions to be answered. In the subsequent sections, we will discuss the applicability of each of these methods for measuring activity, growth, or biomass formation, emphasizing the kind of information they provide and their limitations. For information on the technical aspects of the measurements, the reader is referred to the literature cited.

3.1. Biological Oxygen Consumption Rate and Diurnal Oxygen Curve Analysis

In aerobic mineralization processes, oxygen is the terminal electron acceptor in the mineralization reaction. As the oxygen concentration is also an important ecological factor in many environments, the biological oxygen consumption (BOC) measurement is among the most widely employed measures of heterotrophic activity in water pollution research (Gaudy, 1972), as well as in aquatic ecology (Sorokin and Kadota, 1972; Straskrabova, 1979). In contrast to BOD (biochemical oxygen demand) measurements, when determining BOC, the samples may not be filtered, aerated, or incubated at a fixed temperature (e.g., 20°C) and for prolonged periods, but should immediately be carefully siphoned into acid-cleaned bottles, and incubated at *in situ* temperature for as little time as possible. The BOC of water samples can be measured with a precision of 0.01 mg/liter day (0.3 μmol/liter/day) without expensive instrumentation (Bryan *et al.*, 1976). Moreover, oxygen is a natural substrate in the mineralization process, so nothing has to be added to the sample. But because of the length of the incubation time required, the BOC method is too insensitive to be applied to oligotrophic or low-temperature environments. In this case, the sample may be concentrated (Pomeroy and Johannes, 1968). Though the latter authors found no indications of losses of activity resulting from the concentration step, Holm-Hansen *et al.* (1970) found a drastic decrease of ATP in oceanic samples measured according to this procedure.

Parallel to the well-known light- and dark-bottle technique for determining primary production (Vollenweider, 1974), the oxygen consumption in the dark may be expressed in terms of organic carbon mineralization, when a respiratory quotient is measured or assumed, and oxidation of reduced inorganic compounds (such as NH_4^+) is small. Extreme values would be 0.7 and 1.0 when fatty acids or sugars, respectively, are mineralized. Ogura (1972) and Zsolnay (1975) calculated values of 0.85 and 0.75, respectively, for the easily decomposable fraction of surface seawater samples.

Several problems arise when BOC values are used to calculate bacterial biomass production. First, it is impossible to separate respiration by bacteria from respiration by other small organisms such as unicellular algae, microflagellates, and bacteriovorous protozoa. Moreover, successful separation prior to the incubation would disrupt the interactions between these groups of organisms, possibly with a significant influence on the activities measured. Second, the amount of cell material formed per mole of oxygen consumed depends on the organism, the nature of the organic subtrate(s), and the growth rate as principal factors (for reviews, see Stouthamer, 1977, 1979, and Payne and Wiebe, 1978). Consequently, the relationship between the rate of oxygen consumption and the rate of biomass production cannot be considered constant (see also, Sorokin, 1978).

A different approach is the diurnal oxygen curve analysis introduced by Odum and coworkers (see Odum, 1956, 1960). In common with other oxygen-rate measurements, in this method, the respiration by different groups of organisms cannot be separated, but it offers the possibility of making a comparison between rates measured in containers and *in situ* rates. In principle, the oxygen concentration and saturation value within a water body is followed during at least 24 hr. During the night, the oxygen concentration decreases because of the respiration of all aerobic organisms; during the day, it may increase as a result of primary production. Corrections have to be made for exchange of gaseous oxygen with the air, depending on undersaturation or oversaturation, wind, turbulence, and temperature. Moreover, one must ensure that the measurements are repeated in the same body of water, and the exchange with water bodies of different oxygen content has to be accounted for. For more details, the reader should refer to the review by Hall and Moll (1975). In systems where physical processes dominate dissolved oxygen dynamics, it may be impossible to use the diurnal oxygen curve analysis (Lingeman, 1980), but in other systems it may be used with some success, at least for primary production estimates (Kelly *et al.*, 1974; Schurr and Ruchti, 1977; Kemp and Boynton, 1980).

A great difficulty in applying this technique to measurements of mineralization is that daytime respiration has been assumed to equal nighttime respiration. This in itself would be an important point to be tested. Several studies (Kelly et al., 1974; Hall and Moll, 1975; and literature cited therein) have shown that respiration rates in the first hours after sunset are much higher than during the subsequent period. This is supported by the results of Sieburth et al. (1977), who measured strong diurnal variations in heterotrophic growth activity with higher activity at daytime. Though Kemp and Boynton (1980) probably underestimated in situ community respiration, because it was calculated from diurnal oxygen curves, they still found it to be 1.5–4 times higher than the respiration rates measured in bottles at the same time. This information from dirunal oxygen curve analyses at least indicates that bottle effects are more than just effects of additional surface available, and that considerable dirunal fluctuations exist in heterotrophic activity (c.f. Section 4).

3.2. Heterotrophic $^{14}CO_2$ Uptake

During growth, bacteria constantly use a part of the TCA-cycle intermediates to biosynthesize amino acids and other monomers of cell constituents. To compensate for these losses of TCA-cycle intermediates, anaplerotic CO_2 fixation takes place, in which primarily pyruvate or phosphoenolpyruvate is carboxylated to oxaloacetate (Kornberg, 1966).

If there were some fixed relations between biosynthesis and anaplerotic CO_2 fixation, bacterial biomass production could be measured without knowing the biomass or growth rate of the population. On the basis of this assumption, Russian workers (Romanenko, 1963, 1964, 1965; Sorokin, 1964) have developed a measurement closely resembling the Steemann Nielsen method for primary production measurements (Steemann Nielsen, 1952). Samples are placed in the dark and incubated with $^{14}CO_2$ for up to one day, depending on the activity expected. According to Sorokin (1971, 1973), the result of preincubation in the dark for several hours is that the $^{14}CO_2$ incorporation by algae becomes insignificant, though other researchers (Overbeck and Dalay, 1973; Jordan and Likens, 1980) doubt this. The bacterial biomass production rate can then be calculated using the Steeman Nielsen formula and a conversion factor from dark CO_2 fixation to heterotrophic biomass production. Sorokin found conversion factors from 0.03 to 0.10 (1964), and of 0.05 (1971). Romanenko (1964) measured 0.06–0.07. Overbeck (1972, 1979) found much larger variations (0.004–0.30) in the Pluss See (FRG) and in bacterial cultures.

In several studies on lakes (Kuznetsov and Romanenko, 1966; Anderson and Dokulil, 1977; Overbeck, 1979), the calculated heterotrophic (bacterial) production was comparable to, or higher than, algal primary production. Jordan and Likens (1980) compared the bacterial production in the water of an oligotrophic lake, using four different methods. From a total carbon budget of the lake and a dissolved organic carbon budget, bacterial production of 5.6 and 5.4 g $C/m^2/yr$, respectively, were calculated. From ^{35}S-sulphate uptake rates (assuming a C/S ratio of 50), the production calculated was 6.5 g, and from dark $^{14}CO_2$ fixation, it was 20 or 34 g C/m^2/yr (assuming 6% heterotrophic CO_2 fixation and bacteria to be responsible for 56 and 100% of dark CO_2 uptake, respectively). They concluded that the heterotrophic dark $^{14}CO_2$ fixation rates yielded much too high values. Sorokin (1978), reviewing the Russian work in this field, calculated that in the tropical part of the Pacific Ocean, the ratio between bacterial and primary (algal) production ranged from 1.3 (shelf waters) to 4.8 (neritic waters). He assumed that the extra organic matter required to enhance such a ratio originated from temperate zones of the ocean. In these areas, part of the primary production is not remineralized, but descends to deeper water, where it is largely preserved, because of the lower temperatures and higher pressure. In the oceanic water circulation, this deep water is transported to the tropical regions and, by upwelling, brought to the surface again. Sorokin supposed that there, at higher temperatures, the organic matter originating from the colder regions largely supports the observed growth of bacterial communities (Sorokin, 1971, 1973, 1978). However, this explanation raises other questions. Skopintsev (1973) has argued that the organic matter in the deep ocean water is too refractory to serve as an important food source for bacteria.

On the basis of the spatial distribution of the dissolved organic carbon (DOC), Banse (1974) has questioned the conclusion of high bacterial production rates in the tropical ocean. From the production rates and growth efficiency of the bacterial populations presented by Sorokin (1971), he calculated the amount of DOC mineralized. This should have given rise to a significant concentration gradient from the area of upwelling to the area where the high production rates were measured. This gradient was not observed. Neither could vertical transport from below provide the bacteria in the photic layer with DOC fast enough. Moreover, Banse compared oxygen consumption rates in deep ocean water (0.1–0.2 mg O_2/m^3/day, calculated by Sorokin (1971) from the heterotrophic $^{14}CO_2$ fixation rates) with other published values, which were 10 times lower. Banse concluded that the bacterial production rates, calculated with the $^{14}CO_2$ fixation method, were about one order of magnitude too high.

As bacterial production appears to be unrealistically high in most field measurements reported using this method, we wonder if indeed it is only the conversion factor from dark CO_2 fixation to heterotrophic biomass production under *in situ* conditions that differs considerably from the value established from laboratory cultures. Apart from the unknown variation in this conversion factor (Overbeck and Daly, 1973), dark CO_2 fixation by algae, as well as bottle effects as discussed above, may have a serious influence, especially in oligotrophic waters, as incubation times have to be relatively long for samples from those environments.

Waters in contact with anaerobic environments may contain active populations of nitrifiers, reduced sulfur oxidizers and (or) methylotrophs that incorporate much more CO_2 in cell material (Sorokin, 1964; Sen Gupta and Jannasch, 1973; Overbeck, 1976). In such environments, the $^{14}CO_2$ incorporation in the dark cannot be a measure of heterotrophic activity.

3.3. Uptake of Labeled Organic Substrates

3.3.1. The Kinetic Approach

In 1962, Parsons and Strickland published a new method of measuring heterotrophic activity that was based on their observation that in heterotrophic communities, the uptake of small amounts (μg/liter) of ^{14}C-labeled glucose or acetate followed Michaelis-Menten enzyme kinetics. As neither the nature nor the total amount of utilizable substrate is known in a sample of seawater, they called the V_{max} of assimilation of the added substrate "a relative heterotrophic potential." With the aid of this technique, Hobbie and Wright (1965) and Wright and Hobbie (1966) presented evidence that low molecular weight organic substrates are taken up by bacteria and algae in different ways. They found that active uptake by bacteria largely dominated at low (natural) substrate concentrations, irrespective of the relative abundance of algae and bacteria, and that uptake through diffusion by algae dominated at much higher substrate concentrations. From this kinetic approach, they concluded (Wright and Hobbie, 1966) that bacteria may prevent substantial heterotrophy of algae in nature.

The Parsons and Strickland technique yielded much new information, showing, for example, that bacteria are very effective in capturing dissolved organic substrates. Moreover, it could be demonstrated that relatively small additions (i.e., small as compared with substrate concentrations in routine physiological experiments) of 1–10 μmol glucose or amino acids have a rapid and considerable effect on aquatic bacterial communities (Vaccaro, 1969; Williams and Gray, 1970). Furthermore, some estimates of natural substrate concentrations could be made at a time when chemical determinations were not yet available (Wright and Hobbie, 1966; Vaccaro and Jannasch, 1966; Allen, 1968).

A valuable extension of the technique was made by Williams and Askew (1968) and Hobbie and Crawford (1969). They quantified the fraction of substrate that was respired to $^{14}CO_2$ subsequent to uptake and showed that the respiration of labeled substrates during the incubation was frequently a significant fraction of the total amount of substrate taken up. The same type of kinetics was found to be applicable to both the fraction incorporated and the fraction respired (Hobbie and Crawford, 1969). V_{max} values for both processes were additive, whilst the affinity constants (K) were the same. The results indicated that the amounts of substrate incorporated by bacteria might lead to a biomass production of roughly the same magnitude (when expressed per m^2) as primary production by algae.

3.3.2. Determination of the Parameters V_{max}, $(K + S_n)$, T_t, and Percentage Respired

For convenience, the Michaelis-Menten curve may be transformed into a straight line, using a Lineweaver-Burke transformation. This enables an easy statistical analysis of the results as well as the calculation of some interesting parameters. V_{max}, the relative heterotrophic potential, is a measure of the part of the bacterial community that is actively taking up the added substrate, and it can be used to differentiate the trophic status of different habitats (Crawford et al., 1973). Review tables (Sepers, 1977; Gocke, 1977a; Hoppe, 1978) comparing V_{max} values for different substrates and from different habitats show a difference of several orders of magnitude between oligotrophic oceanic waters and coastal or estuarine waters, reflecting in some way the differences in population densities and activities in those environments.

For several reasons, $(K + S_n)$, the sum of the affinity constant K and the natural substrate concentration (another parameter derived graphically from the Lineweaver-Burke plot) is difficult to interpret in terms of community properties. First, it is composed of two terms, both contributing an unknown fraction to the sum (unless S_n is determined separately). Furthermore, for theoretical reasons, it is impossible to determine a discrete K value for a heterogeneous community. Reported values of $(K + S_n)$ for glucose roughly range from five to several hundred nmoles (Sepers, 1977; Hoppe, 1978).

The turnover time (T_t), the time that would be required to totally utilize the substrate if there were no renewal, is a reflection of the dynamics of the substrate considered. Using the additional information on the natural substrate concentration, the flux rate of the substrate can be calculated from $T_t = S_n/V_n$. T_t can be calculated in two different ways. T_t is the intercept with the ordinate in a Lineweaver-Burke plot when the straight line of S versus S/V is extrapolated to $S_a = 0$. This will be referred to as the kinetic approach. T_t can also be measured by adding trace amounts $(S_a < S_n)$ of labeled substrate and following its utilization: $T_t =$ incubation time/fraction utilized.

This measurement, introduced by Williams and Askew (1968), will be referred to as the tracer approach. Both approaches were compared on a theoretical basis by Wright (1974) and through actual measurements by Gocke (1977b). Wright (1974) considered the assumptions to be made in each approach. In the tracer approach, it has to be assumed that T_t is not affected by the addition of the labeled substrate, so, in $T_t = S/V$, either S (and hence V) should not be affected significantly, or V should increase proportionally with the increase in S. By approximation, this is true in the first, almost linear part of the Michaelis-Menten curve and was confirmed by the results of Williams and Askew (1968). In the kinetic approach, the critical assumption is that the uptake follows Michaelis-Menten kinetics, and the straight line in the Lineweaver-Burke plot is also rectilinear below the concentration traject where the measurements were done. Researchers using the kinetic approach have frequently been unable to observe Michaelis-Menten kinetics, especially in oligotrophic oceanic waters (Vaccaro and Jannasch, 1967; Vaccaro, 1969; Hamilton and Preslan, 1970; Barvenick and Malloy, 1979) and, consequently, have been unable to calculate turnover times for the substrates used. Furthermore, the applicability of Michaelis-Menten kinetics to heterogeneous bacterial communities has been questioned. Williams (1973a), using a model calculation, showed that the heterogeneity in the natural community, and especially in K values, could introduce significant deviations from linearity in the Lineweaver-Burke plot, especially at low, natural substrate concentrations. This was confirmed experimentally by Gocke (1977b), who observed the largest differences in estimated T_t values in oligotrophic waters: 33% on average.

The ratio of the amount of substrate incorporated into cell material to the amount respired, as measured using U-^{14}C substrates, indicates which fraction of the organic carbon flow through the bacterial community may become available to higher trophic levels. At low added substrate concentrations (nmole-range) and short incubation times (1–4 hr), the percentage that is respired is fairly constant in a given sample (Hobbie and Crawford, 1969; Williams, 1973b, Gocke, 1976). Apparently, the proportion between incorporated and respired substrate carbon is not disrupted during the experiments by the substrate concentrations added.

Results from estuarine, coastal, and offshore marine environments have been reviewed by Hoppe (1978). It is surprising that, for environments so widely differing in trophic status and for such different substrates, the variations in percentage respired are relatively small. For glucose and amino-acid mixtures, the substrates most frequently used, the percentage of total uptake respired at incubation times of up to 4 hr is generally between 20 and 40%.

From the amounts of uniformly labeled substrates incorporated and respired, an apparent growth yield (Y_c) can be calculated if it is assumed that the fraction incorporated at the end of the incubation time represents growth:

$$Y_c = \frac{\text{Amount incorporated}}{\text{Amount incorporated} + \text{amount respired}}$$

Using this value and the daily or annual flux of the substrate investigated ($V_n = S_n/T_t$), its contribution to the food web via bacterial biomass production may be quantified (Andrews and Williams, 1971). Williams (1973b) compared Y_c values of samples from a variety of marine environments (0.71 ± 0.10 of the carbon from glucose and 0.77 ± 0.08 from an amino acid mixture) with yield values reported for cultures (Forrest, 1969; Payne, 1970). He noted a striking difference: the relatively slow-growing natural population appeared much more efficient than fast-growing laboratory cultures, though for the latter, it is known that the growth efficiency (yield) is highest at high growth rates (Pirt, 1975). The reasons for this discrepancy are still unclear. Probably during the relatively short incubation times (up to 4 hr), no isotope equilibrium is established, so that relatively little of the newly assimilated substrate is respired. Moreover, natural bacterial communities grow on complex mixtures of substrates, which may enhance their growth yield (Payne and Wiebe, 1978).

3.3.3. Validity of the Kinetic Parameters V_{max}, $(K + S_n)$ and T_t

To date, the validity of the assumption that uptake and utilization of organic substrates by heterogeneous natural communities can be described by simple enzyme kinetics has never been confirmed. Attempts to correlate the kinetic parameters with those calculated independently from continuous culture data (Sepers and Van Es, 1979) have failed completely. Moaledj and Overbeck (1980) applied uptake measurements of glucose and acetate to nine strains of oligocarbophylic bacteria (requiring very low concentrations of organic substrates) and concluded that five of them did not show Michaelis-Menten uptake kinetics at all. Nevertheless, field data frequently fit the enzyme kinetics model.

The empirical nature of the relation between substrate concentration and utilization rate has considerable implications for the interpretation of the parameters V_{max}, $(K + S_n)$ and T_t. Wright (1973) has listed the basic assumptions to be made when applying enzyme kinetics of natural communities. In fact, they can be reduced to the assumption that the natural community should behave like a single enzyme converting one substrate under optimum conditions in the absence of interfering substrates. Williams (1973a) and Gocke (1977b) showed that heterogeneity of the community considerably influenced measured T_t (and $K + S_n$) values. Interactions between uptake of various carbohydrates have been reported by Wood (1973), who suggested that multiple substrate systems are operating. This means that the kinetic parameters for each substrate are, to an unknown degree, dependent on the concentration of many other natural substrates. Krambeck (1979) discussed the deviations that could be expected if bacterial communities reflected enzyme chains and networks rather than single enzymes. She concluded that the overall K is a function of all V_{max} values of the intermediate enzymes in the reaction chain. As V_{max} of each single step is directly proportional to the amount of enzyme present, K is dependent on the amount of enzymes induced in the recent history of the cells and can be altered by the organisms. This clearly shows the limited comparability with single enzyme kinetics. An example of changed affinity (K) of bacterial cells at changed environmental conditions was given by Law and Button (1977). Genetic adaptation of K_s (half-saturation constant for growth) may play a role in other organisms, as indicated by the observations of Jannasch (1968), who noticed a strong increase of K_s in bacterial strains isolated from seawater in continuous culture at low substrate concentrations. After cultivation on rich media, their K_s values had increased considerably.

The validity of the calculated T_t value, determined by either the kinetic or the tracer approach, is affected by observations (Burnison and Morita, 1974; Dawson and Gocke, 1978; Gocke et al., 1981) that suggest considerable differences between the chemically determined concentrations of dissolved glucose and amino acids and the concentrations available to bacteria. The most direct evidence for this was obtained from the observation by Gocke et al. (1981) that in various biotopes, the $(K + S_n)$ values calculated for glucose were only between 6 and 31% of the chemically determined S_n values.

3.3.4. Actual Uptake Rate Measurement

After the addition of very small amounts of labeled substrate, the actual rate of substrate utilization can be estimated in a tracer approach by measuring the assimilation and respiration of the added substrate at short time intervals. The natural substrate concentration must be known in order to calculate the specific activity. The added substrate should have a concentration that is much lower, in order to prevent enzyme induction. The total incubation time must be kept short, in order to prevent exhaustion of the substrate or changes in the specific activity. Both enzyme induction and exhaustion of the labeled substrate result in a nonlinear time course of uptake. This technique has mainly been applied to sediment samples (Harrison et al., 1971; Meyer-Reil, 1978b), where incubation times of only a few minutes are required.

3.4. Rate of Nucleotide Incorporation into RNA and DNA

A special application of labeled organic substrates is in uptake measurements of specific nucleic acid precursors that are incorporated into RNA and/ or DNA at a rate that can be related to the cells' growth rate. DNA is only synthesized in growing cells, at a rate proportional to total biomass formation, as total DNA content stays fairly constant, and DNA does not turn over (Maaløe and Kjeldgaard, 1966). Therefore, the rate of DNA synthesis excellently reflects growth rate of the organisms.

Brock (1967) followed DNA synthesis in a study on growth rates of *Leucothrix mucor* during epigrowth on seaweeds. During incubation of samples with ^{3}H-thymidine, he took subsamples at regular intervals and prepared autoradiograms in order to determine the fraction of cells that had accumulated the label in DNA. In prior experiments, Brock had ascertained that at a constant growth rate, the accumulation of labeled cells was linear with time and was proportional to growth rate. One percent of the cells had scored positively in 0.002 generation. He used this relationship as a constant for estimating the generation time of *L. mucor* in natural samples. He found generation times of approximately 11 h in a wide variety of habitats. As he pointed out (Brock, 1967), this technique is only applicable to organisms that can be recognized microscopically and for which the relation between growth rate and rate of accumulation of labeled cells can be determined independently.

Since Brock's studies, several measurements based on the rate of nucleotide incorporation into RNA and DNA have been published (Thomas et al., 1974; Straat et al., 1977; LaRock et al., 1979; Karl, 1979, 1981; Karl et al., 1981a,b; Fuhrman and Azam, 1980; Moriarty and Pollard, 1981). Karl (1979, 1981) stressed several requirements that have to be fulfilled before the incorporation of labeled nucleotides can be used to calculate growth or biomass formation. The most important ones are that (1) almost all the growing cells must take up and metabolize the label, (2) the rate of nucleic acid synthesis must be directly coupled with growth and (3) the intracellular specific activity of the immediate nucleic acid precursor must be known. Karl and coworkers (Karl, 1979; Karl et al., 1981a,b) argued that these requirements are met to a satisfactory degree when measuring the incorporation of $(2-^3H)$ adenine into stable nucleic acids (predominantly tRNA and rRNA) of microbial communities. Metabolic regulation results in a strong reduction of de novo synthesis of adenine if adenine is available in the external medium and uptake is directly linked to nucleic acid synthesis. As both mRNA and its direct precursor ATP have very short turnover times, the isotope equilibrium within the ATP pool will be reached very quickly. However, a time series analysis of the rate of ^3H-RNA synthesis is essential, as during the first stages of the incubation, a possible ATP-pool compartmentalization may influence the linearity of the rate of labeling (Karl et al., 1981a).

In samples from the Caribbean Sea, the fastest rate of RNA synthesis was observed at 450 m depth, coinciding with the minimum dissolved oxygen concentration (Karl, 1979). Karl et al. (1981b) showed that RNA synthesis is a sensitive measure of the effect of several environmental factors in growth activity (i.e., light versus dark incubation, inorganic and organic nutrient additions). However, it is difficult to interpret these results, as the overall activity of the whole community is measured and its different components have quite different activities and specific growth rates.

According to Fuhrman and Azam (1980), methyl-^3H-thymidine, if added at concentrations of 11 nmole or less, is taken up almost exclusively($> 90\%$) by bacteria. They argued that to assess bacterial production, it is preferable to measure DNA snythesis (from methyl-^3H-thymidine) rather than RNA synthesis, for two main reasons: the unknown specific activity of the precursor in bacteria, and the absence of a simple relation between RNA synthesis and growth rate during unbalanced growth.

Fuhrman and Azam assumed that the necessity to know the specific activity of the precursor is much less critical for labeled thymidine incorporation into DNA than it is for ATP incorporation into RNA (Karl, 1979; Karl et al., 1981a,b). The rapid turnover of mRNA dilutes its labeled precursors, but the extent of dilution cannot be quantified because bacterial nucleic acid precursors cannot be determined separately in a total microbial community. Inclusion of precursor pools from other organisms could lead to underestimates of specific activity of one to two orders of magnitude and hence to overestimates of bacterial growth rates. Fuhrman and Azam assumed that isotope dilution of labeled thymidine in bacterial intracellular pools was not significant because, in contrast to RNA, DNA turnover does not take place to a significant degree. This assumption leads to a conservative estimate of bacterial growth rates, as *de novo* synthesis of unlabeled thymidine may continue to some extent. Experiments with different concentrations of labeled thymidine produced a doubling to tripling of estimated growth rate as the concentration of isotope added was increased 20–100-fold. On the other hand, Karl (1981), Karl *et al.* (1981a,b), and Moriarty and Pollard (1981) stressed the importance of measuring the actual specific activity of the nucleotide precursor. Moriarty and Pollard (1981) came to the conclusion that it cannot be assumed that the addition of a large amount of thymidine will lead to a cessation or even a considerable lowering of the *de novo* synthesis. This conclusion was based on results of Rosenbaum and Zamenhof (1972), who observed that in cultures of *E. coli*, only 35–63% of external thymidine was utilized for DNA synthesis, even at the extremely high concentration of 1 g thymidine per liter. Moriarty and Pollard (1981) used an isotope-dilution method to correct for the effect on labeled thymidine incorporation of intracellular and extracellular thymidine pools as well as *de novo* synthesis in seagrass bed sediment.

Fuhrman and Azam (1980) reported that with methyl-^{3}H-thymidine, 80% of the cold TCA insoluble fraction was incorporated into DNA. Labeling of the methyl group was supposed to circumvent labeling of RNA, as during conversion of thymidine to uridine or cytidine, the methyl group is lost. On the other hand, Karl (1981, unpublished results) showed that methyl-^{3}H-thymidine is not incorporated exclusively into DNA. Considerably more label may appear in other cold TCA insoluble fractions, especially RNA and protein. Of course, this seriously interferes with the estimate of DNA synthesis. Because of the variable fate of this label (depending on habitat, incubation time, and the concentration of the added substrate; Karl, 1981, unpublished results), it is absolutely necessary to measure the specific activity of the immediate nucleic acid precursor, the nucleotide triphosphate.

A further reason put forward by Fuhrman and Azam (1980) in favor of DNA synthesis is that, to estimate growth rates from nucleic acid synthesis, the concentration of RNA or DNA present must also be known. At present, the bacterial fraction of total RNA and DNA in natural samples cannot be determined; therefore, some value has to be assumed. The range of reported concentrations per cell is much smaller for DNA than it is for RNA (Maaløe and Kjeldgaard, 1966), mainly because RNA content is largely dependent on growth rate. The concentration of bacterial DNA, therefore, can be estimated with less uncertainty from microscopic or other biomass determinations than can RNA concentration. The dependence of the cellular RNA content on growth rate further gives the complication that, during shifts from one growth rate to another, the rate of RNA synthesis is disproportional to the growth rate.

The almost complete incorporation of thymidine in organisms < 1 μm indicates that at the concentrations applied, the rates measured reflected the growth of the bacterial part of the total community (Fuhrman and Azam, 1980). For comparison, Fuhrman and Azam measured the increase in cell numbers and biomass in 3-μm filtered samples (see Section 3.5) and found a reasonable agreement between both estimates of biomass production.

Recently, Karl (1981) has extended the measurement of 2-^{3}H adenine incorporation to enable the simultaneous assessment of RNA and DNA synthesis rates in microbial communities. The measurements included determining the specific activity of ATP. He showed that dATP and ATP, the precursors of DNA and RNA, were in isotopic equilibrium within minutes. The rates of incorporation of the labeled nucleotide into the DNA and RNA of a fast-growing batch culture of *Serratia marinorubra* were constant and at a ratio of 0.02 as long as the nucleotide was present in the growth medium. For surface water from the Kaneohe Bay (Hawaii), the ratio of DNA/RNA synthesis rates varied by 400% over a daylight period. In contrast to the relatively constant rates of RNA synthesis, the rates of DNA synthesis consistently increased, suggesting that cell growth (RNA synthesis) and multiplication (DNA synthesis) were not closely linked throughout the day (Karl, 1981). In this procedure, the advantages of determining rates of DNA synthesis as a preferable measure of growth rate are combined with the additional information on rates of RNA synthesis, which may give a good reflection of the physiological status of the organisms being investigated (Karl et al., 1981a,b).

Nucleotide incorporation (either 2-^{3}H-adenine or methyl-^{3}H-thymidine) into nucleic acids, especially DNA, appears to be a promising future technique for assessing bacterial growth rates and biomass production in natural waters. However, as yet unsettled problems are the determination of specific activity of precursors within the bacterial part of the community and the question whether all cells that contribute to the measured precursor concentrations are indeed the same as those incorporating the label. Using microautoradiography, Ramsay (1974) studied bacteria colonizing *Elodea canadensis* (pond weed) leaves and reported that the percentage of cells taking up ^{3}H-thymidine was always less than the percentage of bacteria labeled with ^{3}H-glucose. The ratio ^{3}H-thymidine: ^{3}H-glucose count was not constant but tended to increase as the percentage of the total community metabolizing ^{3}H-glucose increased. Hoppe (1976) found the ratio ^{3}H-thymidine: ^{3}H-amino acid mixture counts to be 0.81 in Kiel Bay samples. Many more cells were labeled with thymidine than with glucose. More physiological work is needed, especially on organisms starved of energy and nutrients, in order to verify whether those organisms react to the labeled nucleic acid precursor in the same way as fast-growing laboratory cultures.

3.5. Enzyme Activity

The determination of enzymatic activities offers an approach for studying decomposition processes in natural samples. Whereas some of the enzymes are released extracellularly into the surrounding medium, other enzymes are located within the cells and may be released by mechanical breakage or cell lysis.

Packard and coworkers (Packard, 1971; *et al.*, 1971) introduced the enzymatic approach to estimate oxygen utilization in phytoplankton samples from the activity of the respiratory electron-transport systems (ETS). This enzyme system can be measured specifically and highly sensitively by the reduction of a colorless tetrazolium salt to a colored formazan. In a number of subsequent publications, the close relationship between ETS activity and respiration rate has been established for various samples (e.g. Kenner and Ahmed, 1975a,b; Olanczuk-Neyman and Vosjan, 1977; Devol and Packard, 1978; Christensen *et al.*, 1980). However, extrapolation to *in situ* oxygen consumption depends upon a constant ratio between ETS activity measured at optimum velocity of electron transfer *in vitro* and respiratory activity *in situ*.

Various enzymes have been detected as secretion products of culturable marine bacteria (Corpe and Winters, 1972; Corpe, 1974) and/or as "free-dissolved" enzymes in natural waters and sediments (Kim and ZoBell, 1974): phosphatase, amylase, β-glucosidase, and proteolytic enzymes. The importance of these enzymes, especially for the turnover of particulate organic matter in the sea, has already been emphasized by Corpe and Winters (1972). By microbial enzymatic decomposition, products of the primary production (e.g., cellulose) enter the secondary production and become available to higher trophic levels. Furthermore, microbial decomposition processes play an important role in the digestion in marine animals (e.g., Goodrich and Morita, 1977a,b; Meyer-Reil and Faubel, 1980).

In addition to microbial cells (bacteria, fungi), dead and lysing plant and animal cells also contribute to the liberation of enzymes (Meyer-Reil, 1981). Given the energy available in bacterial communities, it seems unlikely that microorganisms continuously excrete enzymes. Furthermore, it is probably too easy to think in terms of extracellular enzymes free in the water. According to studies done by Burns (1980), enzymes persist in the soil and retain their activity by the formation of humic-enzyme complexes, which may react as "starter" enzymes, inducing the production of microbial enzymes in the presence of sufficient concentrations of suitable substrates.

Various enzyme systems have been assayed using natural untreated samples as well as homogenized extracts. As substrates, dye-labelled substances (e.g., Hide Powder Azure; cf. Little *et al.*, 1979) have been used; these are stable, insoluble, and sensitive to certain groups of enzymes (e.g., proteolytic enzymes in the case of Hide Powder Azure as a substrate). Alternatively, the remineralization of natural ^{14}C-labeled substrates (e.g., ^{14}C cellulose; Federle and Vestal, 1980) can be determined as an indirect measurement of microbial enzymatic decomposition.

3.6. Increase of Cell Density in Incubators

3.6.1. Incubation of Prefiltered Samples in Bottles

This method, developed by Rasumov (1947) and Ivanov (1955); both cited in Sorokin and Kadota (1972), is widely employed by Eastern European limnologists and marine microbiologists (e.g., Godlewska-Lipova, 1969,1970; Rodina, 1972; Maksimova, 1976; Sorokin, 1980). In principle, a water sample is filtered through a 3–6 μm filter in order to remove protozoa and other organisms that graze on bacteria, and the filtrate is incubated at simulated *in situ* conditions, allowing approximately one doubling of the number of cells. At t_0 and t_1, the numbers of bacteria N_0 and N_1 are counted microscopically. The generation time g can then be calculated from

$$g = \frac{(t_1 - t_0) \log 2}{\log N_1 - \log N_0}$$

However, the generation time will be seriously overestimated if no allowance is made for the fraction of nongrowing cells. If, for example, this fraction is 50% of the total bacterial community, a doubling of the total number has to be produced by only half of the initial number. Consequently, this latter fraction must have much shorter generation times to produce the same amount of new cells, compared with a situation where all the original cells would grow. A correction of the constant (2) in the numerator has been proposed (Kuznetsov and Romanenko, 1963, cited in Godlewska-Lipowa, 1969). At an assumed fraction of 90% of growing cells, the constant was reduced to 1.8.

However, this appears to be a correction in the wrong place. The correction should be applied in the numbers of cells present at the beginning and the end of the incubation. It is, however, very difficult to say whether the fraction of inactive cells (for example, determined by microautoradiography and epifluorescence microscopy combined; see Section 2.2) is equal to the fraction of nongrowing cells in the original sample and during the incubation. As long as the reasons and kinetics of inactivation and cell death in natural habitats are obscure, a proper correction for the fraction of inactive cells cannot be made.

The true (averaged) growth rate or generation time of the growing cells cannot be determined in this way whenever there is a considerable fraction of inactive cells. This drawback may be partly overcome by following the increase over time of incorporated radioactivity from heterotrophic $^{14}CO_2$ fixation, instead of the increase of cell number (Romanenko, 1969, cited in Sorokin and Kadota, 1972). Moreover, techniques that follow the progeny of single growing cells developing to microcolonies may be used.

Apart from bottle effects that influence every method in which samples are incubated, the measured generation time in filtered samples may be altered by the removal of predators, particles, and part or all of the phytoplankton. Significant fractions of dissolved organic matter are released by algae (Lancelot, 1979; Mague et al., 1980), ciliates (Seto and Tazaki, 1971), copepods (Lampert, 1978; Copping and Lorenzen, 1980), and probably other grazers (Johannes and Satomi, 1967). Moreover, grazers may play an essential role in nutrient regeneration (Johannes, 1965,1968; Pommeroy, 1970,1974), especially in oligotrophic waters. Once the grazers are removed, this regeneration is seriously affected, and bacterial growth may decline, as a result of the exhaustion of inorganic nutrients. Organic particles, present in the original sample, may serve directly as a food source for bacteria colonizing them (Corpe and Winters, 1972), and particles in general may provide adsorption sites where substrate concentrations are higher than in the surrounding water (Paerl, 1980; Marshall, 1980) and bacterial activity is higher (Jannasch and Pritchard, 1972; Wangersky, 1977; see Section 4.3). The effect on growth rate of the removal of particles and grazers by filtration is hard to quantify.

The biomass production rate of the original bacterial community can be derived from the calculated generation time (g) when the conversion factor (c) from cell number to biovolume and biomass is measured concomitantly. In natural habitats, the number of bacteria remains fairly constant when grazers are present (Gak, 1971; cited in Sorokin and Kadota, 1972), and, therefore, the biomass production P may be calculated from the number of cells present in the original (unfiltered) sample (N):

$$P = c \cdot N \cdot \mu \cdot t$$

where μ represents the specific growth rate (hr^{-1}). As $\mu = \ln2/g$, it follows that $P = 0.693 \cdot c \cdot N \cdot t/g$.

3.6.2. Incubation in Diffusion Chambers

In order to avoid substrate exhaustion during incubation, several devices have been developed using a semipermeable connection between the sample being investigated and the water it originates from (Lloyd and Morris, 1971; Kunicka-Goldfinger, 1973; Baskett and Lulves, 1974; Meyer-Reil, 1975; Sieburth *et al.,* 1977). Toth (1980) made *in situ* incubations of lakewater samples in dialysis bags and glass bottles. He noted a considerably lower (45%) bacterial production as well as a lower (40%) phytoplankton production rate in the bottles during 6-hr incubations. However, the increase in bacterial numbers in the free water was much lower than that in dialysis bags over the corresponding time period. As grazing by zooplankton was responsible for only a small part of this difference, he attributed the resulting difference to higher mortality in the free water (67% of production for bacteria; Toth, 1980). Unfortunately, Toth gave no evidence to discount the possibility that the slower increase in cell numbers in the free water and the glass bottle resulted from the stimulation of enclosure in dialysis bags.

Bacteria tend to adhere to and grow on dialysis membranes of regenerated cellulose (Varga *et al.,* 1975). For that reason, Sieburth (1976) recommended 0.1 μm polycarbonate (Nuclepore) filters as the ideal barriers for caging bacterioplankton communties.

3.6.3. Incubation in an Open System

In addition to incubation in dialysis culture, another possibility of avoiding the depletion of substrates and the accumulation of end products is to provide the incubated sample directly with a continuous flow of fresh natural water. For this purpose, Jannasch (1969) used continuous cultures that were inoculated with selected marine bacteria and fed with sterilized unsupplemented seawater. The bacteria had to grow at the expense of the supplied natural (dissolved) organic and inorganic nutrients in order to maintain a population in the culture vessel. (For information on the theoretical background to continuous culture techniques, see Pirt, 1975; Veldkamp, 1976,1977). The growth rate that heterotrophic bacteria are able to achieve depends on the concentration and composition of the nutrients that are in the shortest supply, normally the utilizable organic substrates. Jannasch selected the dilution rates of the cultures to be somewhat higher than the maximum growth rates of the isolates under the given conditions, and, therefore, the numbers of cells within the culture vessel slowly fell. From the dilution rate of the culture vessel and the (slower) washout rate of the bacterial population, the growth rate or generation time of the latter could be determined.

Jannasch (1969) found that on unsupplemented, filter-sterilized coastal water, several organisms grew (at 24°C) with generation times of 50–170 hr. This is exceptionally slow compared with the range of results from other estimates (Table VI; see Section 4.4). The isolates Jannasch used may not have been well adapted to the organic matter composition and concentrations. Moreover, the generation times may have been overestimated because of the exhaustion of the growth medium. During bacterial growth in unsupplemented seawater, the concentrations of utilizable carbon sources must have declined as the potential producers of fresh organic matter (algae, protozoa, zooplankton) were removed. The continuous input of fresh filtered seawater cannot fully counterbalance the utilization. Consequently, in the system described above, the observed generation times should decrease with decreasing cell densities applied in the culture vessel. This apparently happened in one of the experiments (Jannasch, 1969, Fig. 3), where population effects dependent on cell density (Jannasch, 1963) further complicated the outcome.

Apart from these problems, an accurate determination of the washout rate of the bacterial population requires several countings. Moreover, continuous cultures still appear to be uncommon tools for the majority of microbial ecologists. This may also have hindered further application and development of this technique in field measurements.

3.7. Frequency of Dividing Cells

Krambeck (1978) was the first to note that, in bacterial communties in the epilimnion of the Pluss See (FRG), approximately 7% of the cells were in the stage of cell division at sunrise compared with only 4–5% during the rest of the light period. This fluctuation correlated with the growth activity of the cells. Hagström et al. (1979) assessed experimentally (in continuous culture) a positive correlation between the frequency of dividing cells (FDC) and growth rate. A statistically significant relationship between FDC and growth rates was also observed by Newell and Christian (1981). With this correlation and measurements of FDC in water samples, it is possible to calculate the growth rate of natural bacterial communities, without incubating the samples. This is certainly an important advantage. Furthermore, Larsson and Hagström (1981) found a significant positive correlation between mean cell volume and FDC of a mixed bacterial population. This offers the possibility of assessing both the growth rate and the rate of biomass production.

Several problems with this very elegant approach are discussed by Newell and Christian (1981). One problem is that relations between FDC, growth rate, and mean cell volume have to be established in continuous culture, a condition which is selective towards actively growing cells. Besides, in natural environments, the actively growing cells may be a small percentage of the total community (Meyer-Reil, 1977) and, hence, the number of dividing cells may be related to the wrong basis. This problem can only partly be overcome by simultaneous determination of the number of metabolically active cells, as it is not yet clear which fraction of the active cells is indeed growing. Moreover, this relation may vary between environments and as a result of fluctuating environmental conditions. Another problem is the detectability of division stages in very small cells. Nevertheless, this approach deserves rapid further development as the conditions of not requiring an incubation or time series measurement is a great advantage. Moreover, in the future, it might also be extended to unicellular eucaryotic organisms.

3.8. Incubation Effects

As mentioned above, one of the most serious difficulties when assessing bacterial activity is that bacteria respond very sharply to changes in their environment. This can be illustrated by the close coupling between the diurnal variations in bacterial activity and the diurnal rhythm of phytoplankton activity (which in its turn is dependent on the discontinuous input of solar radiation; Sieburth *et al.*, 1977; Meyer-Reil *et al.*, 1979; V. Straskrabova and J. Fuksa, unpublished results). Changes in the cells' environment may also be expected to occur when a water sample is enclosed in a container for the purposes of measuring heterotrophic activity and bacterial growth. Some aspects of bottle effects have already been mentioned. The organisms have more access to surface per unit of volume, and this can have a stimulatory effect. On the other hand, the influx of growth rate-limiting substrate is interrupted and products of metabolism accumulate, which may result in a fall in metabolic activity. The possible effects of removing larger organisms before the incubation has been considered in Section 3.5. The effects mentioned so far can be minimized by using short incubation times (a few hours or less) and large containers.

A further aspect, frequently overlooked, is the influence of light. Most samples are taken during daytime from euphotic habitats and are subsequently incubated in the dark. Thus, some widespread phenomena, such as photoassimilation of organic substrates in algae and light-dependent production of exudates, are prevented. Results obtained by Williams and Yentsch (1976) indicated that in the light, the uptake of amino acids is slightly higher than in the dark, whereas Azam and Holm-Hansen (1973) measured light uptake rates of glucose that were only 30% of dark uptake. Spencer (1979), in a more experimental approach, size-fractionated samples from the Pluss See before or after incubation in the light and in the dark. He observed a 50–109% stimulation of glucose uptake (concentration 1.4–8.6 μg/liter) in the light, with 76–98% of the total activity being taken up in the size fraction < 3 μm and $\pm 80\%$ in the size fraction < 1 μm. From the results of pre- and postfiltration experiments and the inducibility of the stimulatory effect (after about 10 min in the light), he concluded that photoheterotrophy is a bacterial phenomenon induced by algal photosynthetic activity, possibly the appearance of algae release products. This shows that one should be aware of the effects that incubation in the dark may have on the heterotrophic activity in samples from the euphotic zone.

4. Discussion

So far, we have discussed different methods for counting numbers of bacteria and estimating biomass, and for assessing bacterial activities. We will now consider some selected results that together may give an idea of the place of bacteria within the total community of the marine aquatic ecosystem.

4.1. Bacterial Numbers and Biomass

Some results of bacterial counts are summarized in Table III. To avoid distortions from the different methods used, only counts obtained by epifluorescence microscopy after the samples were stained with acridine orange (AODC) are included in the table. The results obtained by Ferguson and Rublee (1976) and Ferguson and Palumbo (1978) are probably too low by a factor of 2–3 because of the use of cellulose acetate filters (Ferguson and Palumbo, 1979). Moriarty (1979) counted only freely suspended bacteria but reported that about 50% of cells over coral reefs were attached to particles.

Despite tidal (Wright, 1978; Wilson and Stevenson, 1980), diurnal (Meyer-Reil *et al.*, 1979), and seasonal variations (Goulder, 1977; Meyer-Reil, 1977; Zimmermann, 1977; Wilson and Stevenson, 1980), there are some general trends. Bacterial numbers decrease by roughly two orders of magnitude from estuarine interiors ($>50 \times 10^8$ cells/liter) to tidal inlets and coastal regions (10–50×10^8 cells/liter), and further to euphotic offshore and oceanic waters (0.5–10×10^8 cells/liter). In deep waters, a gradient with depth is observed from 0.5–10×10^8 at the surface to less than 0.1×10^8 cells/liter in the aphotic zone, although high numbers may be present in intermediate layers and near the bottom. Nevertheless, even in the most impoverished water masses studied thus far, each ml of water contains several thousands of bacteria.

The difference between counts with epifluorescence and other light microscopic techniques may be illustrated by the results obtained by Wiebe and Pomeroy (1972) who used phase contrast microscopy. They found less than 10^6 bacterial cells/liter in 40 samples from euphotic ocean water and 0.01–1×10^8 cells/liter in 42 samples from continental shelf waters off the Georgia coast. On the other hand, Sorokin (1978), summarizing much of the Russian work with bright field microscopy after erythrosin staining, reported ranges from 20×10^8 to 100×10^8 cells/liter for polluted estuaries and lagoons to 0.4–2×10^8 cells/liter for the surface layers of oligotrophic seawater and freshwater, which is almost as high as the range apparent from Table III.

In our opinion, epifluorescence microscopy currently yields the most reliable determination of bacterial number and of biovolume of samples from natural environments. But as biovolume determinations are rather time consuming, relatively few data on bacterial biomass are determined with this method (Table III). Biomass values range from 1 to 5 μg C/liter for offshore surface water to 200 μg C/liter in the inner part of an eutrophic estuary. Sorokin (1978) summarized values, based on erythrosin-stained preparations, which cover the same range as the data in Table III. He reported generalized ranges of 10–100 μg C/liter for surface waters in polluted estuaries and lagoons and ranges of 1 to 5 μg C/liter in oligotrophic surface waters. Again there is a clear gradient from estuarine to coastal and offshore environments. The biomass gra-

Table III. Bacterial Numbers and Derived Biomass Values in Marine Habitats as Determined by Epifluorescence Microscopy on 0.1–0.2 μm Nuclepore Filters (Unless Otherwise Stated)

Area	Sampling depth (m)	Number of stations	Number of cells × 10^8/liter	Bacterial biomass μg C/liter	Reference	Remarks
Coastal and Estuarine waters						
U.S. east coast	0–20	3	5.5–6.8	5.2–6.0	Ferguson and Rublee, 1976	0.45-μm cellulose acetate filters
U.S. east coast	0–50	7	21		Ferguson and Palumbo, 1979	Ferguson and Palumbo, 1979
U.S. east coast	5	1	18		Johnson and Sieburth, 1979	
U.S. east coast	5	3	23	63.3	Sieburth et al., unpubl. data	
U.S. East Coast: Essex estuary					Wright, 1978	
Inner part	1	3	68			
Inlet	1	1	49			
Outside	1	5	29			
U.S. East coast Newport River estuary	0–2	4	61		Palumbo and Ferguson, 1978	0.45 μm cellulose and acetate filters
U.S. East coast North inlet High marsh creeks	0.2	3	78		Wilson and Stevenson, 1980	Annual averages
Primary tidal channel	0.2	1	23			
English estuaries					Goulder, 1977	0.45 μm cellulose acetate filters
Humber summer	Surface	1	64			
Humber winter	Surface	1	133			
Tyne winter	Surface	1	264			

(Continued)

Table III (*Continued*)

Area	Sampling depth (m)	Number of stations	Number of cells $\times$ 10^8/liter	Bacterial biomass μg C/liter	Reference	Remarks
Wear autumn	Surface	1	94			
Tees summer	Surface	1	47			
Esk summer	Surface	1	101			
North Sea						
Elbe estuary, low salinity, inner part	1	2	115	200	Saltzmann, 1980	
Baltic Sea						
Kiel Fjord	2	2	28	13.0	Zimmermann, 1977	Annual averages
Kiel Bight	2	3	14	5.9		
Baltic Sea	1–2	1	22–31	16–26	Meyer-Reil *et al.*, 1979	Range gives diurnal variation
	2–240	3	4.3–8.4	1.7–3.2	Dawson and Gocke, 1978	No obvious gradient with depth
Southern California Bight	1–30	3	7–19	4–12	Fuhrman *et al.*, 1980	
Southern California Bight	1–65	29	4 (0.6–20.8)		Fuhrman *et al.*, 1980	Stations 1–100 km offshore
South African coast			63		Hobbie *et al.*, 1977; Schleyer, 1981;	
Australian coast, over coral reefs	Surface	2	2–5		Moriarty, 1979	Only freely suspended cells counted
Antarctica						
McMurdo Sound	Surface	5	6.5	5.4	Fuhrman and Azam, 1980;	
East						
West		2	0.7			

Offshore and oceanic waters	Depth	No.			Reference	
N.W. Atlantic, continental shelf	0–600	2	6.7		Ferguson and Palumbo, 1979	
N.W. Atlantic, continental shelf	5	7	10.4	5.3	Sieburth et al., unpubl. data	
Sargasso Sea	5	1	4.2		Johnson and Sieburth, 1979	
	50	2	2.3			
	100	3	2.0			
Sargasso Sea	3–400	1	0.2–0.5		Liebezeit et al., 1980	
	900–1500	1	0.1	0.7		
North Central Pacific	1	1	1.4		Carlucci and Williams, 1978	
	75	1	0.5			
	500	1	0.2			
	1500	1	0.07			
	5550	1	0.05			
Deep waters of Africa	Surface		16		Hobbie et al., 1977	
	4200		0.3			
Antartic water under Ross Ice Shelf	66–200	1	0.1	0.1	Azam et al., 1979	Station covered with ice (360 m thick)

dient, which becomes obvious by comparing the cell sizes of bacteria in samples from nearshore to offshore environments (c.f. Fig. 2), will be more pronounced compared with that for bacterial number, as long rods are predominantly found in estuarine and near-coastal water, especially in summer (Meyer-Reil, 1977; Zimmermann, 1977). Though minibacteria (length > 0.3 μm) are often a large fraction of total number, their contribution to total biomass is usually small.

4.2. The Active Fraction

A long-standing question is what fraction of the total bacterial community is active. Less than 1% is able to grow on agar plates as colony-forming units (CFU), but, as discussed in Section 2.3 there may be several reasons for this. Microscopic techniques in combination with autoradiography or other methods that enable the direct observation of individual cells give a better estimation of active bacterial cells. The capability tested is the active uptake and metabolism of organic or inorganic substrates under close to natural conditions. Some results are summarized in Table IV. Because of differences in the methods, the data are certainly not completely comparable, but it may be concluded that roughly between 6 and 60% of the cells, with an average around 30%, show some kind of activity, with no apparent difference between the types of habitats investigated to date.

As only cells that show activity under the given conditions (method applied, kind of substrate, concentration, etc.) can be scored positively and obligate autotrophs will always score negative on organic substrates, the fraction of active cells thus determined is a conservative estimate. Although many bacteria in natural habitats are very versatile with respect to their ability to utilize a wide variety of organic substrates (Sepers, 1979; Konings and Veldkamp, 1980), microautoradiographic studies (Ramsay, 1974; Hoppe, 1976) have shown that the number of positive scores is largely dependent upon the substrates used. But if approximately 30% of the cells are active, then the discrepancy between the total number of bacteria and CFU is not a result of a very high fraction of dead cells.

Large temporal and spatial variations have been observed in the ratios of CFU to total counts and active cell counts to total counts (e.g., Jannasch and Jones, 1959; Sorokin and Kadota, 1972; Hoppe, 1977; Meyer-Reil, 1978a; Meyer-Reil et al., 1979). These fluctuations are much greater than can be explained by methodological errors.

Meyer-Reil (1978a) observed a 200-fold variation in viable count to total count ratios in samples taken above sandy sediment on 12 stations in the Kiel Fjord and Kiel Bight area, whereas the ratios of counts of active cells (with glucose as substrate) to total counts varied only 20-fold and the total count varied twofold. There was no apparent relation between the number of active

Table IV. Active Fraction of Total Bacterial Number

Location	Temp. °C	Substrate	Total counts determined	% Active (range)	Reference	Remarks
Freshwater lake	4–8	Glucose Thymidine	Light microscopy	17–45 8–27	Ramsay, 1974	Epigrowth bacteria on *Elodea canadensis* leaves
Rhode River estuary		Organic substrates	Light microscopy	63–85	Faust and Corell, 1977	
Seawater below Ross Ice Shelf, Antarctica	−2		AODC[b]	up to 50	Azam et al., 1979	Samples taken 66–200 m below 360 m-thick ice
Kiel Bight		Amino acid mixture	AODC	9–60 av: 28	Hoppe, 1976,1977	Year-round values
Kiel Fjord		Amino acid mixture		18–52 av: 30		
Kiel Bight	18–20	Glucose	AODC	2.3–50.5 av: 24	Meyer-Reil, 1978a	Active and total cell counts in same microscopic field
Kiel Fjord	18–20	Glucose	AODC	4.5–56.2 av: 38		
Kiel Fjord	15	Glucose	AODC	11–15	Meyer-Reil, et al., 1979	(a) Active and total cell counts in same microscopic field; (b) range gives diurnal variation
Baltic Sea	15	Glucose	AODC	4–6		
Kiel Fjord		Tetrazolium salt	AODC	19–23	Iturriaga and Rheinheimer, 1975	
Kiel Bight	0.5–2	Tetrazolium salt	AODC	12	Zimmerman et al., 1978	(a) Active and total cell counts in same microscopic field; (b) lower limit of detection of active cells; 0.4 μm
Kiel Fjord	4	Tetrozolium salt	AODC	6–7		
Salt marsh creek, California Surface layer (300μm)		Tetrazolium salt	AODC	16.0 5.1	Harvey and Young, 1980	
Subsurface Pacific Ocean		Yeast extract + naladixic acid	AODC	5–10	Kogure et al., 1979	

[a]Consult text for a description of the methods.
[b]AODC: acridine orange direct counts with epifluorescence microscopy.

cells and the number of viable cells. Apparently, against a background of relatively stable numbers of total bacterial cells, under certain environmental conditions, a small but strongly variable part of the living bacterial community is active and capable of growing out to visible colonies on enriched media, whilst under other conditions they are not. Although the reasons for the low CFU as compared with total counts are not entirely clear, determining the CFU gives valuable information for describing water bodies of different trophic levels (Sorokin and Kadota, 1972) as the CFU reflect an aspect of bacterial communities not revealed by other methods of investigation. The large spatial fluctuations in the CFU at relatively constant total counts apparently reflect sharp physiological responses of the bacterial cells to not yet adequately measured environmental parameters.

Another long-standing question is whether the active fraction of the total bacterial community is free-floating in the water or associated with particles. Wangersky (1977) came to the conclusion that the particle content of seawater controls the total bacterial utilization of dissolved and particulate organic matter. He argued that freely suspended bacteria "see only the envelope of water which immediately surrounds them and which travels with them wherever they go." In contrast, bacteria attached to particles that move either more slowly or faster than the surrounding water are continuously exposed to new water (Wangersky, 1977). It has, however, to be borne in mind that this new water will probably already contain many thousands of bacteria per ml (Table III). The second argument (Wangersky, 1977) is more convincing: organic molecules are adsorbed by the surfaces of particles, thus providing attached bacteria with enhanced substrate concentrations (for reviews, see Marshall, 1976,1980).

Harvey and Young (1980) investigated the proportion of actively respiring cells in a salt marsh creek (Palo Alto, California). They reported that in subsurface water, the majority (62%) of actively respiring cells were attached, though only 22% of the total cell number were attached cells. In the 300 μm surface film, this was even more extreme: virtually all respiring bacteria were attached to particles. Jannasch and Pritchard (1972) observed a distinctly higher activity when isolates from the turbid river Nile were grown on a medium of peptone (0.5 mg/liter) plus particulate material, instead of on a medium with peptone only. In continuous culture enrichment experiments with seawater, two different strains became dominant in the presence and absence of kaolinite particles. These experiments clearly showed that particles stimulate the activity of the organisms being studied, although this argument cannot be used vice versa, i.e., that active bacteria must be particle associated.

On the other hand, dissolved organic substrates, added at close-to natural concentrations to seawater samples, are predominantly assimilated by organisms that can pass through a 3 μm filter (Table V), presumably free-floating bacteria. Much of the organic matter produced as exudates by algae is dissolved, and approximately half of the total bacterial production results from

Table V. Size Fractionation of Heterotrophic Uptake Activity on Dissolved Substrates

Type of water	Pore size of filter (μm)	Type of filter[a]	Substrate	% of total incorporation passing filter	Reference
Oceanic	1.2	c	Glucose	50	Williams, 1970
	3	c	Amino acids mixture	68	
Oceanic	3	c	Glucose	30–68	Azam and Holm-Hansen, 1973
Continental shelf	3	p	Glucose	>80	Hanson and Wiebe, 1977
Estuarine and coastal	3	p	Glucose	<10	Hanson and Wiebe, 1977
Below Ross Ice shelf, Antarctic	1	p	Thymidine, adenine	11–71	Azam et al., 1979
Coastal Antarctic	0.6	p	ATP	71–88	Hodson et al., 1981
Various	1	p	Thymidine	>90	Fuhrman and Azam, 1980
Coastal	3	p	Glucose	>90	Berman, 1975
Coastal	1	p	Glucose	90	Azam and Hodson, 1977
	5	p	Acetate, serine	100	
Turbid near shore	1	p	Glucose, acetate	51–60	Paerl, 1980
Coastal, over subtidal reef	1	p	Glucose	82	Schleyer, 1981
	3	p		93	

[a] c = cellulose-based filter; p = polycarbonate filter.

the utilization of these dissolved exudates (Derenbach and Williams, 1974; Wolter, 1980; Larsson and Hagström, 1981). These combined data indicate that a considerable part of the total organic matter that flows through the bacterial part of the food web consists of dissolved organic substrates and that free-floating bacteria are its predominant utilizers. In fact, the question of whether particle-associated or free-floating bacteria are the most active, may largely depend on the temporarily and spatially varying input of both dissolved and particulate organic matter to the body of water being investigated. The discussion on this question, which started over 50 years ago (see review by ZoBell, 1946; Jannasch and Pritchard, 1972; Sieburth et al., 1974, and unpublished data; Wangersky, 1977; Stevenson, 1978; Schleyer, 1981), would be much more fruitful if data were available on the growth rate or biomass production rate of both groups of bacteria determined with a method not dependent upon the addition of any substrate. In that respect, the frequency of dividing cells (Section 3.7) of both groups may give valuable information. Newell and Christian (1981) observed up to three times higher FDC values for bacteria associated with particles in some nearshore water samples, but in other samples essentially equivalent values were found for free and attached or associated bacteria.

4.3. Heterotrophic Bacterial Activity

From heterotrophic uptake experiments with labeled substrates, it has been concluded that bacteria are the predominant consumers of organic substances at low, natural concentrations (Wright and Hobbie, 1966). This conclusion has been confirmed by microautoradiographic studies (Munro and Brock, 1968; Hoppe, 1976) and experiments in which organisms were size-fractioned after incubation (Table V). It is evident that, with respect to dissolved organic matter, heterotrophic activity is predominantly in the size class of free-living bacteria and possibly bacteria that are loosely associated with larger particles. The relatively high percentages of activity retained on 3-μm filters in early studies (Williams, 1970; Azam and Holm-Hansen, 1973) may be ascribed to the use of cellulose acetate filters. In turbid coastal and estuarine waters, it has been observed that there is a tendency for relatively more bacterial cells to be attached to particles (see Section 2); this is reflected in a higher fraction of heterotrophic activity retained on 3-μm filters (Hanson and Wiebe, 1977; Paerl, 1980). This pattern is in accordance with that for lake systems, where in turbid and eutrophic regions, 50–60% of the heterotrophic activity was found in organisms <1 μm; in all other cases, 70–90% of several substrates was incorporated within this size fraction (Paerl, 1980).

Wangersky (1978) has criticized the design of experiments from which it was concluded that utilization of dissolved organic substrates is predominantly bacterial. He argued that they are "heavily biased in favour of the bacteria" as samples are mostly taken from the euphotic zone and incubated in the dark with selected organic substrates. This not only prevents photoassimilation by algae, but also means that the bacterial community is provided with a substrate that limits its growth, whilst the same substrate is not limiting for algae. The results obtained by Mitamura and Saijo (1975) and Wheeler *et al.* (1977) indicate that under nitrogen limitation, algae can use a significant part of nitrogenous organic substrates.

However, though many algae, protozoans, and metazoans can utilize organic substrates as an additional or a sole source of carbon and energy (for reviews, see Droop, 1974; Neilson and Lewin, 1974; Sepers, 1977; Conover, 1978; Stephens, 1981), the possession of high-affinity uptake systems for a wide variety of organic substrates (Sepers, 1979) and their high surface-to-volume ratio undoubtedly make heterotrophic bacteria more suitable for uptake at natural (nanomolar) concentrations than any of the eucaryotic organisms in the sea.

Heterotrophic bacteria in aquatic environments simultaneously exhibit various types of activity, e.g., the hydrolysis of macromolecules, the uptake of small dissolved molecules, inorganic nutrients, and oxygen, which together result in the production of biomass and new cells. It should be kept in mind that different methods analyze different specific aspects of microbial activity.

These aspects, though related to each other, are not necessarily closely linked. When, for example, nitrogen-poor detritus is the main carbon source, bacteria will take up ammonia or nitrate, whereas ammonia will be excreted when protein-rich detritus is decomposed, though in both cases the growth rate can be the same.

Bacterial activity is directly dependent on a number of chemical and physicochemical parameters: temperature, pH, pressure, salinity, and ion composition. This dependence has been the subject of numerous studies. But under equal macroenvironmental conditions there may be microenvironments of high activity surrounded by low activity. The surfaces of particles may be such environments, and anyone who has ever observed decaying algal cells under the microscope will agree that such cells are spots of extremely high bacterial activity. Unfortunately, there are almost no data on the small-scale distribution of heterotrophic activity. Such knowledge would certainly help in finding the ultimate factors that regulate the heterotrophic activity within natural communities. On a larger scale, this activity will primarily be regulated by the rate at which utilizable organic matter becomes available, as most communities are severely energy limited.

Heterotrophic bacterial communities generally keep the concentrations of dissolved sugars, amino acids, and lower fatty acids in the nanomole range. The turnover times of these substrates, as calculated from uptake kinetics experiments (Section 3.3), is in the order of hours to a few days in the euphotic zone in the temperate climates (Gocke, 1977a; Sepers, 1977; Hoppe, 1978). So there must be an almost continuous supply of new material to replenish the losses resulting from utilization. The main sources are exudation by algae, algal cell lysis, losses of algal material during zooplankton grazing, zooplankton excretion, hydrolysis of complexed dissolved and particulate organic matter, and, especially in estuarine and coastal waters, external sources. Some of these subjects have recently been reviewed (Williams, 1975; Conover, 1978; Wangersky, 1978; Mague et al., 1980; Gagosian and Lee, 1981; Skopintsev, 1981).

In the last ten years, the production of exudates by phytoplankton and its subsequent utilization has received much attention. Studies from marine environments include those by Derenbach and Williams, 1974; Williams and Yentsch, 1976; Hanson and Wiebe, 1977; Iturriaga and Hoppe, 1977; Wiebe and Smith, 1977a,b; Lancelot, 1979; Larsson and Hagström, 1979, 1981; Bell, 1980; Bell and Sakshaug, 1980, and Wolter, 1980. It is beyond the scope of this chapter to discuss the physiology and kinetics of exudate production by algae. Hellebust (1970) and Mague et al. (1980) give detailed information on this, and Sharp (1977) discusses the overestimates probably made in earlier work.

Most determinations of the flow of exudates from autotrophs to heterotrophs are based on incubation of samples with $^{14}CO_2$, size fractionation of the samples using 3-μm and 0.2-μm filters, and tracing the appearance of dissolved

(<0.2 μm) labeled organic matter and incorporation into microheterotrophs (<3 μm) (Derenbach and Williams, 1974). Several procedures have been proposed to quantify the bacterial uptake of exudates in the presence of primary producers (Berman, 1975; Iturriage and Hoppe, 1977; Wiebe and Smith, 1977a,b). Wolter (1980) has dealt extensively with the procedural problems and has arrived at a procedure with many controls, including *in situ* incubation of unfiltered and 3-μm filtered samples with $^{14}CO_2$, as well as ^{14}C glucose in light and dark bottles for 6 and 24 hr. In samples from the Kiel Fjord (FRG), she observed a large fluctuation in exudate production that paralleled the rate of primary production. The percentage of primary production that was released as exudates fluctuated considerably (range 0–24%, average $\pm15\%$), but this was mainly because of changes in the composition of the phytoplankton over the year: the highest precentages were observed during flagellate blooms. Wolter found that bacterial incorporation was closely linked with exudate production. The percentage of exudates incorporated by bacteria within 6 hr ranged from 0 to 65% and averaged approximately 20%. The exudates of nanoflagellates and dinoflagellates appeared to be taken up more rapidly than those of other algae.

Results of the studies on the utilization of algal exudates give good examples of the dependence of bacterial activity on the availability of utilizable organic substrates.

4.4. Bacterial Growth and Biomass Production in Relation to the Food Web

One way to study an ecosystem and the organisms involved is to unravel the structure of its food web and to determine quantitatively the relations between organisms, their food, and their predators in terms of transfers of carbon or energy. To assess the quantitative contribution of microheterotrophs, especially bacteria, to food webs in the sea it is necessary, as for all other organisms, to have information on biomass and activity, especially specific growth rate or generation time and biomass production.

Some results of measurements of generation time are given in Table VI. In some investigations, the growth of isolated bacteria in unsupplemented seawater was followed (Jannesch, 1967; Carlucci and Shimp, 1974; Carlucci and Williams, 1978). But isolates may not be very representative. As discussed in Section 3.5, Jannasch's (1967) estimates of generation time may be too high. In other investigations, the total microplankton or the total bacterial community was included. Surprisingly, in spite of the completely different methods used for the determination of generation times, comparable results were obtained. Generally, generation times were in the the range of 10–100 hr. In the euphotic zone, the generation time can be considerably less than 12 hr, which would mean more than two doublings per day. The considerable seasonal

Table VI. Estimates of Generation Time (g)

Location	Method (section)	Sampling depth (m)	Temp. (°C)	g (hr)	Reference	Remarks
Several coastal habitats	Autoradiography (3.4)	Euphotic		11–11.5	Brock, 1967	Epigrowth of *Leucotrix mucor* on seaweeds
North Atlantic, coastal	Washout rate in continuous culture (3.6.3)	Surface	24	52–170	Jannasch, 1967	Isolate, growing in filtered, unsupplemented seawater
North Atlantic, coastal	Frequency of dividing cells (3.7)	Subsurface	24–29	8–37	Newell and Christian, 1981	Total bacterial community
Sea of Japan, coastal inlets, June–Oct.	Increase in no. of cells in bottle (3.6.1)	Subsurface		7.4–41.7 av. 24	Vyshkwartsev, 1980	Total bacterial community
Pacific Ocean, coastal	Increase in no. of cells in bottle (3.6.1)	Surface	20	3	Carlucci and Shimp, 1974	Isolate, growing in aged seawater
		Surface	5	25		
North Central Pacific	Increase in no. of cells in bottle (3.6.1)	1	22.6	14	Carlucci and Williams, 1978	Isolates, growing in filtered, unsupplemented seawater at *in situ* pressure
		75	18.8	11		
		500	7.5	67		
		1500	2.5	145		
		5550	1.5	210		
Caribbean Sea (Columbian basin)	RNA synthesis (3.4)	0–700	10–25	2.2–6.5	Karl, 1979	Total plankton highest growth rate at appr. 450 m depth (10°C)
North Atlantic	Increase in ATP conc. in diffusion chamber (3.6.2)	40–80		4 8	Sieburth *et al.*, 1977	Total plankton <3 μm daylight hours, mean of diurnal cycle
Kiel Fjord Summer Winter	Increase in no. of cells in flow system (3.6.2)	2 2		7.7–13.4 15.2–37.2	Meyer-Reil, 1977	Total bacterial community, only cells capable of growth included
Kiel Bight Summer Winter		2 2		11.9–71.4 >114		
Kiel Fjord summer (water overlaying sandy sediment)	¹⁴C-glucose uptake (3.3.4)	Beaches	19	10	Meyer-Reil, unpubl. data	Total bacterial community
Baltic Sea, coastal year-round	Frequency of dividing cells (3.7)	0–20		10–100	Hagström *et al.*, 1979	Total bacterial community

variation of generation time (Meyer-Reil, 1977; Vyshkvartsev, 1980) can partly be ascribed to differences in temperature. In winter, even negative values may occasionally be recorded when the decrease in total number of cells is not compensated for by the increase caused by the growth of cells capable of growth (Meyer-Reil, 1977).

Bacterial biomass production rates are summarized in Table VII. Again, temporal and spatial differences make it difficult to compare the results of different studies obtained with quite different methods. The results in both tables VI and VII should rather serve to give an idea of the order of magnitude. Nevertheless, it is evident that bacterial production per liter is highest in estuarine and coastal waters and some two orders of magnitude lower in the top 1000 m of open ocean water.

Frequently, the measurements of bacterial production were accompanied by determinations of bacterial biomass and algal primary production, thus allowing bacterial production to biomass (P/B) ratios and the percentage of bacterial production relative to primary production to be calculated. This latter percentage indicates the quantitative importance of bacteria in the flow of organic matter and energy through food webs in the sea. The high values resulting from the dark CO_2 fixation measurements (Sorokin, 1971,1978; Vyshkvartsev, 1980) have been discussed in Section 3.2. The other data indicate that bacterial production in the euphotic zone is some 20% of primary production. If the growth efficiency of bacteria is approximately 0.5, this means that some 40% of annual primary production may be utilized by bacteria.

Wolter (1980) estimated the diurnal bacterial biomass production at the expense of algal exudates in surface water of the Kiel Bight to be up to 15.6% of primary production. Larsson and Hagström (1981) calculated an annual exudate release of 12% of primary production, supporting 53% of annual bacterial production at two eutrophic stations in the Northern Baltic. For a control station, the values were 16% and 57%, respectively. Earlier reports on the contribution of phytoplankton exudates to bacterial production showed an average exceeding 20% (Derenbach and Williams, 1974), and an annual average of 45% (Larsson and Hagström, 1979), respectively. These results may partly be explained by the passage of significant amounts of nanoplankton cells through the 3-μm separation filter, which requires correction (Wolter, 1980; Larsson and Hagström, 1981). Occasionally a significant fraction of the nanoplankton may even pass through a 1μm filter (Berman, 1975).

Nevertheless, it may be concluded that in the euphotic zone, exudates are an important source of organic matter for bacteria. Besides this direct transfer of organic matter from algae to bacteria, senescent and decaying algal cells, as well as algal cell constituents leaking into the environment during feeding of zooplankton, will be other important sources of organic matter for heterotrophic bacteria.

If bacteria in the euphotic zone of the sea do indeed grow with generation

Table VII. Estimates of Bacterial Biomass Production

Location	Method (section)	Time of the year	µg C/liter/day	Production rate g C/m²/yr	Daily P/B ratio	Annual bacterial/ primary prod. ratio (× 100%)	Reference	Remarks
Tropical Pacific								
Lagoon of Atoll (24 m)	Heterotrophic CO_2 fixation (3.2)		19		0.3	270	Sorokin, 1971, 1978	Water column considered for calculations: 22 m, 300 m, 5000 m
Neritic waters (2900 m)			2		1.1	480		
Equatorial zone (5000 m)			0.1		0.05	106		
Sea of Japan, coastal inlets	Heterotrophic CO_2 fixation (3.2)	Summer / Winter	40–70 / 2–6	43	1.55 / 0.06	100	Vyshkvartsev, 1980	
	Increase of cell biovolume in bottle (3.6.1)	Summer and fall	18–160 (av. 55)		0.3–1.9 (av. 0.77)			
Caribbean Sea at depth of 450 m	Rate of RNA synthesis (3.4)	Summer	1.9–5.5				Karl, 1979	Zone of production maximum; total community
Antarctic McMurdo Sound	Rate of DNA synthesis (3.4)	Summer	0–2.9			±10	Fuhrman and Azam, 1980	
Californian Coast	Increase of biovolume in bottle (3.6.1)		10–34					
	DNA synthesis (3.4)		0.7–53					
North Atlantic, coastal	Frequency of dividing cells (3.7)	Spring	19–178				Newell and Christian, 1981	
Baltic Sea, coastal	Frequency of dividing cells (3.7)	Year-round					Larsson and Hagström, 1981	
Eutrophied			12–30	50		18		
Control			10–15	38		24		
Kiel Fjord	Increase of cell biovolume in flow system (3.6.2)	July–Oct. (summer–fall)	117–520 (av. 259)	57		29	Meyer-Reil, 1977	
		Nov.–May (winter–spring)	2–96 (av. 70)					
Kiel Bight		July–Oct.	10–57 (av. 25)	9		15		
		Nov.–May	0–8 (av. 2)					
Kiel Fjord (water overlying sandy sediments)	^{14}C glucose uptake (3.3.4)	Summer	av. 150		3		Meyer-Reil, unpubl. data	

times of between 3 and 20 hr, and with a biomass production of 20% of primary production, growth must be counter-balancing comparable rates of loss. A high death rate may be assumed. This implies that within the same sample, rapidly growing cells and rapidly dying cells coexist, pointing to the existence of microenvironments in which more bacterial activity occurs (e.g., within decaying algae cells, fecal pellets), whereas in the bulk of water, the activity may be much lower with a net negative result for the cell number. A second mechanism for counterbalancing a high rate of bacterial production is a comparable rate of predation. Fenchel (1980c) calculated that bacteriovorous ciliates require bacterial concentrations of 10^7–10^8/ml for growth. It is evident from Table III that in general, such high concentrations are only found in estuarine waters. It is therefore unlikely that ciliary suspension feeders are responsible for the control of bacterial populations in offshore waters (Fenchel, 1980c). This may hold especially for free-living bacteria; on the other hand, ciliated protozoa appear to form an essential link in the transfer of energy from bacteria that decompose particulate detritus to zooplankton copepods (Heinle *et al.*, 1977; Sorokin, 1977). Pomeroy (1979) supposed that free-living bacteria in the water could only be effectively removed by mucus-net feeders, but R. T. Wright (personal communication) has observed that *Geukensia demissa*, an estuarine bivalve species, is capable of filtering free-floating natural bacteria.

An explanation for the observations that most particles in the open sea are devoid of bacterial cells and that most cells are freely suspended may be that the grazing pressure on attached cells is so high that they, whilst growing fast, are gathered from the particles up to several times per day. The free floating cells may, in that light, be regarded as a survival mechanism of the species, as they are subject to much lower grazing pressure. Such cells may be waiting for a chance to colonize new particles, senescent algal cells, fresh fecal pellets, etc.; they are helped in finding them by chemotaxis (Gessey and Morita, 1979). This hypothesis would be boosted if it could be shown that the activity of attached bacteria is much higher than that of freely suspended bacteria. Unfortunately, no such data are yet available. As suggested by Pomeroy (1974), it may be hypothesized that the large fraction of free-living cells assimilates approximately half of the available organic matter, but almost completely respires this in the course of time without growing very much. The low respiration percentages calculated from ^{14}C tracer experiments (Section 3.3) would then have to be considered spurious, because they result from the short incubation periods used (a few hours). Another consequence of this hypothesis would be that algal exudates would be largely lost to the food web. If, however, freely suspended bacteria are responsible for a large part of bacterial production, a highly specialized and effective grazing mechanism [presumably mucus-net feeders (Pomeroy, 1979)] has to be supposed in order to maintain the bacterial numbers at existing levels. Unfortunately, though a picture begins to emerge of the productivity of bacteria in marine environments, our knowledge of the fate of the produced bacterial biomass is far less complete.

4.5. Outlook

Today, the importance of bacteria as mineralizers and biomass producers is unquestioned. It is encouraging that determination of important parameters such as bacterial biomass, generation time, and biomass production yield comparable orders of magnitude (Tables III, VI, and VII), in spite of the different techniques applied. Nevertheless, field data on the quantification of these parameters are still scarce. Moreover, it is necessary that existing techniques are further developed and compared under *in situ* incubation conditions.

For a better understanding of the regulation of processes in natural bacterial communities, the scale of field measurements has to be reduced both in time and in space. Short-term (hours to days) and small-scale (μm to m) variations should be studied rather than variations between samples from water masses differing in many ways.

Thus far, attention has mainly been focused on whole bacterial communities, neglecting their diversity. Studies of the physiology of the composing species, and of the differences between species, are needed to enable explanation or prediction of community reactions on environmental changes. Unfortunately, up till now, most physiological work has been carried out with selected organisms which had the capacity to grow relatively fast under steady state conditions at micromolar concentrations of one growth-rate-limiting substrate. These conditions are certainly considerably different from diurnal fluctuations in the nanomolar range of dozens of substrates in a mixed community. This may be reflected in essential differences of characteristics between the organisms studied and those predominating in the field. Consequently, ecophysiological experiments should be done at more close-to-natural conditions.

A number of specific problems become obvious from the foregoing discussion: bacterial decomposition of particulate organic matter, activity and interactions of free-floating bacteria and those attached to particles, diversity and succession of bacteria in natural communities, excretion of products by bacteria, and grazing upon bacteria. These are only some of the most urgent questions in microbial ecology. We feel that the solution of some of the problems is not limited by adequate methods available, but is rather dependent upon asking the right questions and approaching the questions by combining adequate methods.

ACKNOWLEDGMENTS. We are indebted to Prof. H. Veldkamp for his constructive criticism on the manuscript, to Mrs. M.Th. Broens-Erenstein and Mrs. M Pras for typewriting several drafts of the manuscript, and to Mrs. J. Burrough-Boenisch for correcting the English text. This work comprises Publication no. 52 of the project "Biological Research in the Ems-Dollard Estuary" and no. 367 of the Joint Research Program of Kiel University (Sonderforschungsbereich 95 der Deutschen Forschungsgemeinschaft).

References

Abrams, B. I., and Mitchell, M. J., 1980, Role of nematode-bacterial interactions in hetero-trophic systems with emphasis on sewage sludge decomposition, *Oikos* **35**:404–410.

Allen, H. L., 1968, Acetate in fresh water: natural substrate concentrations determined by dilu-tion bioassay, *Ecology* **49**:346–349.

Anderson, R. S., and Dokulil, M., 1977, Assessment of primary and bacterial production in three large mountain lakes in Alberta, Western Canada, *Int. Rev. Ges. Hydrobiol.* **62**:97–108.

Andrews, P., and Williams, P. J. LeB., 1971, Heterotrophic utilization of dissolved organic com-pounds in the sea. III. Measurement of the oxidation rates and concentrations of glucose and amino acids in sea water, *J. Mar. Biol. Assoc. U.K.* **51**:111–125.

APHA, 1971, *Standard Methods for the Examination of Water and Wastewater* (13th ed.), American Public Health Association, Washington D.C.

Azam, F., and Holm-Hansen, O., 1973, Use of tritiated substrates in the study of heterotrophy in sea water, *Mar. Biol.* **23**:191–196.

Azam, F., and Hodson, R. E., 1977, Size distribution and activity of marine microheterotrophs, *Limnol. Oceanogr.* 22:492–501.

Azam, F., Beers, J. R., Campbell, L., Carlucci, A. F., Holm-Hansen, O., Reid, F. M. H., and Karl, D. M., 1979, Occurrence and metabolic activity of organisms under the Ross Ice Shelf, Antartica, at Station J9, *Science* **203**:451–453.

Banse, K., 1974, On the role of the bacterioplankton in the tropical oceans, *Mar. Biol.* **24**: 1–5.

Barvenick, F. W., and Malloy, S. C., 1979, Kinetic patterns of microbial amino acid uptake and mineralization in marine waters, *Est. Coast Mar. Sci.* **8**:241–250.

Baskett, R. C., and Lulves, W. J., 1974, A method of measuring bacterial growth in aquatic environments using dialysis culture, *J. Fish. Res. Bd. Can.* **31**:372–374.

Bell, W. H., 1980, Bacterial utilization of algal extracellular products. 1. The kinetic approach, *Limnol. Oceanogr.* **25**:1007–1020.

Bell, W. H., and Sakshaug, E., 1980, Bacterial utilization of algal extracellular products. 2. A kinetic study of natural populations, *Limnol. Oceanogr.* **25**:1021–1033.

Berman, T., 1975, Size fractionation of natural aquatic populations associated with autotrophic and heterotrophic carbon uptake, *Mar. Biol.* **33**:215–220.

Bowden, W. B., 1977, Comparison of two direct-count techniques for enumerating aquatic bac-teria, *Appl. Environ. Microbiol.* **33**:1229–1232.

Brock, M. L., and Brock, T. D., 1968, The application of microautoradiographic techniques to ecological studies, *Mitteil. Int. Ver. theor. Angew. Limnol.* **15**:1–29.

Brock, T. D., 1967, Bacterial growth rate in the sea: direct analysis by thimidine autoradiogra-phy, *Science* **155**:81–83.

Bryan, J. R., Riley, J. P., and Williams, P. J. LeB., 1976, A Winkler procedure for making precise measurements of oxygen concentration for productivity and related studies, *J. Exp. Mar. Biol. Ecol.* **21**:191–197.

Burnison, B. K., and Morita, R. Y., 1974, Heterotrophic potential for ammino acid uptake in a naturally eutrophic lake, *Appl. Microbiol.* **27**:488–495.

Burns, R. G., 1980, Microbial adhesion to soil surfaces: consequences for growth and enzyme activities, in: *Microbial Adhesion to Surfaces* (R. C. W. Berkeley, J. M. Lynch, J. Melling, P. R. Rutter, and B. Vincent, eds.), pp. 249–262, Ellis Horwood, Chichester.

Carlucci, A. F., and Shimp, S. L., 1974, Isolation and growth of a marine bacterium in low concentrations of substrate, in: *Effect of the Ocean Environment on Microbial Activities* (R. R. Colwell and R. Y. Morita, eds.), pp. 363–367, Univ. Park Press, Baltimore.

Carlucci, A. F., and Williams, P. M., 1978, Simulated in situ growth rates of pelagic marine bacteria, *Naturwissenschaften* **65**:641–542.

Christensen, J. P., Owens, T. G., Devol, A. H., and Packard, T. T., 1980, Respiration and physiological state in marine bacteria, *Mar. Biol.* **55**:267–276.

Copping, A. E., and Lorenzen, C. J., 1980, Carbon budget of a marine phytoplankton-herbivore system with carbon-14 as a tracer, *Limnol. Oceanogr.* **25**:873–882.

Corpe, W. A., 1974, Periphytic marine bacteria and the formation of microbial films on solid surfaces, in: *Effect of the Ocean Environment on Microbial Activities* (R. R. Colwell and R. Y. Morita, eds.), pp. 397–417, University Park Press, Baltimore.

Corpe, W. A., and Winters, H., 1972, Hydrolytic enzymes of some periphytic marine bacteria, *Can. J. Microbiol.* **18**:1483–1490.

Conover, T. J., 1978, Transformation of organic matter, in: *Marine Ecology, Vol. IV, Dynamics* (O. Kinne, ed.), pp. 221–499, John Wiley & Sons, Chichester.

Coveney, M. F., Cronberg, G., Enell, M., Larsson, K., and Olafsson, L., 1977, Phytoplankton, zooplankton and bacteria-standing crop and production relationships in a eutrophic lake, *Oikos* **29**:5–21.

Crawford, C. C., Hobbie, J. E., and Webb, K. L., 1973, Utilization of dissolved organic compounds by microorganisms in an estuary, in: *Estuarine Microbial Ecology* (L. H. Stevenson and R. R. Colwell, eds.), pp. 169–180, Univ. South Carolina Press, Columbia.

Dalay, R. J., and Hobbie, J. E., 1975, Direct counts of aquatic bacteria by a modified epifluorescence technique, *Limnol. Oceanogr.* **20**:875–882.

Darnell, R. M., 1967, Organic detritus in relation to the estuarine ecosystem, in: *Estuaries* (G. H. Lauff, ed.), pp. 376–382, Publ. A.A.A.S. no. 83, Washington, D.C.

Davis, W. M., and White, D. C., 1980, Fluorometric determinations of adenosine nucleotide derivates as measures of the microfouling, detrital and sedimentary microbial biomass and physiological status, *Appl. Environ. Microbiol.* **40**:539–548.

Dawson, R., and Gocke, K., 1978, Heterotrophic activity in comparison to the free amino acid concentrations in balthic sea water samples, *Oceanol. Acta* **1**:45–54.

Derenbach, J. B., and Williams, P. J. LeB., 1974, Autotrophic and bacterial production: fractionation of plankton populations by different filtration of samples from the English Channel, *Mar. Biol.* **25**:263–269.

Devol, A. H., and Packard, T. T., 1978, Seasonal changes in respiratory enzyme activity and productivity in Lake Washington microplankton, *Limnol. Oceanogr.* **23**:104–111.

Droop. M. R., 1974, Heterotrophy of carbon, in: *Algal Physiology and Biochemistry*, Bot. Mongr. 10 (W. D. P. Stewart, ed.), pp. 530–559, Blackwell, Oxford.

Fallen, R. D., and Pfaender, F. K., 1976, Carbon metabolism in model microbial systems from temperate salt marsh, *Appl. Environ. Microbiol.* **31**:959–968.

Faust, M. A., and Correll, D. L., 1977, Autoradiographic study to detect metabolically active phytoplankton and bacteria in the Rhode River Estuary, *Mar. Biol.* **41**:293–305.

Federle, T. W., and Vestal, J. R., 1980, Lignocellulose mineralization by artic lake sediments in response to nutrient manipulation, *Appl Environ. Microbiol.* **40**:32–39.

Fenchel, T., 1980a, Relation between particle size selection and clearance in suspension-feeding ciliates, *Limnol. Oceanogr.* **25**:733–738.

Fenchel, T., 1980b, Suspension feeding in ciliated protozoa: functional response and particle size selection, *Microb. Ecol.* **6**:1–11.

Fenchel, T., 1980c, Suspension feeding in ciliated protozoa: feeding rates and their ecological significance, *Microb. Ecol.* **6**:13–25.

Fenchel, T., and Harrison, P., 1976, The significance of bacterial grazing and mineral cycling for the decomposition of particulate detritus, in: *The Role of Terrestrial and Aquatic Organisms in Decomposition Processes* (J. M. Anderson and A. MacFadyan, eds.) pp. 285–299, Blackwell, Oxford.

Fenchel, T. M., and Jørgensen, B. B., 1977, Detritus food chains of aquatic ecosystems: the role of bacteria, in: *Advances in Microbial Ecology, Vol. 1* (M. Alexander, ed.), pp. 3–37, Plenum Press, New York.

Ferguson, R. L., and Rublee, P., 1976, Contribution of bacteria to standing crop of coastal plankton, *Limnol. Oceanogr.* **21**:141–145.

Ferguson, R. L., and Palumbo, A. V., 1979, Distribution of suspended bacteria in neritic waters south of Long Island during stratified conditions, *Limnol. Oceanogr.* **24**:697–705.

Field, E. O., Dawson, K. B., and Gibbs, J. E., 1965, Autoradiographic differentiation of tritium and another beta-emitter by a combined color-coupling and double stripping film technique, *Stain Technol.* **40**:295–300.

Fliermans, C. B., and Schmidt, E. L., 1975, Fluorescence microscopy: direct detection, enumeration and spatial distribution of bacteria in aquatic systems. *Arch. Hydrobiol.* **76**:33–42.

Forrest, W. W., 1969, Energetic aspects of microbial growth, *19th Symp. Soc. Gen. Microbiol.*, pp. 65–86, Cambridge Univ. Press, London.

Francisco, D. E., Mah, R. A., and Rabin, A. C., 1973, Acridine-orange-epifluorescence technique for counting bacteria in natural waters, *Trans. Am. Microsc. Soc.* **92**:416–421.

Fuhrman, J. F., 1981, Influence of method on the apparent size distribution of bacterioplankton cells: epifluorescence microscopy compared to scanning electron microscopy, *Mar. Ecol. Progr. Ser.* **5**:103–106.

Fuhrman, J. A., and Azam, F., 1980, Bacterioplankton secondary production estimates for coastal waters of British Columbia, Antartica and California, *Appl. Environ. Microbiol.* **39**:1085–1095.

Furhman, J. A., Ammerman, J. W., and Azam, F., 1980, Bacterioplankton in the coastal euphotic zone: distribution, activity and possible relationships with phytoplankton, *Mar. Biol.* **60**:201–207.

Gagosian, R. B., and Lee, C., 1981, Processes controlling the distribution of biogenic organic compounds in seawater, in: *Marine Organic Chemistry* (E. K. Duursma and R. Dawson, eds.), pp. 91–123, Elsevier, Amsterdam.

Gaudy, A. F., 1972, Biochemical oxygen demand, in: *Water Pollution-Microbiology* (R. Mitchell, ed.), pp. 305–322, John Wiley & Sons, New York.

Geesey, G. G., and Morita, R. Y., 1979, Capture of arginine at low concentrations by a marine psychrophilic bacterium, *Appl. Environ. Microbiol.* **38**:1092–1097.

Gieskes, W. W. C., Kraay, G. W., and Baars, M. A., 1979, Current ^{14}C methods for measuring primary production: gross underestimates in oceanic waters, *Neth. J. Sea Res.* **13**:58–78.

Gocke, K., 1976, Respiration von gelösten organischen Verbindungen durch natürliche Mikroorganismen Populationen. Ein Vergleich zwischen verschiedenen Biotopen, *Mar. Biol.* **35**:375–383.

Gocke, K., 1977a, Heterotrophic activity, in: *Microbial Ecology of a Brackish Water Environment* (G. Rheinheimer, ed.), pp. 61–70, Springer-Verlag, Berlin.

Gocke, K., 1977b, Comparison of methods for determining the turnover times of dissolved organic compounds, *Mar. Biol.* **42**:131–141.

Gocke, K., Dawson, R., and Liebezeit, G., 1981, Availability of dissolved free glucose to heterotrophic microorganisms, *Mar. Biol.* **62**:209–216.

Godlewska-Lipowa, W., 1969, Relationship between the generation time of a group of bacteria in water, and the exposure time and capacity of flasks, *Bull. Acad. Pol. Sci.* Cl. II, Vol. XVIII, **41**:233–237.

Godlewska-Lipowa, W., 1970, Generation time of a group of bacteria in the water of Mazurian lakes, *Pol. Arch. Hydrobiol.* **17**:117–120.

Goodrich, T. D., and Morita, R. Y., 1977a, Incidence and estimation of chitinase activity associated with marine fish and other estuarine samples, *Mar. Biol.* **41**:349–353.

Goodrich, T. D., and Morita, R. Y., 1977b, Bacterial chitinase in the stomachs of marine fishes from Yaquina Bay, Oregon, U.S.A., *Mar. Biol.* **41**:355–360.

Goulder, R., 1977, Attached and free bacteria in an estuary with abundant suspended solids, *J. Appl. Bacteriol.* **43**:399–405.

Hagström, Å., Larsson, U., Hörstedt, P., and Normark, S., 1979, Frequency of dividing cells, a new approach to the determination of bacterial growth rates in aquatic environments, *Appl. Environ. Microbiol.* **37**:805–812.

Hall, C. A. S., and Moll, R., 1975, Methods of assessing aquatic primary productivity, in: *Primary Productivity of the Biosphere* (H. Lieth and R. H. Whittaker, eds.), pp. 19–53, Springer-Verlag, New York.

Hamilton, R. D., 1973, Interrelationships between bacteria and protozoa, in: *Estuarine Microbial Ecology*, (L. H. Stevenson and R. R. Colwell, eds.), pp. 491–497, Univ. South Carolina Press, Columbia.

Hamilton, R. D., and Preslan, J. E., 1970, Observations on heterotrophic activity in the eastern tropical Pacific, *Limnol. Oceanogr.* **15**:395–401.

Hanson, R. B., and Wiebe, W. J., 1977, Direct measurement of dissolved organic carbon release by phytoplankton and incorporation by microheterotrophs, *Mar. Biol.* **42**:321–330.

Harris, G. P., 1978, Photosynthesis, productivity and growth: the physiological ecology of phytoplankton, *Arch. Hydrobiol. Beih.* **10**:1–171.

Harrison, P. G., and Mann, K. H., 1975, Detritus formation from eelgrass *(Zostera marina)*: the relative effects of fragmentation, leaching and decay, *Limnol. Oceanogr.* **20**:924–934.

Harrison, M. J., Wright, R. Y., and Morita, R. Y., 1971, Method for measuring mineralization in lake sediments, *Appl. Microbiol.* **21**:698–702.

Harvey, R. J., 1970, Metabolic regulation in glucose-limited chemostat cultures of *Escherichia coli, J. Bacteriol.* **104**:698–706.

Harvey, R. V., and Young, L. Y., 1980, Enumeration of particle-bound and unattached respiring bacteria in the salt marsh environment, *Appl. Environ. Microbiol.* **40**:156–160.

Heinle, D. R., Harris, R. P., Ustach, J. F., and Flemer, D. A., 1977, Detritus as food for estuarine copepods, *Mar. Biol.* **40**:341–353.

Hellebust, J. A., 1970, The uptake and utilization of organic substances by marine phytoplankters, in: *Organic Matter in Natural Waters* (D. W. Hood, ed.), p. 225, Univ. of Alaska, Occasional Publication no. 1.

Hobbie, J. E., 1976, Acridine orange as an indicator of the activity of bacteria, in: *Abstr. 39th Meet. Amer. Soc. Limnol Oceanogr.* Grafton, Wisconsin.

Hobbie, J. E., and Crawford, C. C., 1969, Respiration corrections for bacterial uptake of dissolved organic compounds in natural waters, *Limnol. Oceanogr.* **14**:528–532.

Hobbie, J. E., and Wright, R. T., 1965, Competition between planktonic bacteria and algae for organic solutes, *Mem. Ist. Ital. Idrobiol. Suppl.* **18**:175–187.

Hobbie, J. E., Holm-Hansen, O., Packard, T. T., Pomeroy, L. R., Sheldon, R. W., Thomas, J. P., and Wiebe, W. J., 1972, A study of the distribution and activity of microorganisms in ocean water, *Limnol. Oceanogr.* **17**:544–555.

Hobbie, J. E., Daley, R. J., and Jasper, S., 1977, Use of nucleopore filters for counting bacteria by fluorescence microscopy, *Appl. Environ. Microbiol.* **33**:1225–1228.

Hodson, R. E., Azam, F., Carlucci, A. F., Fuhrman, J. A., Karl, D. M., and Holm-Hansen, O., 1981, Microbial uptake of dissolved organic matter in McMurdo Sound, Antartica, *Mar. Biol.* **61**:89–94.

Hollibaugh, J. T., Carruthers, A. B., Fuhrman, J. A., and Azam, F., 1980, Cycling of organic nitrogen in marine plankton communties studied in enclosed water columns, *Mar. Biol.* **59**:15–21.

Holm-Hansen, O., 1973, The use of ATP determinations in ecological studies, *Bull. Ecol. Res. Comm.* (Stockholm) **17**:215–222.

Holm-Hansen, O., and Booth, C. R., 1966, The measurement of adenosine triphosphate in the ocean and its ecological significance, *Limnol. Oceanogr.* **11**:510–519.

Holm-Hansen, O., and Paerl, H. W., 1972, The applicability of ATP determination for estimation of microbial biomass and metabolic activity. *Mem. Ist. Ital. Idrobiol.* **29**.149–168.

Holm-Hansen, O., Packard, T. T., and Pomeroy, L. R., 1970, Efficiency of the reverse-flow filter technique for concentration of particulate matter, *Limnol. Oceanogr.* **15**:832–834.

Hoppe, H.-G., 1974, Untersuchungen zur Analyse mariner Bakterienpopulationen mit einer autoradiographischen Methode, *Kieler Meeresfrsch.* **30**:107–116.

Hoppe, H.-G., 1976, Determination of properties of actively metabolizing bacteria in the sea, investigated by means of microautoradiography, *Mar. Biol.* **36**:291–302.

Hoppe, H.-G., 1977, Analysis of actively metabolizing bacterial population with the autoradiographic method, in: *Microbial Ecology of a Brackish Water Environment* (G. Rheinheimer, ed.), pp. 171–197, Springer-Verlag, Berlin.

Hoppe, H.-G., 1978, Relations between active bacteria and heterotrophic potential in the sea, *Neth. J. Sea Res.* **12**:78–98.

Ishida, Y., and Kadota, H., 1979, A new method for enumeration of oligotrophic bacteria in lake water, *Arch. Hydrobiol. Beih.* **12**:77–85.

Iturriaga, R., and Hoppe, H.-G., 1977, Observations of heterotrophic activity on photoassimilated organic matter, *Mar. Biol.* **40**:101–108.

Iturriaga, R., and Rheinheimer, G., 1975, Eine einfache Methode zur Auszählung von Bakterien mit aktivem Electronentransportsystem in Wasser und Sedimentproben, *Kieler Meeresforsch.* **31**:83–86.

Jannasch, H. W., 1963, Bakterielles Wachstum bei geringen Substratkonzentrationen, *Arch. Mikrobiol.* **45**:323–343.

Jannasch, H. W., 1967, Growth of marine bacteria at limiting concentrations of organic carbon in seawater, *Limnol. Oceanogr.* **12**:264–271.

Jannasch, H. W., 1968, Growth characteristics of heterotrophic bacteria in seawater, *J. Bacteriol.* **95**:722–723.

Jannasch, H. W., 1969, Estimations of bacterial growth rates in natural waters, *J. Bacteriol.* **99**:156–160.

Jannasch, H. W., and Jones, G. E., 1959, Bacterial populations in seawater as determined by different methods of enumeration, *Limnol. Oceanogr.* **4**:128–129.

Jannasch, H. W., and Pritchard, P. H., 1972, The role of inert particulate matter in the activity of aquatic microorganisms, *Mem. Ist. Ital. Idrobiol.* **29**:289–308.

Jassby, A. D., 1975, An evaluation of ATP estimates of bacterial biomass in the presence of phytoplankton, *Limnol. Oceanogr.* **20**:646–648.

Javornitsky, P., and Prokesova, V., 1963, The influence of protozoa and bacteria upon the oxidation of substances in water, *Int. Rev. Ges. Hydrobiol.* **48**:335–350.

Jensen, V., 1967, The plate count technique, in: *The Ecology of Soil Bacteria* (T.R.G. Gray and D. Parkinson, eds.), pp.158–170, Liverpool University Press, Liverpool.

Johannes, R.E., 1965, Influence of marine protozoa on nutrient regeneration, *Limnol. Oceanogr.* **10**: 434–442

Johannes, R.E., 1968, Nutrient regeneration in lakes and oceans, in: *Advances in Microbiology of the Sea, Vol.1* (M.R. Droop and E.J.F. Wood, eds.), pp. 203–213, Academic Press, London.

Johannes, R.E., and Satomi, M., 1967, Measuring organic matter retained by aquatic invertebrates, *J. Fish. Res. Bd. Can.* **24**:2467–2471.

Johnson, P.W., and Sieburth, J.McN., 1979, Chroococcoid cyanobacteria in the sea: a ubiquitous and diverse phototrophic biomass, *Limnol. Oceanogr.* **24**:928–935.

Jordan, M.J., and Likens, G.E., 1980, Measurement of planktonic bacterial production in an oligotrophic lake, *Limnol. Oceanogr.* **25**:719–732.

Karl, D.M., 1979, Measurement of microbial activity and growth in the ocean by rates of stable ribonucleic acid synthesis. *Appl. Environ. Microbiol.* **38**:850–860.

Karl, D.M., 1980, Cellular nucleotide measurements and applications in microbial ecology, *Microbiol. Rev.* **44**:739–796.

Karl, D. M., 1981, Simultaneous rates of RNA and DNA synthesis for estimating growth and cell division of aquatic microbial communities, *Appl. Environ. Microbiol.,* **42**:802–810.

Karl, D. M., Winn, C. D., and Wong, D. C. L., 1981a, RNA synthesis as a measure of microbial growth **64**:1–12 in aquatic environments. I. Evaluation, verification and optimization of methods, *Mar. Biol.,* **64**:1–12.

Karl, D. M., Winn, C. D., abd Wong, D. C. L., 1981b, RNA synthesis as a measure of microbial growth in aquatic environments. II. Field applications, *Mar. Biol.,* **64**:13–21.

Kelly, M. G., Hornberger, G. M., and Cosby, B. J., 1974, Continuous automated measurement of rates of photosynthesis and respiration in an undisturbed river community, *Limnol. Oceanogr.* **19**:305–312.

Kemp, W. M., and Boynton, W. R., 1980, Influence of biological and physical processes on dissolved oxygen dynamics in an estuarine system: implications for measurement of community metabolism, *Est. Coast. Mar. Sci.* **11**:407–431.

Kenner, R. A., and Ahmed, S. I., 1975a, Measurements of electron transport activities in marine phytoplankton, *Mar. Biol.* **33**:119–127.

Kenner, R. A., and Ahmed, S. I., 1975b, Correlation between oxygen utilization and electron transport activity in marine phytoplankton, *Mar. Biol.* **33**:129–133.

Kim, J., and ZoBell, C. E., 1974, Occurrence and activities of cellfree enzymes in oceanic environments, in: *Effect of the Ocean Environment on Microbial Activities* (R. R. Colwell, and R. Y. Morita, eds.), pp. 368–385, Univ. Park Press, Baltimore.

Koch, A. L., 1979, Microbial growth in low concentrations of nutrients, in: *Strategies of Microbial Life in Extreme Environments* (M. Shilo ed.), pp. 261–279, Berlin, Dahlem Konferenzen, Verlag Chemie, Weinheim.

Kogure, K., Simidu, U., and Taga, N., 1979, A tentative direct microscopic method for counting living marine bacteria, *Can. J. Microbiol.* **25**:415–420.

Konings, W. N., and Veldkamp, H., 1980, Phenotypic response to environmental change, in: *Contemporary Microbial Ecology* (D. C. Ellwood, N. J. Hedger, M. J. Latham, J. M. Lynch, and J. H. Slater, eds.) pp. 161–191, Academic Press, London.

Kornberg, H. L., 1966, Anaplerotic sequences and their role in metabolism in: *Essays in Biochemistry, Vol. 2,* (P. N. Campbell, and G. D. Greville, eds.), pp. 1–31, Academic Press, London.

Krambeck, C., 1978, Changes in planktonic microbial populations—an analysis by scanning electron microscopy, *Verh. Int. Ver. Limnol.* **20**:2255–2259.

Krambeck, C., 1979, Applicability and limitations of the Michaelis-Menten equation in microbial ecology, *Arch. Hydrobiol. Beih.* **12**:12–23.

Kunicka-Goldfinger, W., 1973, An attempt to measure the growth of indigenous aquatic bacteria by the technique of semicontinuous cultures on membrane filters, in: *Modern Methods in the Study of Microbial Ecology, Bull. Ecol. Res. Comm.* (T. Rosswall, ed.), pp. 311–316, NFR Stockholm.

Kunicki-Goldfinger, W. J. H., 1974, Methods in aquatic microbiology, A story of apparent precision and frustrated expectations. *Pol. Arch. Hydrobiol.* **21**:3–17.

Kuznetsov, S. I., and Romanenko, V. I., 1966, Produktion der Biomasse heterotropher Bakterien und die Geschwindigkeit ihrer Vermehrung im Rybinsk-Staussee, *Verh. Int. Verein theor. angew. Limnol.* **16**:1493–1500.

Kuznetsov, S. I., Dubinina, G. A., and Lapteva, N. A., 1979, Biology of oligotrophic bacteria, *Ann. Rev. Microbiol.* **33**:377–387.

Lampert, W., 1978, Release of dissolved organic carbon by grazing zooplankton, *Limnol. Oceanogr.* **23**:831–834.

Lancelot, C., 1979, Gross excretion rates of natural marine phytoplankton and heterotrophic uptake of excreted products in the Southern North Sea, as determined by short-term kinetics, *Mar. Ecol. Progr. Ser.* **1:**179–186.

LaRock, P. A., Lauer, R. D., Schwarz, J. R., Watanabe, K. K., and Wiesenburg, D. A., 1978, Microbial biomass and activity distribution in an anoxic, hypersaline basin, *Appl. Environ. Microbiol.* **37:**466–470.

Larsson, K., Weibull, C., and Cronberg, G., 1978, Comparison of light and electron microscopic determinations of the number of bacteria and algae in lake water, *Appl. Environ. Microbiol.* **35:**397–404.

Larsson, U., and Hagström, A., 1979, Phytoplankton exudate release as an energy source for the growth of pelagic bacteria, *Mar. Biol.* **52:**199–206.

Larsson, U., and Hagström, A., 1982, Fractionated phytoplankton primary production, exudate release, and bacterial production in a balthic eutrophication gradient, *Mar. Biol.,* **67:**57–70.

Law, A. T., and Button, D. K., 1977, Multiple-carbon-source-limited growth kinetics of a marine coryneform bacterium, *J. Bacteriol.* **77:**115–123.

Lehmicke, L. G., Williams, R. T., and Crawford, R. L., 1979, ^{14}C-Most-Probable-Number method for enumeration of active heterotrophic microorganisms in natural waters, *Appl. Environ. Microbiol.* **38:**644–649.

Lewin, R. A., 1974, Enumeration of bacteria in seawater, *Int. Rev. Ges. Hydrobiol.* **59:**611–619.

Liebezeit, G., Bölter, M., Brown, I. F., and Dawson, R., 1980, Dissolved free amino acids and carbohydrates at pycnocline boundaries in the Sargasso Sea and related microbial activity, *Oceanol. Acta* **3:**357–362.

Lingeman, R., 1980, Analysis and interpretation of the diel and annual oxygen regimes in two aquatic ecosystems Ph.D. thesis, University of Amsterdam,

Litchfield, C. D., Rake, J. B., Zendulis, J., Watanabe, R. T., and Stein, D. J., 1975, Optimization of procedures for the recovery of heterotrophic bacteria from marine sediment, *Microb. Ecol.* **1:**219–233.

Little, J. E., Sjogren, R. E., and Carson, G. R., 1979, Measurement of proteolysis in natural waters, *Appl. Environ. Microbiol.* **37:**900–908.

Lloyd, G. I., and Morris, E. O., 1971, An apparatus for measuring microbial growth or survival in the marine environment, *Mar. Biol.* **10:**295–296.

Maaløe, O., and Kjeldgaard, N. O., 1966, *Control of Macromolecular Synthesis: A Study of DNA, RNA and Protein Synthesis in Bacteria,* W. A. Benjamin, New York.

Mague, T. H., Friberg, E., Hughes, D. J., and Morris, I., 1980, Extracellular release of carbon by marine phytoplankton; a physiological approach, *Limnol. Oceanogr.* **25:**262–279.

Maeda, M., and Taga, N., 1979, Chromogenic assay method for lipopolysaccharide (LPS) for evaluating bacterial standing crop in seawater, *J. Appl. Bacteriol.* **47:**175–182.

Maksimova, E. A., 1976, Annual cycle of bacterioplankton production in pelagic southern Baikal, *Microbiology* **45:**146–149.

Marshall, K. C., 1976, *Interfaces in Microbial Ecology,* Harvard Univ. Press, Cambridge.

Marshall, K. C., 1980, Reactions of microorganisms, ions and macromolecules at interfaces, in: *Contemporary Microbial Ecology* (D. C. Ellwood, N. J. Hedger, M. J. Latham, J. M. Lynch, and J. H. Slater, eds.), pp. 93–106, Academic Press, London.

Mateles, R. I., Ryu, D. Y., and Yasuda, T., 1965, Measurement of unsteady state growth rates of micro-organisms, *Nature* **208:**263–265.

Matin, A., and Veldkamp. H., 1978, Physiological basis of the selective advantage of a *Spirillum* sp. in a carbon-limited environment, *J. Gen. Microbiol.* **105:**187–197.

Meyer-Reil, L.-A., 1975, An improved method for the semicontinuous culture of bacterial populations on nuclepore membrane filters, *Kieler Meeresforsch.* **31:**1–6.

Meyer-Reil, L.-A., 1977, Bacterial growth rates and biomass production in: *Microbial Ecology of a Brackish Water Environment* (G. Rheinheimer, ed.), pp. 223–235, Springer-Verlag, Berlin.

Meyer-Reil, L.-A., 1978a, Autoradiography and epifluorescence microscopy combined for the determination of number and spectrum of actively metabolizing bacteria in natural waters, *Appl. Environ. Microbiol.* **36**:506–512.

Meyer-Reil, L.-A., 1978b, Uptake of glucose by bacteria in the sediment, *Mar. Biol.* **44**:293–298.

Meyer-Reil, L.-A., 1981, Enzymatic decomposition of proteins and carbohydrates in marine sediments: methodology and field observations during spring, *Kieler Meeresforsch Sonderh.* **5**:311–317.

Meyer-Reil, L.-A., and Faubel, A., 1980, Uptake of organic matter by meiofauna organisms and interrelationships with bacteria, *Mar. Ecol. Prog. Ser.* **3**:251–256.

Meyer-Reil, L.-A., Dawson, R., Liebezeit, G., and Tiedge, H., 1978, Fluctuations and interactions of bacterial activity in sandy beach sediments and overlaying waters, *Mar. Biol.* **48**:161–171.

Meyer-Reil, L.-A., Bölter, M., Liebezeit, G., and Schramm, W., 1979, Short-term variations in microbiological and chemical parameters, *Mar. Ecol. Prog. Ser.* **1**:1–6.

Meyer-Reil, L.-A., Bölter, M., Liebezeit, G., Szwerinski, H., and Wolter, K., 1980, Interrelationships between microbiological and chemical parameters of sandy beach sediments, a summer aspect, *Appl. Environ. Microbiol.* **39**:797–802.

Meynell, G. C., and Meynell, E. W., 1965, *Theory and Practice in Experimental Bacteriology*, Cambridge University Press, Cambridge.

Mitamura, O., and Saijo, Y., 1975, Decomposition of urea associated with photosynthesis of phytoplankton in coastal waters, *Mar. Biol.* **30**:67–72.

Moaledj, K., and Overbeck, J., 1980, Studies on uptake kinetics of oligocarbophilic bacteria, *Arch. Hydrobiol.* **89**:303–312.

Moriarty, D. J. W., 1977, Improved method using muramic acid to estimate biomass of bacteria in sediments, *Oecologia* (Berlin) **26**:317–323.

Moriarty, D. J. W., 1979, Biomass of suspended bacteria over coral reefs, *Mar. Biol.* **53**:193–200.

Moriarty, D. J. W., and Pollard, P. C., 1981, DNA synthesis as a measure of bacterial productivity in seagrass sediments, *Mar. Ecol. Prog. Ser.* **5**:151–156.

Morrison, S. J., and White, D. C., 1980, Effects of grazing by estuarine gammaridean amphipods on the microbiota of allochthonous detritus, *Appl. Environ. Microbiol.* **40**:659–671.

Munro, A. L. S., and Brock, T. D., 1968, Distinction between bacterial and algae utilization of soluble substances in the sea, *J. Gen. Microbiol.* **51**:35–42.

Neilson, A. H., and Lewin, R. A., 1974, The uptake and utilization of organic carbon by algae: an essay in comparative biochemistry, **13**:227–264.

Newell, R., 1965, The role of detritus in the nutrition of two marine deposit-feeders, the prosobranch *Hydrobia ulvae* and the bivalve *Macoma balthica*, *Proc. Zool. Soc. London* **144**:25–45.

Newell, S. Y., and Christian, R. R., 1981, Frequency of dividing cells as an estimator of bacterial productivity, *Appl. Environ. Microbiol.* **42**:23–31.

Novitsky, J. A., and Morita, R. Y., 1976, Morphological characterization of small cells resulting from nutrient starvation of a psychrophilic marine vibrio, *Appl. Environ. Microbiol.* **32**:617–622.

Odum, H. T., 1956, Primary production in flowing waters, *Limnol. Oceanogr.* **1**:102–117.

Odum, H. T., 1960, Analysis of diurnal oxygen curves for the assay of reaeration rates and metabolism in polluted marine bays, in: *Waste Disposal in the Marine Environment* (E. A. Pearson, ed.), pp. 547–555, Pergamon Press, Elmsford, N Y.

Odum, E. P., and de la Cruz, A. A., 1967, Particulate organic detritus in a Georgia salt marshestuarine ecosystem, in: *Estuaries* (G. H. Lauff, ed.), pp. 383–388, Publ. A.A.A.S. no. 83, Washington, D.C.

Ogura, N., 1972, Rate and extent of decomposition of dissolved organic matter in surface seawater, *Mar. Biol.* **13**:89–93.

Olanczuk-Neyman, K. M., and Vosjan, J. H., 1977, Measuring respiratory electron-transport-system activity in marine sediment, *Neth. J. Sea. Res.* **11**:1–13.

Overbeck, J., 1972, Experimentelle Untersuchungen zur Bestimmung der Bakteriellen Produktion im See, *Verh. int. Verein theor. angew, Limnol.* **18**:176–187.

Overbeck, J., 1976, Some remarks on the ecology of the CO_2-metabolism of heterotrophic and methylotrophic bacteria, in: *Microbial Production and Utilization of Gases* (H. G. Schlegel, G. Gottschalk, and N. Pfennig, eds.), pp. 263–266, Goltze Verlag, Göttingen.

Overbeck, J., 1979, Dark CO_2 uptake—biochemical background and its relevance to in situ bacterial production, *Arch. Hydrobiol. Beih.* **12**:38–47.

Overbeck, J., and Dalay, R. J., 1973, Some precautionary comments on the Romanenko technique for estimating heterotrophic bacterial production, in: *Modern Methods in the Study of Microbial Ecology* (T. Rosswall, ed.), pp. 342–344, Stockholm.

Packard, T. T., 1971, The measurement of respiratory electron transport activity in marine phytoplankton, *J. Mar. Res.* **29**:235–244.

Packard, T. T., Healy, M. L., and Richards, F. A., 1971, Vertical distribution of the activity of the respiratory electron transport system in marine plankton, *Limnol. Oceanogr.* **16**:60–70.

Paerl, H. W., 1974, Bacterial uptake of dissolved organic matter in relation to detrital aggregation in marine and freshwater systems, *Limnol. Oceanogr.* **19**:966–972.

Paerl, H. W., 1980, Attachment of microorganisms to living and detrital surfaces in freshwater systems, in: *Adsorption of Microorganisms to Surfaces* (G. Bitton and K. C. Marshall, eds.), pp. 375–402, Wiley Interscience, New York.

Paerl, H. W., and Williams, N. J., 1976, The relation between adenosine triphosphate and microbial biomass in diverse aquatic ecosystems, *Int. Rev. Ges. Hydrobiol.* **61**:659–664.

Palumbo, A. V., and Ferguson, R. L., 1978, Distribution of suspended bacteria in the Newport River estuary, North Carolina, *Est. Coast. Mar. Sci.* **7**:521–529.

Parsons, T. R., and Strickland, J. D. H., 1962, On the production of particulate organic carbon by heterotrophic processes in seawater, *Deep Sea Res.* **8**:211–222.

Payne, W. H., 1970, Energy Yield and growth of heterotrophs, *Annu. Rev. Microbiol.* **24**:17–52.

Payne, W. J., and Wiebe, W. J., 1978, Growth yield and efficiency in chemosynthetic microorganisms, *Annu. Rev. Microbiol.* **32**:155–183.

Peterson, B. J., 1980, Aquatic primary productivity and the ^{14}C-Co_2 method: a history of the productivity problem, *Annu. Rev. Ecol. Syst.* **11**:359–385.

Petipa, T. S., Pavlova, E. V., and Mironov, G. N., 1970, The food web structure, utilization and transport of energy by trophic levels in the planktonic communities, in: *Marine Food Chains* (J. H. Steele, ed.) pp. 142–167, Oliver and Boyd, Edinburgh.

Pirt, S. J., 1975, *Principles of Microbe and Cell Cultivation*, Blackwell, Oxford.

Pomeroy, L. R., 1970, The strategy of mineral cycling, *Annu. Rev. Ecol. Syst.* **1**:171–190.

Pomeroy, L. R., 1974, The ocean's food web, a changing paradigm, *Bioscience* **24**:499–504.

Pomeroy, L. R., 1979, Secondary production of continental shelf communities, in: *Ecological Processes in Coastal and Marine Systems* (R. J. Livingston, ed.), pp. 163–186, Plenum Press, New York.

Pomeroy, L. R., and Johannes, R. E., 1968, Occurrence and respiration of ultraplankton in the upper 500 metres of the ocean, *Deep Sea Res.* **15**:381–391.

Postgate, J. R., and Hunter, J. R., 1964, Accelerated death of *Aerobacter aerogenes* starved in the presence of growth-limiting substrates, *J. Gen. Microbiol.* **34**:459–473.

Pugsley, A. P., and Evison, L. M., 1974, A membrane filtration staining technique for detection of viable bacteria in water, *Water Treat. Exam.* **23**:205–214.

Ramsay, A. J., 1974, The use of autoradiography to determine the proportion of bacteria metabolizing in an aquatic habitat, *J. Gen. Microbiol.* **80**:363–373.

Rasumov, A. S., 1932, Interrelation between bacteria and plankton in connection with some problems of water hygiene, in: *Voprosy Sanitarnoj Bakteriologii* Maskva, Izd. Acad. Med. Nauk. SSSR, pp. 30–43, (after Godlewska-Lipowa, 1970).

Raetz, C. R. H., 1978, Enzymology, genetics and regulation of membrane phospholipid synthesis in *Escherichia coli. Microbiol. Rev.* **42**:614–659.

Rodina, A. G., 1972, *Methods in Aquatic Microbiology,* Univ. Park Press, Baltimore.

Romanenko, V. I., 1963, Potential ability of the microflora in water to heterotrophic CO_2 assimilation and to chemosynthesis, *Microbiology* **32**:569–574.

Romanenko, V. I., 1964, Heterotrophic assimilation of CO_2 by bacterial flora of water, *Microbiology* **33**:610–614.

Romanenko, V. I., 1965, Correlation between oxygen and carbon dioxide uptake in peptone-grown heterotrophic bacteria, *Microbiology* **34**:334–339.

Romanenko, V. I., and Dobrynin, E. G., 1978, Specific weight of the dry biomass of pure bacterial cultures, *Microbiology* **47**:220–221.

Rosenbaum, O. D., and Zamenhof. F., 1972, Degree of participation of exogenous thymidine in the overall deoxyribonucleic acid synthesis in *Escherichia coli, J. Bacteriol.* **110**:585–591.

Saltzman, H. A., 1980, Untersuchungen über die Veränderungen der Mikroflora beim Durchgang von Brackwasser durch die Kühlanlagen von Kraftwerken, Ph.D. thesis, University of Kiel, F.R.G.

Schleyer, M. H., 1981, Microorganisms and detritus in the water column of a subtidal reef of Natal, *Mar. Ecol. Prog. Ser.* **4**:307–320.

Schurr, J. M., and Ruchti, J., 1977, Dynamics of O_2 and CO_2 exchange, photosynthesis and respiration in rivers from time-delayed corrections with ideal sunlight, *Limnol. Oceanogr.* **22**:208–225.

Seki, H., 1972, The role of microorganisms in the marine food chain with reference to organic aggregates, *Mem. Ist. Ital. Idrobiol.* **29**:245–259.

Sen Gupta, R., and Jannasch, H. W., 1973, Photosynthetic production and dark assimilation of CO_2 in the Black Sea, *Int. Rev. Ges. Hydrobiol.* **58**:625–632.

Sepers, A. B. J., 1977, The utilization of dissolved organic compounds in aquatic environments, *Hydrobiologia* **52**:39–54.

Sepers, A. B. J., 1979, De Aerobe Mineralisatie van Aminozuren in Natuurlijke Aquatische Milieus, Ph.D. thesis, Groningen, The Netherlands.

Sepers, A. B. J., and Van Es, F. B., 1979, Heterotrophic uptake experiments with ^{14}C-labeled histidine in a histidine-limited chemostat, *Appl. Environ. Microbiol.* **37**:794–799.

Seto, M., and Tazaki, T., 1971, Carbon dynamics in the food chain system of glucose-*Escherichia coli-Tetrahymena vorax, Jpn. J. Ecol.* **21**:179–188.

Sharp, J. H., 1977, Excretion of organic matter by marine phytoplankton: do healthy cells do it? *Limnol. Oceanogr.* **22**:381–399.

Sieburth, J. McN., 1976, Bacterial substrates and productivity in marine ecosystems, *Annu. Rev. Ecol. Syst.* **7**:259–285.

Sieburth, J. McN., 1979, *Sea Microbes,* Oxford University Press, New York.

Sieburth, J. McN., Brooks, R. B., Gessner, R. V., Thomas, C. D., and Tootle, J. L., 1974, Microbial colonization of marine plant surfaces as observed by scanning electron microscopy, in: *Effects of the Ocean Environment on Microbial Activities,* (R. R. Colwell and R. Y. Morita, eds.), pp. 418–432, University Park Press, Baltimore.

Sieburth, J. McN., Johnson, K. M., Burney, C. M., and Lavoie, D. M., 1977, Estimation of in situ rates of heterotrophy using diurnal changes in dissolved organic matter and growth rates of picoplankton in diffusion culture, *Helgol. wiss. Meeresunters.* **30**:565–574.

Skopintsev, B. A., 1973, A discussion of some views in the origin, distribution and composition of organic matter in deep ocean waters (translated from the Russian) *Oceanology* **12**:471–474.

Skopintsev, B. A., 1981, Decomposition of organic matter of plankton, humification and hydrolysis, in: *Marine Organic Chemistry* (E. K. Duursma and R. Dawson, eds.), pp. 125–177, Elsevier, Amsterdam.

Sorokin, Y. I., 1964, On the trophic role of chemosynthesis in water bodies, *Int. Rev. Ges. Hydrobiol.* **49**:307–324.

Sorokin, Y. I., 1971, On the role of the bacteria in the productivity of tropical oceanic waters, *Int. Rev. Ges. Hydrobiol.* **56**:1–48.

Sorokin, Y. I., 1973, Data on the biological productivity of the Western tropical Pacific Ocean, *Mar. Biol.* **20**:177–196.

Sorokin, Y. I., 1977, The heterotrophic phase of plankton succession in the Japan Sea, *Mar. Biol.* **41**:107–117.

Sorokin, Y. I., 1978, Decomposition of organic matter and nutrient regeneration, in: *Marine Ecology, Vol. IV* (O. Kinne, ed.), pp. 501–616, Wiley Interscience, Chichester.

Sorokin, Y. I., and Kadota, H., 1972, Techniques for the assessment of microbial production and decomposition in fresh waters, International Biological Programme No. 23, Blackwell, Oxford.

Spencer, M. J., 1979, Light–dark discrepancy of heterotrophic bacterial substrate uptake, *FEMS Microbiol. Lett.* **5**:343–347.

Steele, J. H., 1974, *The Structure of Marine Ecosystems,* Harvard University Press, Cambridge, Mass.

Steemann Nielsen, E., 1952, The rate of primary production and the size of the standing stock of zooplankton in oceans, *Int. Rev. Ges. Hydrobiol.* **57**:513–516.

Stephens, G. C., 1981, The trophic role of dissolved organic material, in: *Analysis of Marine Ecosystems* (A. R. Longhurst, ed.), pp. 271–291, Academic Press, London.

Stevenson, L. H., 1978, A case for bacterial dormancy in aquatic systems, *Microb. Ecol.* **4**:127–133.

Stouthamer, A. H., 1977, Energetic aspects of the growth of microorganisms, in: *Microbial Energetics* (B. A. Haddock and W. A. Hamilton, eds.), pp. 285–315, 27th Symp. Soc. Gen. Microbiol., Cambridge University Press, London.

Stouthamer, A. H., 1979, The search for correlation between theoretical and experimental growth yields, *Microbial Biochemistry, Vol. 21* (J. R. Quayle, ed.), pp. 1–47, University Park Press, Baltimore.

Straat, P. A., Wolochow, H., Dimmick, R. L., and Chatigny, M. A., 1977, Evidence for incorporation of thymidine into deoxyribonucleic acid in airborne bacterial cells, *Appl. Environ. Microbiol.* **34**:292–296.

Straskrabova, V., 1979, Oxygen methods for measuring the activity of water bacteria, *Arch. Hydrobiol. Beih.* **12**:3–10.

Strügger, S., 1949, Fluorescent microscope examination of bacteria in soil, *Can. J. Res.* **26**:288.

Sullivan, J. D., Valois, F. W., and Watson, S. W., 1976, Endotoxins: the Limulus amebocyte lysate system, in: *Mechanisms in Bacterial Toxinology* (A. W. Bernheimer, ed.), pp. 217–236, John Wiley & Sons, New York.

Taga, N., and Matsuda, O., 1974, Bacterial population attached to plankton and detritus in sea water, in: *Effects of the Ocean Environment on Microbial Activities* (R. R. Colwell and R. Y. Morita, eds.), pp. 443–448, University Park Press, Baltimore.

Thomas, D. R. Richardson, J. A., and Dicker, R. J., 1974, The incorporation of tritiated thymidine into DNA as a measure of the activity of soil micro-organisms, *Soil Biochem.* **6**:293–296.

Torrella, F., and Morita, R. Y., 1981, Microcultural study of bacterial size changes and microcolony and ultramicrocolony formation by heterotrophic bacteria in seawater, *Appl. Environ. Microbiol.* **41**:518–527.

Toth, L. G., 1980, The use of dialyzing sacks in estimation of production of bacterioplankton and phytoplankton, *Arch. Hydrobiol.* **89**:474–482.

Ulitzur, S., Yagen, B., and Rottem, S., 1979, Determination of lipopolysaccharide by a bioluminescence technique, *Appl. Environ. Microbiol.* **37**:782–784.

Väätänen, P., 1977, Effects of composition of substrate and inoculation technique on plate counts of bacteria in the Northern Baltic Sea, *J. Appl. Bacteriol.* **42**(3):437–443.

Vaccaro, R. F., 1969, The response of Natural microbial populations in seawater to organic enrichment, *Limnol. Oceanogr.* **14**:726–735.

Vaccaro, R. F., and Jannasch, H. W., 1966, Studies on heterotrophic activity in seawater based on glucose assimilation, *Limnol. Oceanogr.* **11**:596–607.

Vaccaro, R. F., and Jannasch, H. W., 1967, Variations in uptake kinetics for glucose by natural populations in seawater, *Limnol. Oceanogr.* **12**:540–542.

Van Veen, J. A., and Paul, E. A., 1979, Conversion of biovolume measurements of soil organisms grown under various moisture tensions, to biomass and their nutrient content, *Appl. Environ. Microbiol.* **37**:686–692.

Varga, G. A., Hargraves, P. E., and Johnson, P., 1975, Scanning electron microscopy of dialysis tubes incubated in flowing seawater, *Mar. Biol.* **31**:113–120.

Veldkamp, H., 1976, *Continuous Culture in Microbial Physiology and Ecology,* Meadowfield Press, Durham, pp. 1–68.

Veldkamp, H., 1977, Ecological studies with the chemostat, in: *Advances in Microbial Ecology, Vol. 1* (M. Alexander, ed.), pp. 59–94, Plenum Press, New York.

Vollenweider, R. A., 1974, Primary production in aquatic environments, *International Biological Programme Handbook no. 12* (2d ed.), Blackwell, Oxford.

Vyshkvartsev, D. I., 1980, Bacterioplankton in shallow inlets of Posyeta Bay, *Microbiology* **48**:603–609.

Wangersky, P. J., 1977, The role of particulate matter in the productivity of surface waters, *Helgol. Wiss. Meeresunters.* **30**:546–564.

Wangersky, P. J., 1978, Production of dissolved organic matter in: *Marine Ecology, Vol. IV* (O. Kinne, ed.), pp. 115–220, Wiley Interscience, Chichester.

Watson, S. W., Novitsky, T. J., Quinby, H. L., and Valois, F. W., 1977, Determination of bacterial number and biomass in the marine environment, *Appl. Environ. Microbiol.* **33**:940–947.

Wheeler, P., North, B., Littler, M., and Stephens, G., 1977, Uptake of glycine by natural phytoplankton communities, *Limnol. Oceanogr.* **22**:900–910.

Wiebe, W. J., and Pomeroy, L. R., 1972, Microorganisms and their association with aggregates and detritus in the sea: a microscopic study, *Mem. Ist. Ital. Idrobiol.* **29**:325–352.

Wiebe, W. J., and Smith, D. F., 1977a, ^{14}C labeling of the compounds excreted by phytoplankton for employment as a realistic tracer in secondary productivity measurements, *Microb. Ecol.* **4**:1–8.

Wiebe, W. J., and Smith, D. F., 1977b, Direct measurement of dissolved organic carbon release by phytoplankton and incorporation by microheterotrophs, *Mar. Biol.* **42**:213–224.

Williams, P. J. LeB., 1970, Heterotrophic utilization of dissolved organic compounds in the sea. I. Size distribution of population and relationship between respiration and incorporation of growth substrates, *J. Mar. Biol. Assoc. U.K.* **50**:859–870.

Williams, P. J. LeB., 1973a, On the question of growth yields of natural heterotrophic populations, in: *Modern Methods in the Study of Microbial Ecology* (T. Rosswall, ed.), pp. 400–401, Bull. Ecol. Res. Comm., Stockholm.

Williams, P. J. LeB., 1973b, The validity of the application of simple kinetic analysis to heterogeneous microbial populations, *Limnol. Oceanogr.* **18**:159–165.

Williams, P. J. LeB., 1975, Biological and chemical aspects of dissolved organic material in seawater, in: *Chemical Oceanography, Vol. 2* (J. P. Riley and G. Skirrow, eds.), 2d ed., pp. 301–363, Academic Press, London.

Williams, P. J. LeB., 1981, Incorporation of microheterotrophic processes into the classical paradigm of the planktonic food web. *Kieler Meeresforsch. Sonderh.* **5**:1–28.

Williams, P. J. LeB., and Askew, Co., 1968, A method for measuring the mineralization by microorganisms of organic compounds in seawater, *Deep Sea Res.* **15:**365–375.

Williams, P. J. LeB., and Gray, R. W., 1970, Heterotrophic utilization of dissolved organic compounds in the sea. II. Observations on the responses of heterotrophic marine populations to abrupt increases in amino acid concentration, *J. Mar. Biol. Assoc. U.K.* **50:**871–881.

Williams, P. J. LeB., and Yentsch, C. S., 1976, An examination of photosynthetic production, excretion of photosynthetic products and heterotrophic utilization of dissolved organic compounds with reference to results from a coastal subtropical sea, *Mar. Biol.* **35:**31–40.

Wilson, C. A., and Stevenson, L. H., 1980, The dynamics of the bacterial population associated with a salt marsh, *J. Exp. Mar. Biol. Ecol.* **48:**123–138.

Wolter, K., 1980, Untersuchungen zur Exsudation Organischer Substanz und deren Aufnahme durch Natürliche Bakterienpopulationen, Ph.D. thesis, University of Kiel, F.R.G.

Wood, L. W., 1973, Monosaccharide and disaccharide interactions on uptake and catabolism of carbohydrates by mixed microbial communities, in: *Estuarine Microbial Ecology* (L. H. Stevenson and R. R. Colwell, eds.), pp. 181–198, Univ. of South Carolina Press, Columbia.

Wright, R. T., 1973, Some difficulties in using ^{14}C-organic solutes to measure heterotrophic bacterial activity, in: *Estuarine Microbial Ecology* (L. H. Stevenson and R. R. Colwell, eds.), pp. 199–217, Univ. of South Carolina Press, Columbia.

Wright, R. T., 1974, Mineralization of organic solutes by heterotrophic bacteria, in: *Effects of the Ocean Environment on Microbial Activities* (R. R. Colwell and R. Y. Morita, eds.), pp. 546–565, University Park Press, Baltimore.

Wright, R. T., 1978, Measurement and significance of specific activities in the heterotrophic bacteria of natural waters, *Appl. Environ. Microbiol.* **36:**297–305.

Wright, R. T., and Hobbie, J. E., 1966, Use of glucose and acetate by bacteria and algae in aquatic ecosystems, *Ecology* **47:**447–453.

Zimmermann, R., 1975, Entwicklung und Anwendung von Fluoreszenz-und Rasterelektronenmikroskopischen Methoden zur Ermittlung der Bakterienmenge in Wasserproben, Ph.D. thesis, Kiel, F.R.G.

Zimmerman, R., 1977, Estimation of bacterial number and biomass by epifluorescence microscopy and scanning electron microscopy, in: *Microbial Ecology of a Brackish Water Environment* (G. Rheinheimer, ed.), pp. 103–120, Springer-Verlag, Berlin.

Zimmerman, R., and Meyer-Reil, L. A., 1974, A new method for fluorescence staining of bacterial populations on membrane filters, *Kieler Meeresforsch.* **30:**24–27.

Zimmerman, R., Iturriaga, R., and Becker-Birck, J., 1978, Simultaneous determination of the total number of aquatic bacteria and the number thereof involved in respiration, *Appl. Environ. Microbiol.* **36:**926–935.

ZoBell, C. E., 1946, *Marine Microbiology,* Chronica Botanica, Waltham, Mass.

ZoBell, C. E., and Feltham, C. B., 1937, Bacteria as food for certain marine invertebrates, *J. Mar. Res.* **8:**312–327.

Zsolnay, A., 1975, Total labile carbon in the euphotic zone of the Baltic Sea as measured by BOD, *Mar. Biol.* **29:**125–128.

5

Starvation-Survival of Heterotrophs in the Marine Environment

RICHARD Y. MORITA

1. Introduction

Because microbes are the principal catalysts in the ocean, the preservation of the catalytic ability of bacteria is an important aspect of the cycles of matter. As long as the genome is preserved, it will be expressed when the environmental conditions become appropriate, and in many instances, this depends on the availability of suitable energy-yielding substrates.

In any given natural environment, the autochthonous and zymogenic bacteria present may or may not find conditions optimal for growth and reproduction. Generally, it is only a few species, if any, of the bacteria that will find the conditions optimal, whereas most of the bacteria present will find the conditions unsuitable. Although some of the favorable factors are an excess or lack of water, acidity, oxygen, suitable hydrogen acceptor, etc., the dominant one is the lack of specific energy-yielding substrates for the various physiological types of bacteria. Marine bacteria are well suited for starvation-survival studies because they may be exposed to either a large quantity or a lack of energy-yielding substrates; the latter is the situation found in oligotrophic waters beyond the continental shelf and in the deeper portions of the ocean. In actuality, it is the lack of energy in most environments which is the question that should be addressed. Even in eutrophic environments, energy sources for some

RICHARD Y. MORITA ● Department of Microbiology and School of Oceanography, Oregon State University, Corvallis, Oregon 97331.

of the physiological types of bacteria are lacking. In other words, no environment can be optimal for all the various physiological types of bacteria present.

In this chapter, I will concentrate on the lack of energy sources and the possible processes by which bacteria can perpetuate the genome until the right energy source plus the other necessary conditions present themselves. For lack of a better word, I have used the term "starvation-survival" to indicate the process of survival in the absence of energy yielding substrates.

The charge, given to me when accepting the assignment of writing this chapter, was that the presentation stress new ideas, generalizations, and principles and not be simply an annotated bibliography. Therefore, I apologize to the many investigators who have addressed the survival question of nonmarine bacteria for not citing their contributions, which are too numerous to be included in this presentation. In order to fulfill my charge, I have chosen to employ a rather chauvinistic presentation with the objective that the paper will stimulate others to work on the problem of starvation-survival. Hopefully, sometime in the future, some of the ideas presented here may be validated by the data obtained by other investigators.

Bacterial survival has been the subject of many papers, and there is a wide range of survival mechanisms to enable bacteria to persist in soil (Gray, 1976). However, many publications deal both with short-term survival, generally a day to a few weeks, and survival under specific conditions. The three longest survival times for vegatative cells under laboratory controlled conditions, other than from this laboratory, are for *Nocardia corallina*, 20 days (Robertson and Batt, 1973); *Arthrobacter globinformis,* 56 days (Luscombe and Gray, 1974); and *A. crystallopoietes,* 100 days (Boylen and Ensign, 1970). Spore forming bacteria will not be discussed, and I will restrict myself to the survival of vegetative marine heterotrophic bacteria because (1) much of the marine environment is oligotrophic, and (2) masses of water can have residence times for as long as a millenium. Although many prokaryotic cells form survival states such as cysts, endospores, conidia, etc., the focus of this presentation will be on those organisms that do not create specific resting cell states. A good discussion of specific survival states may be found in Sudo and Dworkin (1973).

All microbial forms can survive without energy-yielding substrates, but the period of starvation-survival may be very short to very long. The concept of feast or famine was put forth by Koch (1971), but it is a rare event for marine bacteria to be in a feast situation and much more likely for them to be in a famine situation. At any one time in the marine environment, even the nearshore environment, bacteria are generally in a less than optimal environment in relation to energy-yielding compounds; hence they are slow growing.

The ability of one cell to survive in any given environment is the key to studies on survival (Morita, 1980a). That one cell in the environment will express itself in terms of multiplication and functions when conditions become appropriate.

2. Background Information

"Survival of the species" is very important in evolutionary biology, but its importance in microbial ecology should also be stressed. Survival of various bacterial species permits those species to function when certain environmental changes occur, and environmental conditions are always in a state of flux. Today in marine microbiology we pay great attention to the active bacteria, and the dormant ones are neglected. In soil microbiology, there is a growing realization that many of the organisms isolated from soil by conventional methods were present in the dormant state, and there is evidence that the resting bacteria outnumber the active ones (Gray and Williams, 1971). The same situation could prevail in marine microbiology because relatively rich media are generally employed for the isolation of marine bacteria. This situation is well illustrated by the slide culture method employed by Torrella and Morita (1981) for marine organisms.

In order to preserve cultures in the laboratory, it is common practice to place them in nutrient medium and store them in the refrigerator so that one does not have to transfer them frequently. This is an example of survival of the culture over a period of time, in which the mechanism of survival is to lower the metabolism of the organisms so that they virtually cease to function; a dormant state. In this dormant state, the organisms can survive mainly because there are no uncontrolled metabolic processes taking place that will destroy the cell. The reason for the dormant state may be that uptake of the substrate at temperatures below the minimum for growth is nil. Goodrich and Morita (1977) found that, at temperatures below the minimum for growth for a multiple amino acid auxotrophic *Escherichia coli,* uptake of amino acids virtually stopped. A good example of survival at low temperature (4 °C) and high hydrostatic pressure for *E. coli, Streptococcus faecalis, Clostridium perfringens, Vibrio parahaemolyticus,* and other sewage bacteria was documented by Baross *et al.* (1975). Approximately 90% of the sewage isolates survived over 400 hr when exposed to 500 and 1000 atm. In this case, the low temperature and hydrostatic pressure were additive in decreasing the metabolic activity of the organisms.

In this chapter, I will restrict myself to addressing the ability of vegetative organisms to survive long periods of time in the absence of an energy-yielding substrate. Starvation-survival is a special type of dormancy brought about by the lack of specific exogenous energy-yielding substrates for specific species of heterotrophic bacteria, if the definition of Sussman and Halvorson (1966) is employed. This definition states that dormancy is any rest period or reversible interruption of the phenotypic development of the organisms. Furthermore, starvation-survival is an exogenous type of dormancy. Morita (1968) postulated that many marine bacteria are probably in a transient state, waiting until they encounter favorable environmental conditions in order to express themselves. This point was reiterated by Sieburth *et al.* (1974). Dormancy is an important physiological adaptation contributing to the survival of bacteria (Stevenson, 1978) and is known to occur in many marine organisms.

"It is a fact of microbial life that the majority of the bacteria can survive for considerable periods, in the absence of nutrients. The survival characteristics of starved bacteria depend upon the organism and such factors as the growth phase from which they are taken, growth rate, nutritional status, population density, biological history, and the nature of the environment" (Dawes, 1976). The 26th Symposium of the Society for General Microbiology (see, for example, Gray and Postgate, 1976) should be consulted concerning the survival of vegetative cells.

Marine heterotrophic bacteria are more amenable to the study of starvation-survival than soil organisms, mainly because the marine environment does not present the bacteria with water problems (especially dessication), lack of ions, drastic changes in pH, etc. Furthermore, marine bacteria are subjected generally to a mesotrophic energy-yielding environment in the nearshore environment. With the currents of the ocean [including divergence (upwelling) and convergence], the marine bacteria associated with water masses will eventually be in an environment where there are little or no energy-yielding substrates. However, the main difficulty with starvation-survival studies is the time element, especially when one considers the residence time of certain water masses in the oceans (Morita, 1980a).

It is questionable whether we can duplicate environmental conditions, which are often unstable and generally not optimal for the growth of the microorganisms. The concept of "feast and famine" existence (Koch, 1971) represents the two extremes in terms of energy-yielding nutrients. Somewhere in between is the condition that most bacteria experience in nearshore waters, but in the open ocean they experience the famine condition.

Examples of long-term survival studies are not common. Lipman (1931) claimed that he was able to isolate living bacteria in soil samples 64 years old and in protected adobes. Living bacteria were even found in Precambrian limestone. He suggested that life is not a state but a pattern of activity followed by dormancy. Living bacteria have been isolated from small pockets of brine found in crystallized salt in the Permian salt mines of Kansas (Reiser and Tasch, 1960).

The following are examples of long-term survival of various soil bacteria. Bollen (1977) found that sulfur bacteria survived in soil for at least 54 yr. Bosco (1960) found that 65 nonspore formers survived for 16 to 48 yr, Garbosky and Giambiagi (1962) that nitrifying bacteria survived for 5 yr; and Jensen (1961) that *Rhizobium* survived for 5 yr. *Pseudomonas capacia* survived in benzalkonium chloride for 14 yr (Geftic *et al.,* 1979). In these examples, one does not know the end point of the survival period of the bacteria and, therefore, the true survival period may be much longer than indicated.

As stated previously, only one viable surviving cell per stressed environment is required for the continuation of the species, not a few cells per ml, liter, etc. An example of this situation is the isolation of obligate halophiles from seawater by Rodrigues-Valera *et al.* (1979) when it was previously thought that the extremely halophilic bacteria were incapable of surviving in less than 10% NaCl. It took 5 liters of seawater to demonstrate the presence of 10 to 35 viable extreme halophiles.

In the marine environment, viable bacteria have been detected in deep red clay taken from the open ocean environment (Morita and ZoBell, 1955). The deposition rate of red clay is 2.3 mm per thousand years (Arrhenius, 1952). Thus, these bacterial cells have been in a survival state (dormant) for many years (Morita and ZoBell, 1955), because the organic content of red clay is very low. Long-term survival of a bacterial culture was demonstrated by Novitsky and Morita (1977). Jones and Rhodes-Roberts (1981) demonstrated that nitrogen-limited cells survive better than glucose-limited cells, and the two strains of marine bacteria they employed showed stable cell numbers after a 40-day period. Other investigators have suggested that some marine bacteria are inactive or dormant in the marine environment (Jannasch, 1967; Wright, 1973).

There are many other reports in the marine literature concerning the appearance and disappearance of specific physiological types of bacteria, but with no mention of mechanisms of survival. For instance, the luminous bacterium, *Photobacterium fisheri,* is present only in the winter months in the Mediterranean Sea (Yetinson and Shilo, 1979; Shilo and Yetinson, 1979). The winter appearance of this organism appears to be related to its temperature for growth. An over-summering survival mechanism must be present in *Ph. fischeri.* Although not mentioned as a mechanism for survival of the species, Ruby and Morin (1979) found luminous bacteria as part of the enteric population of 22 fish. These luminous bacteria multiplied on the fecal material when defecation took place. This latter case is an example of creating large numbers of cells in the hopes that one will survive. The over-wintering of *Vibrio parahaemolyticus* has been associated with chitin particles (Kaneko and Colwell, 1978). Therefore, survival of many species, particularly psychrophilic or mesophilic, of marine bacteria need only be for the unfavorable part of one year.

3. Organic Matter in Oligotrophic Marine Water

The Dahlem Conference (Shilo, 1979) addressed the question of life under low nutrient conditions. As I have stated above, microbes can survive in a low nutrient environment by a process of starvation-survival. This author agrees with Koch (1979) in his minority report that "the survival mechanism during the transit of a propagule to new habitat is an important part of the biology of any inhabitant of an extreme environment even if studies to date have not uncovered any mechanisms such as, or akin to, sporulation." Survival and time were stressed as being very important in microbial ecology (Morita, 1980b). Each microbial species has its lower threshold of nutrient availability, which governs its ability to reproduce, remain viable, die, or to produce a survival state akin to sporulation. Bacteria are isolated and maintained on nutrients (energy-yielding substrates, growth factors, and necessary salts) at levels far greater than those found in nature. When the oligotrophic waters of the ocean are taken into consideration, the amount of organic matter is extremely low. Below 200—300 m, the particulate organic carbon (POC) and the dissolved organic carbon (DOC) ranged between 3 and $10\mu g$ C/liter and from 0.35 to 70 mg C/liter, respectively (Menzel and Ryther, 1970). It should be recognized that the POC and the DOC are arbitrarily separated by use of the filter (generally a Whatman GF/C filter with an effective retention of $1.2\mu m$, which will permit the ultramicrobacteria to pass through easily). Both the POC and DOC in the ocean have been found to be rather resistant to decay (Menzel, 1967,1970; Menzel and Goering, 1966; Menzel and Ryther, 1970). When the DOC was concentrated and allowed to incubate in the presence of the indigenous microflora, there was no change in the amount of DOC from deep water after incubation for 1 or 2 months; yet viable bacteria were present (Barber, 1968). Barber (1968) concluded that the deep water DOC is a mixture of compounds that is extremely resistant to microbial degradation. Hydrolyzed POC serves as a good substrate for bacteria (Seki et al., 1968; Gordon, 1970). How nutritive DOC and POC in the oligotrophic waters are for the indigenous bacteria is not known (Morita, 1979a,b).

Each species of bacteria differ in their ability to compete for the limited nutrients available. Those bacteria that have a high affinity for uptake of various substrates probably have a lower threshold level for nutrients. The affinity constants for various substrates by various heterotrophic bacteria are reviewed by Tempest and Neijssel (1978).

The concentration of amino acids in the oligotrophic waters of the deep sea is in the nanomole range (Menzel and Goering, 1966; Lee and Bada, 1975). However, the dissolved amino acids detected in seawater could be the constituents of intact bacterial cells that simply passed through the glass fiber used during the initial water processing step (Bada and Lee, 1977).

Recognizing that the possible energy-yielding organic substrates in oligotrophic marine water are low (0.05 μM carbon per liter), Strickland (1971) stated that in cold water below 1000 m, the average division time of living material is very long, being many tens of days. The oxygen consumption in the deep sea (a few μl per year) is trivial compared to the amount of oxygen present. The oxidation of DOC in the deep sea was calculated to be about 0.1– 0.2% per year (Craig, 1971). Carlucci and Williams (1978), employing a heterotrophic low nutrient bacterium, calculated the generation time to be 210 hr (8.75 days), which agrees with the calculations of Craig (1971). If the low nutrient bacteria (oligotrophic bacteria) are capable of functioning at this rate, then the usual heterotrophs are probably not functional at all. The occurrence of oligotrophic bacteria in aquatic environments only reinforces the concept that each species has its own lower limit of energy-yielding substrate before the starvation-survival process takes place.

The organic matter in the oceans itself is old. Employing radioactive dating, the apparent age of the DOC in two samples taken from 1980 and 1920 m in the Northwest Pacific Ocean was 3740 ± 300 yr (Williams *et al.*, 1969). If one takes into consideration the residence time of deep water, the age may range from 230 to 950 years (Broecker, 1963).

All of the foregoing points to a very low rate of bacterial activity in the oligotrophic waters of the deep sea. However, it should be noted that when bacteria are in the presence of an energy-yielding substrate (eg., in the stomach or on the surface of a macroorganism), metabolic activity may be reasonably fast, i.e., sufficient to measure their growth rate, etc.

4. Starvation-Survival

4.1. Viability Curves or Patterns

The survival characteristics of starved bacteria will depend upon the organism and such factors as the growth phase from which they are taken, growth rate, nutritional status, population density, biological history of the organism, and the nature of the starvation environment (Dawes, 1976). *In situ* starvation-survival studies on bacteria are impossible to perform with our present methodology. Use of the chemostat for growth of organisms for starvation-survival is not practical. The pragmatic solution at the present time is to grow the bacteria in nutrient medium in batch culture, suspend the organisms (after they have been washed several times) in organic-free mineral salts solution, and then place them in a menstruum that has no energy-yielding substrates.

Using the above approach, Novitsky and Morita (1977) determined starvation-survival on a marine vibrio designated Ant-300. It was found that during the first week of the exposure of the cells to a starvation menstruum, the number of viable cells increased while the optical density of the suspended culture decreased, indicating that the cells had become smaller. After the first week, the total cells remained at a high level while the viable cell count decreased, indicating that the nonviable cells did not lyse. The age of the culture plays an important part in the survival pattern mainly in the first 4 weeks of starvation-survival (Fig. 1). One of the characteristics of the starvation-survival pattern is that upon exposure to a starvation menstruum, there is an increase in the number of cells.

The increase in the number of genomes at the onset of starvation is very common in plants and animals, and the number of bacterial cells also increases

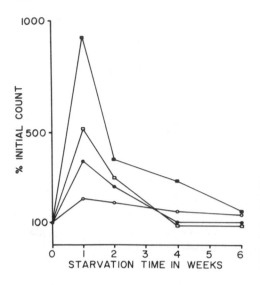

Figure 1. Starvation viability of Ant-300 starved from various times during exponential growth. Viability was determined by plate counts and expressed as a percentage of the initial count. The cells were starved after 65 hr (○), 80 hr (●), 95 hr (□), and 115 hr (■) of growth. After Novitsky and Morita (1976).

upon initial starvation-survival (Fig. 2). The next question is whether or not there is a selection of the cells which will survive and those which will not. The increase in the number of cells upon initial starvation-survival may be a possible strategy for the species to perpetuate itself (Novitsky and Morita, 1978a). This increase may be up to 400% of the initial number of cells placed in the starvation menstruum. The increase in the number of cells depends upon the initial concentration of cells placed in a starvation-survival situation; the lower the number of cells placed in the starvation menstruum, the higher the percentage of cells which survive (Fig. 3).

Ant-300 is the only organism that we have well documented in terms of starvation-survival, but we have tested 16 others. Most of these organisms were isolated during the NORPAX cruise (May to July, 1980) from the open sea, and a few could be tentatively described as facultative chemototrophs (J. A. Baross, personal communication). Eleven of these isolates showed an initial increase (over 100%) in cell numbers, indicating a survival strategy as seen in Ant-300 (Novitsky and Morita, 1978a). The two other patterns exhibited were a decrease in viability until a constant viability was attained and an increase in viable cell count until a constant viable cell count was again attained. An initial increase in cell numbers during the process of starvation-survival was also noted by Tabor *et al.* (1981) for isolates designated GS1 and GS17.

The viability of starved cells (starved 1 week in natural or artificial seawater prior to pressurization) under pressure was 2 to 3 times greater than the 1 atm controls after 10 weeks of incubation. Because Ant-300 was isolated from the Antarctic Convergence, the possibility remains that starvation plus

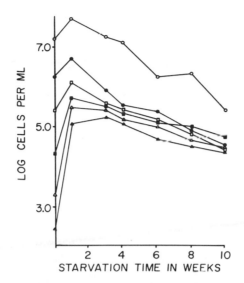

Figure 2. Viability of Ant-300 suspended at various initial concentrations. Log-phase Ant-300 cells were washed and suspended in phosphate buffered salt mixture. The culture was then serially diluted to produce the various cell densities. Viability was determined by plate counts. Initial cell concentrations (cells/ml), 1.77×10^7 (o), 1.77×10^6 (●), 2.45×10^5 (□), 2.08×10^4 (■), 2.01×10^3 (△), and 2.70×10^2 (▲). After Novitsky and Morita (1977).

the hydrostatic pressure aids in the survival of this species from the time of convergence to the time of divergence (Novitsky and Morita, 1978b). Starvation may also play a role in survival of the ultramicrocells obtained by Tabor *et al.* (1981) in the deep sea.

   In starvation-survival studies of a *Pseudomonas* sp., Kurath (1980) noted that the viable population dropped to 0.1% of the initial population after 25 days and stabilized thereafter. However, the number of respiring cells

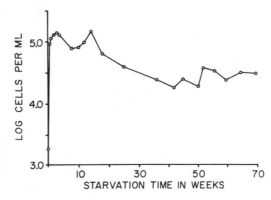

Figure 3. Long-term starvation survival of Ant-300 starving at a low cell density. Viability was determined by plate counts. After Novitsky and Morita (1978a).

accounted for 1% of the population as determined by the INT method (Zimmermann *et al.,* 1978). The difference between the viable and respiring cell counts suggests that there is a subpopulation of cells which is capable of respiring but not capable of reproducing. This phenomenon has been observed by Hoppe (1976), who found that in natural water samples, the actively metabolizing bacteria account for 20–60% of the total number present, and are from 10 to 1000 times as numerous as the viable bacteria.

Bacteria can lose the ability to reproduce but remain biologically completely functional as individuals (Postgate, 1976). However, what percentage of the microbial population in the environment carry out their ecological functions without being able to reproduce?

4.2. Miniaturization of Bacterial Cells

The term "ultramicrobacteria" or "ultramicrocell" as suggested by Torrella and Morita (1981) will be used to designate what might be called by others "dwarf," "minibacteria," or small cells. The use of the term "dwarf" may be confused with cells from a dwarf colony, and minibacteria could be confused with minicells of *E. Coli* which are anucleated. More specifically, the term was used to designate bacteria having a diameter length less than 0.3 μm.

At low nutrient levels, continued bacterial cell division takes place in the absence of growth, and this multiplication results in ultramicrocells (Henrici, 1928; Rahn, 1932). Bisset (1952) stated that nearly all bacteria have a resting cell state or specialized distributive stage, which he termed "microcysts," and certain microorganisms produced microcysts in response to nutrient limitations. Harrison and Lawrence (1963) described starvation-resistant mutants in *Aerobacter aerogenes;* this genetically distinct population produced cells much smaller than the faster growing wild-type strain. Coccoidal ultramicrocells were observed by Casida (1977) in soil. Ultramicrocells in seawater that were capable of passing through a Millipore filter (0.45 nm) were demonstrated by Oppenheimer (1952) and Anderson and Heffernan (1966). Ultramicrocells have been reported in seawater by Zimmermann and Meyer-Reil (1974), Daley and Hobbie (1975), Morita (1977), Zimmermann (1977), Watson *et al.* (1977), Kogure *et al.* (1979), Torrella and Morita (1981), Tabor *et al.* (1981), and MacDonell and Hood (1982). Tabor *et al.* (1981) reported the existence of ultramicrocells (filterable bacteria) in water samples taken from the deep sea. A marine *Pseudomonas* sp. also produced ultramicrocells when starved (G. Kurath and R. Y. Morita, unpublished data). When 16 isolates initially obtained from open ocean seawater beyond the continental shelf were examined for starvation-survival in the laboratory, ultramicrocells were noted in all isolates (P. Amy and R. Y. Morita, unpublished data). It should be noted that the resulting ultramicrocells were not necessarily coccoid (F. Torrella and R.

Y. Morita, unpublished data; P. Amy and R. Y. Morita, unpublished data; Johnson and J. McN. Sieburth, 1978).

Jannasch (1958) presented indirect evidence for the existence of ultra-microcells. Ultramicrocells were produced as a result of starvation by Novitsky and Morita (1976). Employing the same method as Novitsky and Morita (1976), Tabor *et al.* (1981) were also able to produce small ultramicrocells of isolates GS1 and GS17. Marshall *et al.* (1971) reported that ultramicrobac-teria were primary colonizers of glass surfaces immersed in the marine envi-ronment. These ultramicrobacteria appear to be replaced by normal size bac-teria after 12 to 24 hr. Such a situation could easily result as evidenced by the microculture slide technique employed by Torrella and Morita (1981). This microculture technique indicated that some of the ultramicrobacteria did develop into normal size bacteria while others remained as ultramicrocells. If ultramicrocells can be formed by starvation, then it is only logical to assume that they will form normal size bacteria when suitable nutrients are supplied. This does happen with all the various starvation induced ultramicrocells that we have studied in the laboratory (P. Amy and R. Y. Morita, unpublished data). Marshall (1979) suggested that the ultramicrobacteria present in oli-gotrophic marine waters may respond to the nutrients concentrated at the sur-face by growing into normal size bacteria.

Ultramicrocells of *Alcaligenes, Flavobacterium, Pseudomonas,* and *Vibrio* spp. taken from the deep sea were identified by Tabor *et al.* (1981). Viable ultramicrocells from estuarine waters that passed through a 0.2 μm polycar-bonate membrane filter were identified as *Aeromonas, Vibrio, Pseudomonas,* and *Alcaligenes* (MacDonell and Hood, 1982).

The occurrence of ultramicrobacteria in the ocean is now a recognized fact, but what percentage of the observed ultramicrobacteria are due to star-vation-survival is an open question. One should take into consideration not only the heterotrophs but also the chemoautotrophs and chemoheterotrophs.

According to Dow and Whittenbury (1980), the form(s) of bacteria reflect a function related to their existence and persistence in the natural hab-itat. According to Poindexter (1979), there are morphological adaptations to low nutrient environments. Miniaturization of bacterial cells may then reflect the lack of an energy-yielding substrate for a specific species of marine bacte-ria. This miniaturization increases the cell's surface/volume ratio; hence its ability to scavenge energy-yielding substrates from its environment will be increased (see Section 5).

4.3. Energy of Maintenance and Endogenous Metabolism

The energy need for cellular activity other than the production of new cellular material is referred to as the "energy of maintenance," (Pirt, 1965). For a review of the concept for the "energy of maintenance," the contributions by the various authors to the 26th Symposium of the Society for General Microbiology (see, for example, Gray and Postgate, 1976) should be consulted. Prolonged viability without nutrients is dependent upon the regulation of endogenous metabolism in accordance with the energy of maintenance requirements (Thomas and Batt, 1969). Nearly all of the data obtained thus far concerning endogenous metabolism (and energy of maintenance) during survival have been obtained during short-term periods, usually in terms of hours rather than months and years.

There is evidence that some microorganisms do not synthesize any specific reserve compounds and therefore are entirely dependent on protein and RNA degradation for their endogenous metabolism and energy of maintenance. Dawes and Senior (1973) discuss the various types of energy reserve polymers in microorganisms and their publication should be consulted.

Dawes (1976) raises the following questions concerning endogenous metabolism: (1) "Does survival bear any direct or indirect relationship to endogenous metabolism? (2) Is endogenous metabolism wholly catabolic or do available precursors and energy liberated in the process support some degree of macromolecular synthesis? (3) If so, are cellular components whose loss is particularly likely to result in death selectively reformed from dispensable materials? (4) Does the possession of specialized reserve materials exert a sparing action on the degradation of protein and RNA, and does it confer longevity? (5) Can death be attributed to the loss of any specific cellular constituent, or to the energetic state of the cell?" In terms of survival of the species in oligotrophic ocean waters, we are concerned with the preservation of the genome rather than a large number of cells. However, when dealing with survival of a culture for short periods of time (eg., ca. 50 hr), there are answers to some of these questions (Dawes, 1976).

Foster (1947) states that endogenous metabolism occurs simply because the organism cannot help it, and therefore it bears no relationship, direct or otherwise, to the period of survival. Monod (1942) stated that there was no energy of maintenance, and Bauchop and Elsden (1961) arrived at a similar conclusion. Jones and Rhodes-Roberts (1981) stated that "no peculiar physiological property such as nitrogen scavenging, ability to survive at the expense of intracellular polybetahydroxybutyric acid or protein, abnormally low cellular protein content, low maintenance energy requirements or a low adenylate charge state fully accounts for the starvation resistance of these marine bacteria." The data to support the concept of energy of maintenance come from short-term survival studies (eg., generally much less than 50 hr). However, when one considers the oligotrophic marine waters with their old organic matter (Williams et al., 1969), which is resistant to decomposition (Barber, 1968), and the residence time of the water (Broecker, 1963), the concept of energy of maintenance may have no bearing on the long-term survival of marine bacteria.

Novitsky and Morita (1977) addressed the question of endogenous metabolism during starvation-survival and found that it was reduced over 99% during the first week of starvation; they calculated that it would take 58 weeks to reduce the cellular carbon by one-half (Fig. 4). However, the actual measurement indicated that one-half of the cellular carbon is lost after only 39 days, a value close to the time when the viability of Ant-300 has been reduced to 50% of the initial number of cells. After a period of 7 days starvation, the endogenous respiration was reduced to 0.0071% per hr and remained at that level for 28 days. However, it should be remembered that total cellular carbon was measured for both viable and dead cells because it was impossible to separate one from the other. It is possible that, after 28 days of starvation-survival, a sharp decrease in the endogenous respiration may occur and that true dormancy might set in. There is evidence that endogenous respiration is not coupled with long term starvation-survival but with rapidly dying populations (Burleigh and Dawes, 1967; Dawes and Holms, 1958; and Postgate and Hunter, 1962). It is difficult to determine whether a decrease in endogenous respiration is actually a survival mechanism or occurs as a result of, or previous to, cell death. In starvation-survival, the induced endogenous respiration reduction may act to prevent macromolecular degradation, so that cells can immediately utilize a substrate when available.

No cryptic growth could be demonstrated in starvation-survival cells of Ant-300 (Novitsky and Morita, 1978a) or in a *Pseudomonas* sp. Kurath (1980). These data, coupled with the survival of Ant-300 for more than four years, indicate that in long-term survival without an energy-yielding substrate present, the organism does not require any energy of maintenance. The starvation-survival studies for a period of over four years were performed in a

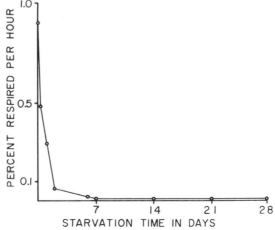

Figure 4. Endogenous respiration of starving Ant-300 cells. Respiration was determined at various times during starvation by the amount of $^{14}CO_2$ evolved from a previously labeled starving culture. Values are expressed as a percentage of the total cellular carbon respired per hour. After Novitsky and Morita (1978a).

screw-capped flask so that there could not be an exchange of possible energy-yielding substrate coming into contact with the contents of the flask.

Nevertheless, oligotrophic waters of the ocean do contain a small amount of organic matter that is resistant to bacterial degradation. It is possible that this material is utilized at a very slow rate to provide energy of maintenance or that the bacteria are utilizing oxidizable compounds other than the organic compounds as a source of maintenance energy. For example, it has been documented that methane oxidizers will participate in the first step in nitrification (Hutton and ZoBell, 1949; Whittenbury et al., 1976; and Hutchinson et al., 1976).

4.4. Changes in Cellular Components

Koch (1971) stated that "the micro-organisms have not only been selected for ability to grow under chronic starvation, but also for ability to respond quickly to unannounced and irregular windfalls of food." Therefore, they must be able to quickly utilize any energy-yielding substrate which implies a good mechanism to transport the substrate into the cell. However, what physiological state must the organisms be in in order to fully utilize the substrate quickly? Koch (1971) further hypothesizes that the bacteria must have evolved a very rapid rate of translation, even if it is at the expense of a high cost of maintaining elevated levels of mRNA, tRNA, amino acids, and energy supplies. The bacteria would also be expected to have an abundance of RNA polymerase. The synthesis of all these macromolecules is an expensive energy consuming process. In addition, there is a great need to make stable kinds of RNA as soon as possible when energy is available because they are part of the means for reproduction. In a carbon-limited state of slow but balanced growth of cells, RNA, tRNA, ribosomal protein, and possibly other proteins needed for protein synthesis are synthesized but not fully utilized (Koch, 1979). Furthermore, Koch (1971) suggested that the "extra" RNA in slow-growing bacteria is not functioning to make proteins, but there is a high selective value to the cell in having unused protein synthesizing machinery ready to function. The question is: What is the level of these macromolecules in starvation-survival cells?

When protein, RNA, and DNA content was measured in Ant-300 cells undergoing starvation-survival, it was found that the RNA dropped initially, but at the end of a 6 week period, the level was quite high. Both the protein and DNA levels also dropped rather drastically initially. These data were obtained on the total number of cells (dead and viable) in the suspension. On a respiring cell basis (evidenced by the INT method of Zimmermann et al., 1978), the amount of protein, RNA, and DNA increased after the initial drastic drop and steadily increased between the 1st and 6th week. On a viable cell basis, the same pattern was observed but the amount of the macromolecules per unit viable cell was higher than on a respiring cell basis (P. Amy, C. Pauling, and R. Y. Morita, unpublished data).

When Ant-300 cells are grown in a complete medium and then exposed after several washings to a starvation menstruum, it appears that the energy they have in reserve is utilized to produce more RNA, DNA, and protein in the viable cells. This suggests that the machinery in starvation-survival cells, which is energy expensive to synthesize, is present. Thus, one may conclude that when starvation-survival cells are challenged with a suitable energy-yielding substrate, the machinery is ready to utilize the substrate immediately, and the energy is not needed to synthesize the energy expensive protein synthesizing macromolecules. This appears to be the case, because starvation-survival cells when challenged with glutamate immediately utilize it (P. Amy and R. Y. Morita, unpublished data).

Karl(1980) stated that cells grown in carbon limited chemostats maintain normal ATP pools and high EC^A (energy charge of the adenylate pool) values. In starvation-survival studies of a *Pseudomonas* sp., Kurath (1980) demonstrated that the ATP per viable cell is higher after 10 days of starvation than at zero time.

Beneckea natriegens, when grown with limited nitrogen, converts the excess glucose in the medium to glycogen, and glycogen-rich cultures survived longer than glycogen-poor cultures during starvation (Nazly *et al.,* 1980). However, their experiments were only 50 hr. During the 50 hr experiment, very little protein or RNA was degraded as the primary endogenous source of energy. RNA was the preferred endogenous source of energy for maintenance of starving *B. natriegens.*

All the data thus far obtained indicate that the cells which have the ability to survive long periods of time in the absence of a proper energy-yielding substrate also have the expensive protein synthesizing machinery necessary for immediate use of any exogenous substrate that the cells encounter in the environment. As Koch (1971), dealing with slow growing cultures, stated: "we believe they should have evolved very rapid rates of translation, even if it is at the expense of a high cost of maintaining elevated levels of mRNA, tRNA, amino acids, and energy supplies. We would also expect the bacteria to have an abundance of RNA polymerase. All this is so that the ribosome, which on proration is the heaviest and therefore the most expensive part of the protein-synthesizing machinery, will be used at highest efficiency and therefore lowest total cost to the cell. Evolution would also select for very efficient activating enzymes for they too represent a costly item in the cell's economy." The foregoing does not answer the question as to how or why the cell diverts its energy towards the synthesis of the macromolecules necessary for protein synthesis. The lack of exogenous energy or exogenous essential amino acids may initiate the control mechanism in starvation-survival cells to produce the protein-synthesizing machinery, but this has yet to be proven.

During the process of starvation-survival of Ant-300, it was noted that chemotaxis of the organism did not develop until it was starved. Freshly harvested cells of Ant-300 do not exhibit active chemotaxis towards arginine (Geesey and Morita, 1979). Chemotaxis towards other amino acids does not occur unless Ant-300 cells are starved for at least 48 hr (F. Torrella and R. Y. Morita, unpublished data). During starvation-survival, the cells, through the chemotaxis process, are seeking energy-yielding substrates to keep them actively metabolizing. This chemotaxis response during the starvation-survival process is an indication that cells without an exogenous energy source appear to change their metabolic systems in order to seek an energy source, and if none is found, then the formation of the macromolecular synthesizing mechanisms appears to be the prime factor preparing the cell for long-term survival.

5. Substrate Capture

Koch (1971) recognized that bacteria must have an efficient transport mechanism to take up carbon and nitrogen sources, growth factors, or trace elements when confronted with an environment which has these compounds at very low concentrations. The same situation confronts the marine bacterium when it is taken by currents, etc., into the oligotrophic waters of the open and deep ocean. However, it should be recognized that the microbes themselves have contributed to the formation of the oligotrophic situation by utilizing what available organic compounds were originally present. In the process of decreasing the organic content of the water mass, less and less organic material is available for the heterotrophic bacteria to use as an energy source. Hence, the microorganism must be able to utilize the small amounts of organic matter until it reaches a point that it probably no longer supports the reproduction of the cells. Likewise, they must be able to utilize energy-yielding substrates when they become available.

In oligotrophic waters, the microorganism must first "see" the substrate and then affix it to its surface before it can be transported into the cell. The ability of a microorganism to affix the substrate to its surface we term "substrate capture." The basis for substrate capture is the organism's binding substances (eg., binding proteins) that permit the organism to capture the substrate on its surface. Substrate capture is not transport. Transport involves the substrate moving from the external to the internal portion of the cell. However, in order that substrates can be transported into the cell, they must first be located on the outside of the cell. The capture of substrate on the surface of the cells becomes very important in oligotrophic waters, where the substrates are in low concentration. Hence, transport of the substrate into the cell is against a concentration gradient. The question then is: How does a bacterium capture its substrate in a drastically decreased nutrient environment, in oligotrophic waters, or in oligotrophic waters where the nutrients are increasing slowly?

The concept of substrate capture is closely aligned with the active transport of substrates into the cells. It is not surprising, therefore, that the intense interest in bacterial transport systems has revealed gram-negative bacteria which bind ions, sugars, amino acids, and vitamins. To date, osmotic shock-releasable proteins have been isolated which bind sulfate (Pardee, 1968), phosphate (Medveczky and Rosenberg, 1969), galactose (Anraku, 1967; and Boos, 1969), L-arabinose (Hogg and Englesberg, 1969), valine (Piperno and Oxender, 1966), glutamine (Weiner and Heppel, 1971), glutamate and asparate (Willis and Furlong, 1975), phenylalanine (Verisek, 1972), histidine (Ames and Lever, 1970), cystine (Berger and Heppel, 1972), and arginine, lysine, and ornithine (Rosen, 1973).

Although most of the binding proteins that have been characterized have been isolated from extracts of E. coli, they appear to be present in other organisms as well. Histidine and sulfate binding proteins have been isolated from Salmonella typhimurium (Ames and Lever, 1972; Pardee, 1968).

Binding proteins may also participate in the formation of substrate pools or reservoirs in the periplasmic region of the cell. Langridge et al. (1970) estimated that there are as many as 10^4 binding proteins per cell for sulfate in S. typhimurium. Extensive studies have indicated that binding proteins do not possess enzymatic activity, nor do the substrates undergo chemical change during complex formation (Wilson and Holden, 1969). In addition, reversible binding of substrate to protein is a common feature of these systems (Wilson and Holden, 1969).

The binding proteins appear to be loosely bound to the cell because mild osmotic shock causes their release into the culture menstruum. Although osmotic shock-releasable binding proteins have been assigned a "periplasmic" location (the regions between the cytoplasmic membrane and the cell wall), loose association of the proteins with the membrane or cell wall may exist (Nakane et al., 1968).

Kinetic studies have provided evidence for the involvement of substrate-binding proteins in active transport. Osmotic shock causes a substantial decrease in the active transport of those substrates which bind shock-releasable proteins (Anraku, 1967; Rosen and Heppel, 1973; Wilson and Holden, 1969). In certain cases, a marked stimulation in transport occurs upon addition of purified binding protein to shocked cells (Wilson and Holden, 1969). In addition, the dissociation constants (K_d) for substrate-protein binding are similar to the K_m values for substrate transport. Piperno and Oxender (1966) determined for the LIV-binding protein, which binds leucine, isoleucine, and valine, K_d values of 1.1 and 2.2 μM. Furthermore. compounds acting as noncompetitive inhibitors to the transport of various amino acids also interfere with the respective amino acid-binding protein reaction (Barash and Halpern, 1971).

Many of the transport systems involving binding proteins operate at very low substrate concentrations. For instance, arginine specific transport in E. coli has a K_m of 2.6×10^{-8} M (Rosen, 1973), histidine specific transport in S.

typhimurium exhibits a K_m of 2.0 $\times$ 10^{-8} M (Ames and Lever, 1970), and galactose (Pmg system) transport has a K_m of 5.0 $\times 10^{-7}$ M (Rotman and Radojkovic, 1964). Kalckar (1971) proposed that the galactose-binding protein, which participates in the high affinity galactose transport system of *E. coli,* acts as a scavenging mechanism during low external galactose concentrations.

Not all substrates that are transported by permeases or that bind shock-releasable binding proteins elicit a chemotactic response (Mesibov and Adler, 1972). L-glutamine, histidine, isoleucine, leucine, methionine, phenylalanine, tryptophan, tyrosine, and valine are transported by *E. coli,* yet do not produce a tactic response. To date, only two receptors have been identified with amino acids (Mesibov and Adler, 1972). The aspartate chemoreceptor exhibits taxis toward aspartate and glutamate, and the serine chemoreceptor recognizes serine, cysteine, alanine, and glycine.

Due to the vast amount of information available on *E. coli* and a few other well-characterized bacteria, it is not surprising that these organisms have been employed to investigate the mechanisms by which bacteria accumulate nutrients. However, the sequestering of nutrients is of paramount importance to other organisms as well, especially marine bacteria. MacLeod and coworkers (Sprott and MacLeod, 1974; Thompson and MacLeod, 1971,1974) have characterized various transport systems in a marine pseudomonad species, but their studies have not dealt directly with substrate capture.

The foregoing illustrates that while numerous substrate receptors have been isolated and described for enteric bacteria, there is virtually no information available on substrate-binding components in marine bacteria. The binding of substrate in marine bacteria was investigated by Griffiths *et al.* (1974) who demonstrated that glutamate was bound loosely to the cell surface of Ant-300 and that a reduced salinity could affect the retention of the glutamate. A loosely bound amino acid pool, shown to exist in the same organism, could be released when the salinity was decreased (Geesey and Morita, 1981). Furthermore, differential centrifugation and fluorescent antibody studies indicate that the binding sites are located on or near the surface of the intact cell. In addition, as shown by equilibrium dialysis, membrane fragments of Ant-300 have the ability to bind a relatively large amount of arginine.

6. Chemotaxis

Several binding proteins, in addition to participating in active transport, are also involved in chemotaxis. Chemotaxis is a response by an organism which results in a directed movement toward a particular nutrient. A bacterium is capable of migrating from an area of no nutrients to an area in which a particular attractant is concentrated. The cellular component that recognizes a specific attractant is termed a "chemoreceptor." The galactose-binding protein serves as the recognition component for galactose chemoreception (Hazelbauer and Adler, 1971). Chemoreceptors for ribose and maltose also correspond to shock-releasable binding proteins (Askamit and Koshland, 1972; Kellerman and Szmecman 1974).

Adler (1969) demonstrated that the functioning of chemoreceptors does not require metabolism or general transport of the compound. It only requires recognition of the attractant by a specific binding protein. Galactose- and maltose-binding proteins each have mutationally separate sites for ligand binding, for coupling to transport, and for interaction with chemoreception (Hazelbauer, 1975).

Bell and Mitchell (1972) have shown that marine bacteria are capable of chemotaxis toward extracellular products of marine algae. They further demonstrated that in mixed populations derived from natural seawater samples, those bacteria exhibiting taxis toward the algal products were selected over nonchemotactic bacteria. Coral mucous (excreted in copious amounts) also elicits a chemotactic response from the indigenous bacteria of coral reefs (R. Y. Morita, unpublished data).

Ant-300 does display chemotaxis towards arginine (Geesey and Morita, 1979), other amino acids, alcohols, organic acids, and carbohydrates (F. Torrella and R. Y. Morita, unpublished data). The chemotactic ability is dependent upon the duration of the starvation period, with a period of 2 or 3 days being the optimum chemotactic period (F. Torrella and R. Y. Morita, unpublished data). Ant-300 cells freshly harvested from rich medium display a poor chemotactic response to various substrates. However, the beneficial effect of chemotaxis on starvation-survival depends on the species in question, because some marine bacteria rapidly loose their chemotactic response upon starvation.

According to Carlile (1980), motility has been neglected by microbial ecologists; active movement can be guided by means of a sensory system, so that the organism can select a suitable environment. Both active and passive movement are complementary and help bring the organisms to the most favorable position for growth, genetic recombination, or dispersal (Carlile, 1980).

Both chemotaxis and substrate capture play important roles in the starvation-survival of cells. Substrate capture is the mechanism by which the organism has the ability to concentrate substrates against a concentration gradient. As stated previously, chemotaxis is observed in the early stages of starvation-survival of Ant-300. What role it plays in the recovery process is not known. Torrella and Morita (unpublished data) have shown that 48-hr starved Ant-300 cells which are energized with one substrate (as the energy source) display very active chemotaxis towards another substrate. The extrapolation of this finding could be made to a situation where carbon compounds resulting from the decomposition of phytoplankton energizes the cell so that it can actively seek out an essential amino acid or other nitrogenous compound.

7. ATP and Energy Charge

Upon starvation, a high energy-charge ratio is maintained for a few hours (Niven *et al.,* 1977). An energy-charge of about 0.95 was demonstrated with *Beneckea natriegens* when growing in nutrient-rich medium; it decreased slowly from about the 13th hour, but was still 0.6 at 50 hr when the total adenylate content was less than 10% of the maximum level and the viability was only 6% of the original culture (Nazly *et al.,* 1980).

Cells grown in carbon-limited chemostats maintain normal ATP pools and therefore have a high energy charge (Karl, 1980). When a marine *Pseudomonas* sp. undergoes starvation-survival, there is an initial drop in the level of ATP per viable cell (Kurath, 1980). At the initiation of starvation-survival, the amount of ATP per viable cell is approximately 10^{-9} μg and drops to 10^{-10} μg in seven days of starvation. After the seventh day, the value increases and reaches its highest value on the 25th day (ca. 0.5×10^{-8} μg ATP/viable cell). According to Geesey and Morita (1979) and Glick (1980), starvation-survival cells maintain an energized membrane as shown by the effect of DNP on amino acid uptake.

Thus, it appears that starvation-survival cells have the energy necessary to utilize any substrate that can be captured by the binding substances on the surface of the cell. In all likelihood, this energy is used for active transport of certain substrates into the cell. It is also possible that the available energy might be used to synthesize extracellular enzymes needed to degrade whatever particulate matter the cell adheres to.

8. Substrate Uptake

Geesey and Morita (1979) demonstrated that Ant-300 cells display different affinities when exposed to nanomolar and micromolar concentrations of arginine. The Lineweaver-Burk plot (Fig. 5) revealed a bimodal relationship for arginine uptake from 0.034 to 0.59 μM. Assuming two simultaneous reactions, K_t, values of 1.7×10^{-8} and 4.5×10^{-6} M and V_{max} values of 12 and 51 pmol/min/5×10^7 bacteria, respectively, were obtained. For details refer to Geesey and Morita (1979). For Ant-300, a high-affinity uptake can be operable when the amino acid concentration level in the open ocean is low (Lee and Bada, 1975). Geesey and Morita (1979) suggest that Ant-300 possesses two independent arginine transport systems as has been reported in other bacteria (Reid *et al.,* 1979; Rosen, 1973). High-affinity uptake at low nutrient concentrations is discussed by Jannasch (1979) and Matin (1979), but these low nutrient concentrations are still higher than the starvation-survival threshold. According to Koch (1979), the cell's transport mechanisms are present in excess in *E. coli* cells starved for a few hours.

When 7-week starvation-survival cells of Ant-300 were challenged with nutrients, there was an immediate 100-fold increase in cell numbers (Novitsky and Morita, 1978a). Starvation-survival cells of Ant-300 were found to bind 113% of the glutamic acid, 56.7% of the proline, 30.6% of the arginine, and 23.4% of the leucine bound by nonstarved cells (Glick, 1980). Ant-300 cells killed by a hydrostatic pressure have also been shown to bind a small amount of glutamate (M. A. Glick and R. Y. Morita, unpublished data). Because the starved cells are much smaller (cocci of a 0.5 μm diameter) than nonstarved cells (1 × 4 μm rods), Novitsky (1977) calculated the volume of the starved cell to 0.09 × the volume of nonstarved cells. On this basis, 30-day starved cells are binding 30 times as much glutamic acid as fresh cells, as well as 8.4 times as much arginine, 15.6 times as much proline, and 6.4 times as much leucine per unit volume. Matin and Veldkamp (1978) stated that a high surface-to-volume ratio contributed to the efficient scavenging capacity of a *Spirillum* sp. in a low nutrient environment. Hoppe (1976) stated that "the small size of a majority of non-culturable bacteria together with their high substrate affinity may enable these bacteria to live in water bodies containing threshold concentrations of nutrients." Just what level threshold means is not known.

The uptake of glucose by freshly harvested cells and 25-day starvation-survival *Pseudomonas* cells (per viable cell) was found to be 27 times greater in the starvation-survival cells, but with a starvation-survival time of 40 days, the value dropped to 7.7 times (Kurath, 1980). The same trend was found in the uptake of glutamate by freshly harvested cells and starvation-survival cells.

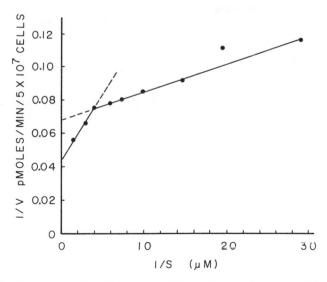

Figure 5. Double-reciprocal plot of initial rates of L-arginine uptake by cells of Ant-300. After Geesey and Morita (1979).

Glick (1980) and Amy and Morita (unpublished data) demonstrated that starvation-survival cells take up substrate immediately. This is in contrast to the work of Williams and Gray (1970), who noted two separate increases in the rates of amino acid uptake by marine populations. The second rate increase followed the first by from 20 to 36 hr. This was thought to be the result of the activation of the inactive cells or of growth. If the pattern of quick activation characterized by Ant-300 is typical for dormant bacteria, then the biphasic uptake seen by Williams and Gray (1970) was probably caused by growth rather than activation. Starvation-survival cells can also take up substrate under hydrostatic pressure (Yorgey, 1980).

From the foregoing data presented, starvation-survival cells have the ability to take up substrate immediately when it becomes available. In fact, they are more capable of doing so than nonstarved cells.

9. Conclusions

I propose that in order for a marine bacterial population to be biologically "fit" for starvation-survival, the following processes greatly aid in its survival in the absence of an energy-yielding substrate:

1. All metabolic processes are reduced to a dormant state or nearly dormant state.
2. Upon starvation many species will increase their number.
3. In the starvation-survival process, any reserve energy material of the cell is used for preparing the cell for survival.
4. All metabolic mechanisms upon starvation go to the formation of specific proteins, ATP, and RNA so that the cell, when it comes in contact with a substrate, will utilize it immediately and not have to expend the initial amounts of energy for the synthesis of RNA and proteins. Both RNA and protein synthesis are high energy-consuming processes. The high ATP level per viable cell is used primarily for active transport of substrates across the membrane.
5. The miniaturization of cells permits the cell to be more efficient to scavenge what little energy there is in the environment and also permits the cell's survival against other adverse environmental factors.

Most of the above points have yet to be proven experimentally.

It is recognized that 1% or less of a culture survives for long periods of time without exogenous energy-yielding substrates. The question still remains as to what dictates which few cells will survive. Is there communication between the cells in starvation-survival or is there a random or genetic selection of the cells that will survive?

ACKNOWLEDGMENT. The author is indebted to his students (P. Amy, M. A. Glick, G. Kurath, and P. S. Yorgey) and his associates (C. Pauling and F. Torrella) for use of unpublished material. This work was supported, in part, by NSF grant OCE 8108366, and is published as technical paper no. 5988, Oregon Agricultural Station.

References

Adler, J., 1969, Chemoreceptors in bacteria, *Science* **166**:1588–1597.

Ames, G. F., and Lever, J., 1970, Components of histidine transport: histidine binding proteins and his P protein, *Proc. Natl. Acad. Sci. U.S.A.* **66**:1096–1103.

Ames, G. F., and J. Lever, 1972, The histidine-binding protein J is a component of histidine transport. *J. Biol. Chem.* **247**:4309–4316.

Askamit, R., and Koshland, D. E., 1972, A ribose binding protein of *Salmonella typhimurium,* *Biochem. Biophys. Res. Commun.* **48**:1348–1352.

Anderson, J. I. W., and Heffernan, W. P., 1965, Isolation and characterization of filterable marine bacteria, *J. Bacteriol.* **90**:1713–1718.

Anraku, Y., 1967, The reduction and restoration of galactose transport in osmotically shocked cells of *Escherichia coli, J. Biol. Chem.* **242**:793–800.

Arrhenius, G., 1952, Sediment cores from the East Pacific. I. Properties of the sediment and their distribution. *Rep. Swed. Deep-Sea Exped.* 1947–1948.

Bada, J. F., and Lee, C., 1977, Decomposition and an alteration of organic compounds in seawater, *Mar. Chem.* **5**:523–534.

Barash, H., and Halpern, Y. S., 1971, Glutamate-binding protein and its relation to glutamate transport in *Escherichia coli* K-12, *Biochem. Biophys. Res. Commun.* **45**:681–688.

Barber, R. T., 1968, Dissolved organic carbon from deep waters resists microbial oxidation, *Nature* **220**:274–275.

Baross, J. A., Hanus, F. J., and Morita, R. Y., 1975, Survival of human enterics and other sewage microorganisms under simulated deep-sea conditions, *Appl. Environ. Microbiol.* **30**:309–318.

Bauchop, T., and Elsden, S. R., 1961, The growth of microorganisms in relation to their energy supply, *J. Gen. Microbiol.* **23**:457–469.

Bell, W., and Mitchell, R., 1972, Chemotactic and growth responses of marine bacteria to algal extracellular products, *Biol. Bull.* **143**:265–277.

Berger, D. A., and Heppel, L. A., 1972, A binding protein involved in the transport of cystine and diaminopimelic acid in *Escherichia coli, J. Biol. Chem.* **247**:7684–7694.

Bisset, K. A., 1952, *Bacteria,* B. and S. Livingstone, Edinburgh.

Bollen, W. B., 1977, Sulfur oxidation and respiration in 54-year soil sample, *Soil. Biol. Biochem.* **9**:405–410.

Boos, W., 1969, The galactose binding protein and its relationship to the β-methylgalactoside permease from *Escherichia coli, Eur. J. Biochem.* **10**:66–73.

Bosco, G., 1960, Studio della sensibilita, In vitro algi antibiotica de parte di microorganismi isolate in epoca preantibiotica, *Nuovi. Ann. Igiene Microbiol.* **11**:227–240.

Boylen, C. W., and Ensign, J. C., 1970, Long-term starvation survival of rod and spherical cells of *Arthrobacter crystallopoietes, J. Bacteriol.* **103**:569–677.

Broecker, W., 1963, Radioisotopes and large-scale organic mixing, in: *The Sea,* Vol. 2 (M. N. Hill, ed.), pp. 88–108, Wiley-Interscience, New York.

Burleigh, I. C., and Dawes, J. R., 1967, Studies on the endogenous metabolism and senescence of starved *Sarcina lutea, Biochem. J.* **102**:236–250.

Carlile, M. J., 1980, Positioning mechanisms—the role of motility, taxis and trophism in the life of microorganisms, in *Contemporary Microbial Ecology,* (D. C. Ellwood, N. J. Hedger, M. J. Latham, J. M. Lynch, and J. H. Slater, eds.), pp. 54–74, Academic Press, London.

Carlucci, A. F., and Williams, P. M., 1978, Simulated in situ growth rates of pelagic marine bacteria, *Naturwissenschaften* **65:**541–542.

Casida, L. W., Jr., 1977, Small cells in pure cultures of *Agromyces ramosus* and in natural soil. *Can. J. Microbiol.* **23:**214–216.

Craig, H., 1971, The deep metabolism: oxygen consumption in abyssal ocean water, *J. Geophys. Res.* **76:**5078–5086.

Daley, R. J., and Hobbie, J. E., 1975, Direct count of aquatic bacteria by a modified epifluorescent technique. *Limnol. Oceanogr.* **20:**875–881.

Dawes, E. A., 1976, Endogenous metabolism and the survival of starved prokaryotes, *Symp. Soc. Gen. Microbiol.* **26:**19–53.

Dawes, E. A., and Holms, W. H., 1958, Metabolism of *Sarcina lutea.* III. Endogenous metabolism, *Biochim. Biophys. Acta* **30:**278–293.

Dawes, E. A., and Senior, P. J., 1973, The role and regulation of energy reserve polymers in micro-organisms, *Adv. Microb. Physiol.* **10:**135–266.

Dow, C. S., and Whittenbury, R., 1980, Prokaryotic form and function, in: *Contempory Microbial Ecology* (D. C. Ellwood, J. N. Hedger, M. J. Latham, J. M. Lynch, and J. H. Slater, eds.), pp. 391–417, Academic Press, London.

Foster, J. W., 1947, Some introspections of mold metabolism, *Bacteriol. Rev.* **11:**166–191.

Garbosky, A. J., and Giambiagi, N., 1966, The survival of nitrifying bacteria in soil. *Plant and Soil* **17:**271–278.

Geesey, G. G., and Morita, R. Y., 1979, Capture of arginine at low concentrations by a marine psychrophilic bacterium, *Appl. Environ. Microbiol.* **38:**1092–1097.

Geesey, G. G., and Morita, R. Y., 1981, Relationship of cell envelope stability to substrate capture in a marine psychrophilic bacterium, *Appl. Environ. Microbiol.* **42:**533–540.

Geftic, S. G., Heymann, H., and Adair, F. W., 1979, Fourteen-year survival of *Pseudomonas cepacia* in a salts solution preserved with benzalkonium chloride. *Appl. Environ. Microbiol.* **37:**505–510.

Glick, M. A., 1980, Substrate capture, uptake, and utilization of some amino acids by starved cells of a psychrophilic marine *Vibrio,* M.S. thesis, Oregon State University, Corvallis.

Goodrich, T. D., and Morita, R. Y., 1977, Low temperature inhibition on binding, transport, and incorporation of leucine, arginine, methionine, and histidine in *Escherichia coli, Zeit. Alleg. Mikrobiol.* **17:**91–97.

Gordon, D. C., 1970, Some studies on the distribution and composition of particulate organic carbon in the North Atlantic Ocean, *Deep-Sea Res.* **17:**233–243.

Gray, T. R. G., 1976, Survival of vegetative microbes in soil, *Symp. Soc. Gen. Microbiol.* **26:**327–364.

Gray, T. R. G., and Postgate, J. R., eds., 1976, The survival of vegetative microbes, *Symp. Soc. Gen. Microbiol.* **26:**432 pp.

Gray, T. R. G., and Williams, S. T., 1971, Microbial productivity in soil, *Symp. Soc. Gen. Microbiol.* **21:**255–286.

Griffiths, R. P., Baross, J. A., Hanus, F. J., and Morita, R. Y., 1974, Some physical and chemical parameters affecting the formation and retention of glutamate pools in a marine psychrophilic bacterium, *Zeit. Alleg. Mikrobiol.* **14:**359–369.

Harrison, A. P., and Lawrence, F. R., 1963, Phenotypic, genotypic, and chemical changes in starving populations of *Aerobacter aerogenes, J. Bacteriol.* **85:**742–750.

Hazelbauer, G. L., 1975, Maltose chemoreceptor of *Escherichia coli, J. Bacteriol.* **122:**206–214.

Hazelbauer, G. L., and Adler, J., 1971, Role of galactose binding protein in chemotaxis of *Escherichia coli* toward galactose, *Nature New Biol.* **230:**101–104.

Henrici, A. T., 1928, *Morphological Variation and the Rate of Growth of Bacteria,* Charles C Thomas, Springfield, Ill.

Hogg, R. W., and Englesberg, E., 1969, L-arabinose binding protein from *Escherichia coli* B/r, *J. Bacteriol.* **100:**423–432.

Hoppe, H. G., 1976, Determination and properties of actively metabolizing heterotrophic bacteria in the sea, investigated by means of microautoradiography, *Mar. Biol.* **36:**291–302.

Hutchinson, D. W., Whittenbury, R., and Dalton, H., 1976, A possible role of free radicals in the oxidation of methane by *Methylococcus capsulatus, J. Theor. Biol.* **58:**325–335.

Hutton, W. E., and ZoBell, C. E., 1949, The occurrence and characteristics of methane-oxidizing bacteria, *J. Bacteriol.* **65:**216–219.

Jannasch, H. W., 1958, Studies on planktonic bacteria by means of a direct membrane filter method, *J. Gen. Microbiol.* **18:**609–620.

Jannasch, H. W., 1967, Growth of marine bacteria at limiting concentrations of organic carbon in sea water, *Limnol. Oceanogr.* **12:**264–271.

Jannasch, H. W., 1979, Microbial ecology of aquatic low nutrient habitats, in: *Strategies of Microbial Life in Extreme Environments,* (M. Shilo, ed.), pp. 243–260, Dahlem Konferenzen, Verlag Chemie, Weinheim.

Jensen, H. L., 1961, Survival of *Rhizobium meliloti* in soil culture, *Nature* **192:**682–683.

Johnson, P. W., and Sieburth, J. McN., 1978, Morphology of non-cultured bacterioplankton from estuarine, shelf and open ocean waters, Abstr. Annu. Meet. Amer. Soc. Microbiol., N95, p. 178.

Jones, K. L., and Rhodes-Roberts, M. E., 1981, The survival of marine bacteria under starvation conditions, *J. Appl. Bacteriol.* **50:**247–258.

Kalckar, H. M., 1971, The periplasmic galactose binding protein of *Escherichia coli. Science* **174:**557–565.

Kaneko, T., and Colwell, R. R., 1978, The annual cycle of *Vibrio parahaemolyticus* in Chesapeake Bay. *Microb. Ecol.* **4:**135–155.

Karl, D. M., 1980, Cellular nucleotide measurements and applications in microbial ecology, *Microbiol. Rev.* **44:**739–796.

Kellermann, O., and Szmelcman, S., 1974, Active transport of maltose in *Escherichia coli* K-12: involvement of a periplasmic maltose binding protein. *Eur. J. Biochem.* **47:**139–149.

Koch, A. L., 1971, The adaptive responses of *Escherichia coli* to a feast and famine existence, *Adv. Microb. Physiol.* **6:**147–217.

Koch, A. L., 1979, Microbial growth in low concentrations of nutrients, in: *Strategies of Microbial Life in Extreme Environments* (M. Shilo, ed.) pp. 261–279, Dahlem Konferenzen, Verlag Chemie, Weinheim.

Kogure, K., Simidu, U., and Tage, N., 1979, A tentative direct microscopic method for counting living marine bacteria. *Can. J. Microbiol.* **25:**415–420.

Kurath, G., 1980, Some physiological bases for survival of a marine bacterium during nutrient starvation, M.S. thesis, Oregon State University, Corvallis.

Langridge, R., Shinagawa, H., Pardee, A. B., 1979, Sulfate-binding protein from *Salmonella typhimurium:* physical properties, *Science* **169:**59–61.

Lee, C., and Bada, J. L., 1975, Amino acids in equatorial Pacific Ocean water, *Earth Plant. Sci. Lett.* **26:**61–68.

Lipman, C. G., 1931, Living microorganisms in ancient rocks, *J. Bacteriol.* **22:**183–196.

Luscombe, B. M., and Gray, T. G. R., 1974, Characteristics of *Arthrobacter* grown in continuous culture, *J. Gen. Microbiol.* **82:**213–222.

MacDonnell, M. T., and Hood, M. A., 1982, Isolation and characterization of ultramicrobacteria from a Gulf Coast estuary, *Appl. Environ. Microbiol.* **43:**566–571.

Marshall, K. C., 1979, Growth at interfaces, in *Strategies of Microbial Life in Extreme Environments* (M. Shilo, ed.), pp. 281–290, Dahelm Koferenzen, Verlag Chemie, Weinheim.

Marshall, K. C., Stout, R., and Mitchell, R., 1979, Selective sorption of marine bacteria to sur-
faces, *Can. J. Microbiol.* **17:**1413–1416.
Matin, A., 1979, Microbial regulatory mechanisms at low nutrient concentrations as studied in
chemostat, in: *Strategies of Microbial Life in Extreme Environments,* pp. 323–340 (M.
Shilo, ed.), Dahlem Konferenzen, Verlag Chemie, Weinheim.
Matin, A., and Veldkamp, H., 1978, Physiological basis of the selective advantage of *Spirillum*
sp. in a carbon-limited environment, *J. Gen. Microbiol.* **105:**189–197.
Medveczky, N., and Rosenberg, H., 1969, The binding and release of phosphate by a protein
isolated from *Escherichia coli, Biochim. Biophys. Acta.* **192:**369–371.
Menzel, D. W., 1967, Particulate organic carbon in the deep-sea. *Deep-Sea Res.* **11:**757–765.
Menzel, D. W., 1970, The role of *in situ* decomposition of organic matter on the concentration
of non-conservative properties in the sea, *Deep-Sea Res.* **17:**751–764.
Menzel, D. W., and Goering, J. J., 1966, The distribution of organic detritus in the ocean, *Lim-
nol. Oceanogr.* **11:**333–337.
Menzel, D. W., and Ryther, J. H., 1970, Distribution and cycling of organic matter in the oceans,
in: *Organic Matter in Natural Waters* (D. W. Hood, ed.) pp. 31–54, Institute of Marine
Science Publication, College, Alaska.
Mesibov, R., and Adler, J., 1972, Chemotaxis toward amino acids in *Escherichia coli, J. Bac-
teriol.* **122:**315–326.
Monod, J., 1942, *Recherches sur la Croissance des Cultures Bacteriennes,* Herman, Paris.
Morita, R. Y., 1968, in: *Marine Microbiology* (C. H. Oppenheimer, ed.), p. 97, Proc. 4th Inter-
national Interdisciplinary Conference, New York Academy of Sciences, New York.
Morita, R. Y., 1977, The role of microbes in the marine environment, in: *Ocean Sound Scatter-
ing Prediction* (N. R. Anderson and B. J. Zuhurance eds.), pp. 445–456, Plenum Press,
New York.
Morita, R. Y., 1979a, Current status of the microbiology of the deep-sea, *Ambio Spec. Rep.*
6:33–36.
Morita, R. Y., 1979b, The role of microbes in the bioenergetics of the deep-sea, *Sarsia* **64:**9–
12.
Morita, R. Y., 1980a, Low temperature, energy, survival and time in microbial ecology, in:
Microbiology—1980 (D. Schlessinger, ed.), pp. 323–324, American Society for Microbiol-
ogy, Washington, D.C.
Morita, R. Y., 1980b, Microbial life in the deep-sea, *Can. J. Microbiol.* **26:**1375–1385.
Morita, R. Y., and ZoBell, C. E., 1955, Occurrence of bacteria in pelagic sediments collected
during the Mid-Pacific Expedition. *Deep-Sea Res.* **3:**66–73.
Nakane, P. K., Nichoalds, G. E., and Oxender, D. L., 1968, Cellular localization of leucine-
binding protein from *Escherichia coli, Science.* **161:**182.
Nazly, N., Carter, I. A., and Knowles, C. J., 1980, Adenine nucleotide pools during starvation
of *Beneckea natriegens, J. Gen. Microbiol.* **116:**295–303.
Niven, D. F., Collins, P. A., and Knowles, C. J., 1977, Adenylate energy charge during batch
culture of *Beneckea natriegens, J. Gen. Microbiol.* **98:**95–108.
Novitsky, J. A., 1977, Effects of long term nutrient starvation on a marine psychrophilic vibrio,
Ph.D. thesis, Oregon State University, Corvallis.
Novitsky, J. A., and Morita, R. Y., 1976, Morphological characterization of small cells resulting
from nutrient starvation of a psychrophilic marine vibrio, *Appl. Environ. Microbiol.*
32:617–662.
Novitsky, J. A., and Morita, R. Y., 1977, Survival of a psychrophilic marine vibrio under long-
term nutrient starvation, *Appl. Environ. Microbiol.* **33:**635–641.
Novitsky, J. A., and Morita, R. Y., 1978a, Possible strategy for the survival of marine bacteria
under starvation conditions, *Mar. Biol.* **48:**289–295.
Novitsky, J. A., and Morita, R. Y., 1978b, Starvation induced barotolerance as a survival mechanism of a psychrophilic marine vibrio in the waters of the Antarctic Convergence, *Mar
Biol.* **49:**7–10.

Oppenheimer, C. H., 1952, The membrane filter in marine microbiology, *J. Bacteriol.* **64:**783–786.

Pardee, A. B., 1968, Membrane transport proteins, *Science* **162:**632–637.

Piperno, J. R., and Oxender, D. L., 1966, Amino acid-binding protein released from *Escherichia coli* by osmotic shock, *J. Biol. Chem.* **241:**5732–5734.

Pirt, S. J., 1965, The maintenance energy of bacteria in growing cultures, *Proc. Roy. Soc. Lond. Ser. B* **163:**224–231.

Poindexter, J. S., 1979, Morphological adaptation to low nutrient concentrations, in: *Strategies of Microbial Life in Extreme Environments* (M. Shilo, ed.), pp. 341–356, Dahlem Konferenzen, Verlag Chemie, Weinheim.

Postgate, J. R., 1976, Death in macrobes and microbes, *Symp. Soc. Gen. Microbiol.* **26:**1–18.

Postgate, J. R., and Hunter, J. R., 1962, The survival of starved bacteria. *J. Gen. Microbiol.* **21:**233–306.

Rahn, O., 1932 *Physiology of Bacteria,* Blakiston, Philadelphia.

Reid, K. G., Utech, N. M., and Holden, J. T., 1970, Multiple transport components for decarboxylic amino acids in *Streptococcus faecalis. J. Biol. Chem.* **245:**5261–5272.

Resier, R., and Tasch, P., 1960, Investigation of the viability of osmophile bacteria of great geological age. *Trans. Kansas Acad. Sci.* **63:**31–34.

Robertson, J. B., and Batt, R. D., 1973, Survival of *Norcardia corallina* and degradation of constituents during starvation, *J. Gen. Microbiol.* **78:**109–117.

Rodriguez-Valera, F., Ruiz-Berraquero, F., and Ramos-Cormenzana, A., 1979, Isolation of extreme halophiles from seawater, *Appl. Environ. Microbiol.* **38:**164–165.

Rosen, B. P., 1973, Basic amino acid transport in *Escherichia coli* II. Purification and properties of an arginine-binding protein, *J. Biol. Chem.* **248:**1211–1218.

Rosen, R. P., and Heppel, L. A., 1973, Present status of binding proteins that are released from Gram-negative bacteria by osmotic shock, in: *Bacterial Membranes and Walls* (L. Lieve, ed.), pp. 209–239, Marcel Dekker, New York.

Rotman, B., and Radojkovic, J., 1964, Galactose transport in *Escherichia coli, J. Biol. Chem.* **239:**3153–3156.

Ruby, E. G., and Morin, J. G., 1979, Luminous enteric bacteria of marine fishes: a study of their distribution, densities and dispersion. *Appl. Environ. Microbiol.* **38:**406–411.

Seki, H., Skelding, J., and Parsons, T. R., 1968, Observations on the decomposition of a marine sediment, *Limnol. Oceanogr.* **13:**440–447.

Shilo, M., ed., 1979, *Strategies of Microbial Life in Extreme Environments,* Dahlem Konferenzen, Verlag Chemie, Weinheim.

Shilo, M., and Yetinson, T., 1979, Physiological characteristics underlying the distribution patterns of luminous bacteria in the Mediterranean Sea and the Gulf of Elat, *Appl. Environ. Microbiol.* **38:**577–584.

Sieburth, J., McN., Brooks, R. D., Gessner, R. V., Thomas, C. D., and Tootle, J. L., 1974, Microbial colonization of marine plant surfaces as observed by scanning electron microscopy, in: *Effect of the Ocean Environment on Microbial Activities* (R. R. Colwell and R. Y. Morita, eds.), pp. 318–326, University Park Press, Baltimore.

Sprott, G. D., and MacLeod, R. A., 1974, Nature of the specificity of alcohol coupling of L-alanine transport into isolated membrane vesicles of a marine pseudomonad, *J. Bacteriol.* **117:**1043–1054.

Stevenson, L. H., 1978, A case for bacterial dormancy in aquatic systems, *Microb. Ecol.* **4:**127–133.

Strickland, J. D. H., 1971, Microbial activity in aquatic environments, *Symp. Soc. Gen. Microbiol.* **21:**231–253.

Sudo, S. Z., and Dworkin, M., 1973, Comparative biology of prokaryotic resting cells, *Adv. Microbial Physiol.* **6:**153–224.

Sussman, A. S., and Halvorson, H. O., 1966, *Spores, their Dormancy and Germination,* Harper & Row, New York.

Tabor, P. S., Ohwada, K., and Colwell, R. R., 1981, Filterable marine bacteria found in the Deep Sea: distribution, taxonomy and response to starvation, *Microb. Ecol.* **7:**67–83.

Tempest, D. W., and Neijssel, O. M., 1978, Eco-physiological aspects of microbial growth in aerobic nutrient-limited environments. *Adv. Microb. Ecol.* **2:**105–153.

Thomas, T. D., and Batt, R. D., 1969, Survival of *Streptococcus lactis* in starvation conditions, *J. Gen. Microbiol.* **50:**367–382.

Thompson, J., and MacLeod, R. A., 1971, Function of Na^+ and K^+ in the active transport of α-aminoisobutyric acid in a marine pseudomonad, *J. Biol. Chem.* **246:**4066–4074.

Thompson, J., and MacLeod, R. A., 1974, Potassium transport and the relationship between intracellular potassium concentration and amino acid uptake by cells of a marine pseudomonad, *J. Bacteriol.* **120:**587–603.

Torrella, F., and Morita, R. Y., 1981, Microcultural study of bacterial size changes and microcolony and ultramicrocolony formation by heterotrophic bacteria in seawater, *Appl. Environ. Microbiol.* **41:**518–527.

Verisek, J., 1972, The cooperative character of phenylalanine binding by a protein fraction isolated from baker's yeast membrane, *Biochim. Biophys. Acta* **290:**256–266.

Watson, S. W., Novitsky, T. J., Quinby, H. L., and Valois, F. W., 1977, Determination of bacterial number and biomass in the marine environment, *Appl. Environ. Microbiol.* **33:**940–946.

Weiner, J. H., and Heppel, L. A., 1971, A binding protein for glutamine and its relation to active transport to *Escherichia coli, J. Biol. Chem.* **246:**6933–6941.

Whittenbury, R., Colby, J., Dalton, H., and Reed, H. L., 1976, Biology and ecology of methane oxidation, in: *Microbial Production and Utilization of Gases* (H.G. Schlegel, G. Gottschalk, and N. Pfenning, eds.). pp. 281–292, E. Goltze KG, Gottingen.

Williams, P. J. LeB., and Gray, R. W., 1970, Heterotrophic utilization of dissolved organic compounds in the sea. II. Observations on the response of heterotrophic marine populations to abrupt increases in amino acid concentrations, *J. Mar. Biol. Assoc. U.K.* **50:**871–881.

Williams, P. M., Oeschger, H., and Kinney, P., 1969, Natural radiocarbon activity of dissolved organic carbon in the Northeast Pacific Ocean, *Nature* **224:**256–258.

Willis, R. C., and Furlong, C. E., 1975, Interactions of a glutamate-aspartate-binding protein with the glutamate transport system of *Escherichia coli, J. Biol. Chem.* **250:**2581–2586.

Wilson, O. H., and Holden, J. T., 1969, Arginine transport and metabolism in osmotically shocked and unshocked cells of *Escherichia coli J. Biol. Chem.* **244:**2737–2742.

Wright, R. T., 1973, Some difficulties in using ^{14}C-organic solutes to measure heterotrophic bacterial activity, in: *Estuarine Microbial Ecology* (L. H. Stevenson and R. R. Colwell, eds.), W. Baruch Library in Marine Science, No. 1, University of South Carolina Press, Columbia.

Yetinson, T., and Shilo, M., 1979, Seasonal and geographic distribution of luminous bacteria in the Eastern Mediterranean Sea and the Gulf of Elat, *Appl. Environ. Microbiol.* **37:**1230–1238.

Yorgey, P. S., 1980, The synergistic effect of starvation and hydrostatic pressure on uptake of alpha-aminoisobutyric acid by a psychrophilic marine vibrio, M.S. thesis, Oregon State University, Corvallis.

Zimmermann, R., 1977, Estimation of bacterial numbers and biomass by epifluorescence microscopy and scanning electron microscopy, in: *Microbial Ecology of a Brackish Water Environment* (G. Rheinheimer, ed.), pp. 103–120, Springer-Verlag, Berlin.

Zimmermann, R., and Meyer-Reil, L. A., 1974, A new method for fluorescence staining of bacterial populations on membrane filter, *Kieler Meeresforch.* **30:**24–27.

Zimmermann, R., Iturriaga, R., and Becker-Birch, J., 1978, Simultaneous determination of the total number of aquatic bacteria and the number thereof involved in respiration. *App. Environ. Microbiol.* **36:**926–935.

6

Are Solid Surfaces of Ecological Significance to Aquatic Bacteria?

MADILYN FLETCHER AND K. C. MARSHALL

1. Introduction

Man has long been aware of microbial attachment to solid surfaces in the sea and in freshwaters, as slime layers formed by attached microorganisms and their associated polymers are easily detected by touch. Moreover, such slime layers frequently become a nuisance, or even a serious fouling problem, when they occur on submerged man-made structures, such as ship hulls, platforms, or dams.

However, it is really over the past 10 years that bacterial attachment has received growing and enthusiastic attention, as workers become aware of the significance of attached bacteria and new methods are devised or adapted to focus on the problem. There have been a large number of descriptive studies on microbial colonization and slime layer development on submerged surfaces, particularly in marine environments. These have shown that essentially all non-biological surfaces are eventually colonized by bacteria, and the only types of surfaces that appear to resist colonization indefinitely are certain surface areas of plants or animals, which are apparently protected from colonization (Section 2.3).

Because of the ubiquitous nature of attached bacteria in aquatic environments, they are expected to play a considerable role in the various processes

MADILYN FLETCHER ● Department of Environmental Sciences, University of Warwick, Coventry CV4 7AL, England. K. C. MARSHALL ● School of Microbiology, The University of New South Wales, Kensington 2033, Australia.

carried out by the microbial community, and the purpose of this chapter is to explore the comparative significance of these attached microorganisms. A number of aspects will be considered, bearing in mind the key question: How does the contribution of attached bacteria compare with that of suspended organisms? Other points are (1) whether attachment surfaces select for particular bacterial types with specific capabilities and the physicochemical and physiological processes involved (Sections 2, 3, 4 and 7), and (2) whether attached bacteria are more or less active physiologically than suspended organisms and whether this depends on the types and numbers of attached cells, the nature of the attachment surface, and various environmental conditions, such as nutrient concentration or hydrodynamic features (Sections 5 and 6).

First, some of the terms used should be clarified. When referring to the attachment process, both *attachment* and *adhesion* are somewhat general terms describing the binding of a bacterium to a surface, and the process may involve physiological activity, such as polymer production. *Adsorption* is used in the chemical sense, i.e., the concentration of a substance on a surface, whereas *sorption* is avoided, as this refers collectively to adsorption and absorption. Similarly, *adherence* is not used as this is appropriate to concepts or principles, whereas *adhesion* is relevant to surfaces. Two terms which are frequently confused are *substratum* and *substrate;* the former refers to the surface, and the latter should be used only for nutrients and assimilated material.

2. Characteristics of a Surface

As the substratum is a major component in the attachment process, its characteristics will have an important influence on the adhesion mechanism. The three principal types of substratum characteristics are: (1) physicochemical factors, which include chemical and thermodynamic features, (2) physical factors, such as surface texture, and (3) biological factors, significant when the substratum is another organism.

2.1. Physicochemical Factors

2.1.1. Surface Energy

Probably the most significant physicochemical property of the substratum is the surface free energy, as it comprises all surface forces capable of interaction with forces belonging to adjacent phases, e.g., the liquid medium, or the bacterial surface. Free energy, a thermodynamic concept, is the "available" energy within a given system and may be represented by G, when it is called the Gibbs free energy. The "unavailable" energy in the system is the product of the temperature (T), and the entropy (S), a function which indicates the degree of disorder or randomness in the system. Together, G and TS give the total energy of the system, called its enthalpy and represented by H, so that

$$H = G + TS$$

When a system is not in equilibrium, it has a tendency to undergo a spontaneous change until equilibrium is reached, and this will be accompanied by a decrease in free energy. This can be achieved by one (or a combination) of two processes: (1) a decrease in enthalpy, or (2) an increase in entropy, and such a spontaneous change is described by the following equation, where ΔG will have a negative value:

$$\Delta G = \Delta H - T\Delta S$$

The free energy of a surface is the available energy resulting from surface groups, molecules, or atoms which are able to interact with other groups, molecules, or atoms which may approach the surface. The possible types of interaction are: (1) van der Waals dispersion interactions, i.e., weak charge interactions resulting from the fluctuating distribution and temporary spatial concentration of electrons in molecules and atoms, (2) electrostatic interactions between charged groups, (3) polar interactions between groups with permanent or induced dipoles, and (4) chemical bonding, including ionic, covalent, and hydrogen bonding. The molecules within the bulk of the substratum can interact with surrounding substratum molecules, whereas those at the surface can interact only with those beneath them in the bulk phase. Thus, free energy at the surface will be greater than that of the interior, and the amount of surface free energy is characteristic of the particular substratum material.

The significance of surface free energy for bacterial attachment lies in the tendency for the free energy within the entire system to be reduced by (1) a reduction in free energy at the substratum-liquid boundary (interfacial energy), which is a function of the surface free energies of both the solid and the liquid; or (2) an increase in the entropy or randomness within the system. In the case of (1), this could be achieved through the adsorption of a bacterial surface polymer on the substratum by means of any of the types of interaction mentioned above; it is described by the equation

$$\Delta G = \gamma_{SB} - \gamma_{SL} - \gamma_{BL}$$

where γ_{SB}, γ_{SL}, and γ_{BL} are the interfacial energies of the surface-bacterium, surface-liquid, and bacterium-liquid, respectively. Thus, when ΔG is negative, adhesion is favored. An increase in entropy (2) could occur if a highly structured layer of adsorbed water at a surface was disrupted or displaced through the adsorption of a bacterial surface polymer, thereby facilitating adhesion.

Thus, the surface free energy of a solid surface is an important indicator of the suitability of that surface for bacterial attachment, as surface free energy indicates the tendency for that substratum to enter into various types of interactions spontaneously. Moreover, it is a measure of the degree to which water can be adsorbed on the surface, as water is able to enter into van der Waals, electrostatic, and polar interactions, as well as hydrogen bonding. Hence, higher-energy surfaces ($\gamma_{SV} > 73$ mN/m, the surface tension of water) tend to be hydrophilic, whereas lower energy surfaces tend to be hydrophobic. Accordingly, the hydrophilicity of a potential substratum will affect the adsorption of bacterial surface adhesives (hence attachment) when the displacement of adsorbed water molecules is a prerequisite for adsorption.

However, when analyzing bacterial attachment phenomena in terms of surface free energies, there are three important points which must be borne in mind. First, the thermodynamic interpretation of bacterial attachment, in terms of a decrease in free energy being the "driving force," is relevant only to spontaneous attachment mechanisms resulting from adsorption of cell surface polymers or structures on the substratum; it is not directly applicable to attachment mechanisms utilizing metabolic energy. Second, the system, constant in total energy and material content, must be considered as a whole. Bacterial adsorption may be favored not only by a decrease in free energy at the substratum, but also, for example, by an increase in entropy within the aqueous phase. Free energy changes in the bacterial envelope may also play a role. Finally, when considering the probability of interaction between bacterial and substratum surfaces, it is not sufficient to consider only their surface free energies. It must also be determined whether the interactive groups on the two surfaces are complementary. For example, if a significant component of the surface free energy of a substratum is due to polar groups, this will have little relationship to bacterial attachment if the bacterial surface has few polar groups. For interaction, the two surfaces must exhibit the same types of forces.

A type of interaction which may be particularly important in attachment is electrostatic attraction or repulsion, involving positively or negatively charged groups on bacterial and substratum surfaces. Where the two surfaces bear ionogenic groups of opposite charge, then binding between the surfaces may occur through acid-base type reactions. However, charged groups also give rise to long-range forces, which may be attractive or repulsive depending upon whether the surfaces bear opposite or like net surface charges, respectively.

Since most bacteria bear a net negative surface charge (Harden and Harris, 1953) and there are few positively charged surfaces, the long-range forces between them will tend to be repulsive. The magnitude of the repulsion will depend on the size of the individual net surface charges and will increase with a decrease in distance between the two surfaces. Nevertheless, adsorption can still occur with two like-charged surfaces when the long-range repulsion forces are balanced by van der Waals attraction forces. In this situation, there are two distances of separation between the surfaces at which there is net attraction. These are called the primary and secondary minima (Fig. 1), and they are separated by an area of separation where repulsive forces may be high and impossible to overcome. If a particle (or bacterial) surface cannot get past the repulsion barrier and reach the area of the primary minimum, then weak adsorption can still be possible at the secondary minimum. Such weak adsorption may be the basis for reversible bacterial attachment where cells exhibit Brownian motion and are easily removed by washing (Marshall *et al.*, 1971a).

Repulsion forces depend not only upon the net charge on the two surfaces and the distance between them, but also on the radii of curvature of the surfaces and on electrolyte concentration. Repulsion forces decrease with a decrease in radius of curvature of the approaching particles, so that repulsion should be less with smaller cells or with projections such as fimbriae or prosthecae. However, there are little data on the significance of projections, etc., in natural attachment systems (Curtis, 1979). Probably a significant factor is the influence of electrolyte concentrations, as adsorption of cations and anions counterbalances the net charge of the surface so that repulsion forces between surfaces are reduced (Fig. 2). Thus, electrostatic repulsion may be more significant in freshwater than in seawater. The role of long range forces in particle adsorption and bacterial attachment are treated in detail in a number of papers (cf. Rutter and Vincent, 1980; Tadros, 1980), and only a brief outline of their potential significance has been given here.

Recent evidence suggests that a type of interaction which may be important in some bacterial attachment processes is hydrophobic bonding (Marshall and Cruickshank, 1973; Fletcher and Loeb, 1979; Dahlbäck *et al.*, 1981).

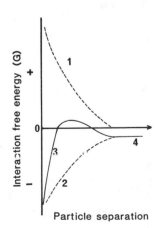

Figure 1. Interaction energy curve obtained by summation of repulsion curve (1) and attraction curve (2). Primary (3) and secondary (4) minima are shown.

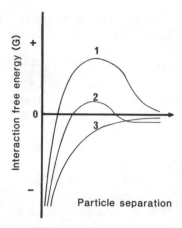

Figure 2. Interaction free energy (G) vs separation as a function of electrolyte concentration at low (1), intermediate (2), and high (3) electrolyte concentration. Adapted from Rutter and Vincent (1980).

Hydrophobic interactions between two nonionic and nonpolar surfaces occur in aqueous solutions. The attraction between the surfaces is weak and due primarily to van der Waals interactions, but because of the strong interaction between the water molecules, the hydrophobic surfaces are excluded from the aqueous phase. In laboratory experiments, many freshwater and marine bacteria (as well as algae and bryozoa; Loeb, 1977) have been found to attach preferentially to hydrophobic surfaces, and only a few relatively nonadhesive strains have attached in slightly higher numbers to hydrophilic surfaces (M. Fletcher, J. J. Bright, and H. P. Pringle, unpublished data). Probably a general preference for hydrophobic surfaces should be expected, as it is difficult to envisage a nonspecific attachment mechanism in which adsorption to a hydrophilic surface is favored over suspension in an aqueous medium.

2.1.2. Methods for Evaluating Surface Free Energy

Contact angle measurements are the most commonly used method of determining surface free energy of solids, and the relationship between a contact angle (θ) of a liquid on a solid and the solid-liquid (γ_{SL}), solid-vapor (γ_{SV}), and liquid-vapor (γ_{LV}) interfacial energies is described by Young's equation:

$$\gamma_{SV} = \gamma_{SL} + \gamma_{LV} \cos \theta$$

One technique for evaluating surface energy is to measure the contact angles of a series of liquids, comprising a range of surface tensions, on a given solid. By plotting $\cos \theta$ against the surface tensions, a straight line is obtained, and its intercept with $\cos \theta = 1$ gives a value called the critical surface tension (γ_C), which is the indirect measure of surface free energy (Zisman, 1964). There are certain drawbacks and inconsistencies with the method, however, as (1) the liquids have to be carefully selected because nonlinear values for surface free energies are obtained if polar liquids are used, and (2) critical surface tensions of liquids are sometimes found to be lower than their measured surface tensions (Wu, 1980). It is also a time-consuming and exacting method, which may make it impractical for a microbiologist investigating bacterial adhesion. Table I gives γ_C values for a number of surfaces, as well as values for surface energy (γ_S), the sum of the dispersion component (γ_S^d) and the dipole-dipole and hydrogen bonding component (γ_S^h). Values for γ_S^d and γ_S^h were determined from contact angles of liquids whose dispersion and polar components were known (Owens and Wendt, 1969).

Another method of evaluating surface free energies is the equation of state approach (Neumann et al., 1974), where an empirically derived equation can be used to solve for the surface free energy of a solid. The only measurable values used in the equation are the contact angle of a liquid, which does not chemically interact with the solid, and the surface tension of the liquid. Although the validity of this approach has not yet been rigorously tested, it offers a possible means of determining surface free energies which could be practically applied by the ordinary experimental biologist, as only one contact angle measurement is required. When critical surface tension values are compared with those derived from the equation of state (Wu, 1980), the latter are usually higher and consistent with figures obtained from liquid homologues (Table II).

When the worker is interested only in the properties of the substratum which are relevant to bacterial attachment, it may not be necessary to determine surface free energies. As water is always the principal liquid phase in attachment processes, then the contact angle of water on various surfaces may give a good indication of their suitability as attachment substrata. Fletcher and Loeb (1979) found a direct correlation between water contact angles on a

Table I. Components of Surface Energy for Various Solids[a]

Surface	m (N/m)			
	γ_S^d	γ_S^h	γ_S	γ_C
Poly(ethylene)	32.0	1.1	33.1	31
Poly(tetrafluoroethylene)	12.5	1.5	14.0	18.5
Poly(vinylidene fluoride)	23.2	7.1	30.3	25
Nylon 6-6	34.1	9.1	43.2	46
Polystyrene	41.4	0.6	42.0	43
Paraffin	25.4	0.0	25.4	23

[a]From Owens and Wendt, 1969.

Table II. The Surface Energies of Solids Determined by the Equation of State Approach and as Critical Surface Tensions $(\gamma_C)^a$

Surface	γ_S (equation of state) $(m, N/m)$	γ_C $(m, N/m)$
Poly(ethylene)	35.9	31
Poly(tetrafluoroethylene)	22.6	18.5
Paraffin wax	32.0	15–22

[a]From Wu (1980).

range of surfaces and the numbers of bacteria which became attached to them. Another approach is to measure the contact angle of a bubble of air against the surface immersed in water, as this is analogous to a bacterium encountering a substratum in a liquid medium (Fletcher and Marshall, 1982; see also Section 3.2.2).

However, one cannot reliably predict surface–surface interactions on the basis of the surface free energy of only one of the interacting surfaces. The surface energies, and the contributing forces and their potential for interaction, of *both* surfaces are important. Thus, a number of attempts have been made to measure the surface free energies of bacteria. One method has been to collect bacteria on a filter and then make contact angle measurements on the lawn of cells. There is still some doubt about the reliability of this method, as some drying of the bacteria is unavoidable, and this would be likely to denature some surface components. However, a value for neutrophil surface energy obtained from contact angle measurement (69.1 mN/m) has since been supported by a value (69.0 mN/m) derived from experiments on the attachment of neutrophils to surfaces of different free energies in media of different surface tensions (Neumann *et al.*, 1979). Certainly, these methods should be tested with a variety of organisms to determine their overall applicability.

Another approach in evaluating cell surface energies is to measure the affinity of the bacteria for different solvents by phase partition chromatography (Gerson, 1980). Biphasic mixtures of dextran and poly(ethylene glycol) have been used to directly measure cell surface hydrophobicity of lymphocytes (Gerson, 1980) and a range of bacteria (Gerson and Akit, 1980; Gerson and Scheer, 1980). Alternatively, the degree of hydrophobicity of bacteria has been evaluated using hydrophobic interaction chromatography (Dahlbäck *et al.*, 1981) or bacterial surface binding of radiolabeled dodecanoic acid (Kjelleberg *et al.*, 1980).

2.2. Physical Factors

Certainly physicochemical factors, such as surface charge or hydrophobicity, will affect the adsorption of bacterial adhesive polymers or structures onto the attachment substratum. In natural environments, however, physical factors, particularly surface texture and hydrodynamic characteristics, may be of considerable importance.

2.2.1. Surface Roughness

Undulations and projections on surfaces can affect electrostatic interactions with bacteria due to their effect on repulsion forces. As mentioned in Section 2.1.1., repulsion forces decrease with a decrease in radii of curvature of the approaching surfaces, so that it may be easier for a bacterium to come into contact with a surface projection than with a comparatively flat plane. However, there is no evidence to suggest that this is a significant factor in bacterial attachment systems. Indeed, the opposite is indicated, as bacteria tend to be found more in crevices than on ridges (Weise and Rheinheimer, 1978), which suggests an overriding importance of hydrodynamic influences.

2.2.2. Hydrodynamic Factors

In fluid systems, there are two types of flow: (1) laminar, i.e., smooth flow without lateral mixing movements, and (2) turbulent flow, where movement at any point within the flow may be erratic and irregular. In natural aquatic environments, however, laminar flow is rare, and most water movements are turbulent. The type of flow has little direct significance to a bacterium attached to a substratum, as at any surface (a substratum or bacterial surface), there is a boundary layer or viscous sublayer, where flow is considerably and increasingly reduced as the surface is neared, so that at the surface itself there is no water flow. Thus, even in the most turbulent waters, the microenvironment of the attached bacterium will be comparatively still. Similarly, a suspended bacterium will be surrounded by a boundary layer. Turbulence, however, will *indirectly* affect bacteria by increasing transfer of the bacteria themselves, nutrients, heat, and momentum between adjacent masses of water.

In situations where flow is laminar, e.g., narrow pipelines, the rate of flow will affect both the rates of deposition of organisms on the surface and of delivery of nutrients contained in the water. Laboratory studies on bacterial attachment in a flowing system showed that, with an increase in flow rate, there was an increase in attachment of slowly grown cells, but a decrease with bacteria grown at faster rates (dilution rates were 0.04/hr and 0.2 or 0.5/hr, respectively). In corresponding experiments, with polystyrene latex spheres, attachment increased with increase in flow rate (Leech and Hefford, 1980; Rutter and Leech, 1980). Thus, attachment of fast growing bacteria may be principally controlled by physiological processes, whereas physicochemical adsorption is the dominant factor with slow growing cells (as with latex particles).

In turbulent systems, such as most natural waters, the attachment of bacteria and the subsequent buildup of a bacterial slime layer, or biofilm, increases with an increase in turbulence. This could be due to (1) preferential attachment by certain species, (2) a microbial physiological response to environmental stress, or (3) the squeezing of loosely bound water from the film by the fluid pressure force (Characklis, 1981b; see also Section 7.1). Turbulence, as well as shear forces associated with laminar flow, will also tend to remove bacteria, particularly those at the surface of slime layers or filamentous forms projecting from these layers, and biofilms are usually sloughed when they reach a certain thickness, generally less than 1000 μm (Characklis, 1981b; see also Section 7.1).

Surface roughness can affect the degree of turbulence near the surface, depending upon the degree of roughness, the width of the flow channel (e.g., a narrow pipeline, wide channel) and the velocity of flow. However, the main significance of roughness for attached bacteria lies in its (1) increasing convective mass transport (e.g., bacteria, nutrients) near the surfaces, (2) increasing the surface area for attachment, and (3) providing bacteria with shelter from shear forces or abrasion (Characklis, 1981b). For example, as sand particles move against one another in turbulent waters, bacteria on the smooth high relief surfaces may be removed by scouring, whereas those in crevices are protected. Scanning electron micrographs of sand grains support this suggestion (Weise and Rheinheimer, 1978).

2.3. Biological Factors

Aquatic plants and animals frequently have attached bacteria on their surfaces, and in many cases, these epiphytic and epizoic bacteria may be able to utilize products secreted by the host. However, some organisms, or parts of organisms, are characteristically free of attached bacteria. For example, scanning electron micrographs of a hydroid showed it to be relatively free from attached bacteria, except for the stalk, which was heavily colonized (Sieburth, 1975). Diatoms are frequently free of attached bacteria until they become senescent or die, while freshwater spermatophytes, e.g., *Rorippa* and *Lemna,* may support a sizeable microbial population of approximately 10^6 bacteria per cm^2 surface area (Hossell and Baker, 1979).

In most cases, it is not known why certain portions of plants and animals remain bacteria-free, but bacterial colonization may be prevented by chemical features of the plant or animal surface, e.g., by a low pH at an algal surface (Sieburth, 1968) or the production of inhibitors by the potential host. For example, a study of the brown alga *Ascophyllum nodosum* (Cundell *et al.,* 1977) showed diverse and abundant colonization by bacteria (short and long rods, *Leucothrix mucor* and flexibacterlike rods), pennate diatoms, and cyanobacteria in areas representing three or four years growth, whereas the apical tips were free of attached microorganisms. Colonization of the new growth was probably prevented to some extent by the secretion of tannins (which can also produce a negative bacterial chemotaxis response; Chet *et al.,* 1975), as well as by desiccation during the tidal cycle. Rapidly growing tissue will also tend to be relatively free from attached bacteria, as they have had insufficient time to colonize young tissue.

Both plants (e.g., large-filamentous algae such as *Spirogyra, Zygonema;* Moss, 1980) and animals (e.g., fish) may sometimes be protected from firm bacterial attachment by the production of surface mucilages which are continually sloughed off the potential host. On the other hand, large numbers of microorganisms are sometimes found within the surface mucilages of some algae or cyanobacteria and are presumably utilizing algal products (Rheinheimer, 1980).

The possibility of lectinlike attachment mechanisms, as found in the *Rhizobium*-legume symbiosis (Dazzo, 1980) has been little explored. To our knowledge, only one such specific attachment mechanism has been found in aquatic environments. This involves the attachment of several types of bacteria to the heterocysts, but not to the vegetative cells, of the cyanobacterium *Anabaena* (Paerl, 1976; Lupton and Marshall, 1981).

Bacteria have been observed on the surfaces of a wide range of animals, including copepods, other crustaceae, worms, and fish (Rheinheimer, 1980). The extent of colonization may be influenced somewhat by exposure of the animal parts to abrasion. For example, the upper portions of exposed exoskeletal surface structures and regularly abraded mouth parts of the marine isopod *Limnoria* were found to be bacteria-free, whereas protected areas between the surface structures were heavily colonized (Boyle and Mitchell, 1980).

The variation in extent of microbial colonization of plants and animals is a clear indication of the complexity of their surface characteristics. Not only do physicochemical and physical factors affect the attachment process, but also the physiological properties of the potential host will have a considerable influence.

3. Conditioning Films

Although the properties of a particular substratum may be defined in terms of surface charge and surface free energy with some degree of precision, such properties may be modified following immersion of the substratum into an aqueous habitat. Spontaneous adsorption of macromolecules at the solid-liquid interface results in the formation of a surface conditioning film. A heterogeneous array of macromolecules, including proteins, glycoproteins, proteoglycans, and polysaccharides, are present in low concentrations in natural waters. These macromolecules may originate as excretory products from living organisms, or as autolytic and decompositional products from dead procaryotes or eucaryotes. Adsorption of these macromolecules at surfaces results in a concentration of these materials at the solid-liquid interface and leads to changes in the apparent charge and free energy characteristics of the surface.

It has been suggested by Baier (1981) that the deviation from the physicochemical predictions of nonbiological adhesion observed in biological systems results from the masking of the solid surface by spontaneously adsorbed macromolecules, particularly by glycoproteins and proteoglycans. Some proteins actually inhibit attachment of bacteria to a variety of substrata (Fletcher, 1976; Ørstavik, 1977). In this section, we shall endeavor to establish the importance of conditioning films in terms of both bacterial adhesion and the potential of the films as nutrient sources.

3.1. Difficulties in Evaluating Adsorbed Films

Surface conditioning films probably range in thickness from about 10 to 20 nm (Baier, 1981). As a consequence, it is difficult to apply routine analytical methods to the qualitative and quantitative determination of macromolecular species adsorbed at solid-liquid interfaces. Alterations to the surface charge of small particulates can be satisfactorily measured as changes in electrophoretic mobilities using a suitable microelectrophoresis apparatus (Santoro and Stotzky, 1967; Neihof and Loeb, 1972,1974). The electrophoretic mobility of particulates in the presence of dilute solutions of macromolecules tends towards the mobility of the macromolecules. Measurements of changes in electrophoretic mobilities have the advantage that they provide a direct measure of modifications to the surface charge of particulates suspended in the aqueous phase.

A variety of techniques have been developed to obtain information on the physicochemical properties of dried conditioning films (Baier and Loeb, 1971; Baier, 1980). By using infrared (IR) transmitting, reflective materials (e.g., germanium prisms) as the solid substratum, it is possible to obtain multiple attenuated internal reflection IR spectra of the conditioning films. Other nondestructive tests that can be carried out on the dried conditioning films include optical thickness by ellipsometric techniques, contact potential from vibrating and ionizing electrode studies, crystal structure from X-ray diffraction patterns, scanning electron microscopy, and critical surface tensions from contact angle measurements (see Section 2.1.2).

It is questionable, however, whether surface free energy data obtained on dried conditioning films provide a true indication of events occurring when the substratum is actually immersed in the aqueous phase. Although a degree of denaturation of macromolecules occurs on immersed substrata, even greater denaturation and, hence, configurational change, will occur on drying. The dried films may not bear any relationship to the conditioned substratum actually encountered by a bacterium when approaching from the bulk aqueous phase. In an attempt to obtain more meaningful information in relation to bacterial adhesion, Fletcher and Marshall (1982) used the contact angles made by bubbles at the substratum surface as a measure of changes in surface free energy resulting from conditioning film formation. This technique has the advantage that the aqueous phase is still present and, in the presence of macromolecules, the behavior of the bubble may reflect that of bacteria approaching the conditioned surface. A difficulty in the technique is that the bubble angles obtained in the presence of macromolecules are determined partly by changes in surface free energy and partly by changes in the surface tension of the aqueous phase. Bubble contact angles did reflect the surface free energies of hydrophilic and hydrophobic polystyrene substrata and, in the presence of dilute protein solutions, gave changes in contact angles that reflected differences in the adhesion of *Pseudomonas* NCMB 2021.

3.2. Modification to Surface Charge and Surface Free Energy

Adsorption of macromolecules to an otherwise clean surface results in a new outer surface that is characterized by the properties of the external molecules of the adsorbed species. Norde and Lyklema (1978) presented a model of a protein adsorbed at a solid-liquid interface with three distinct regions in the adsorbed protein (Fig. 3). Region 1 contains positive groups of the protein that are ion-paired to the negative groups at the solid surface, region 2 is assumed to be devoid of charged groups (by analogy with the interior of globular protein molecules), and region 3 contains charged groups exposed to the aqueous phase. According to Norde and Lyklema, the affinity of a protein for a given negatively charged surface increases with (1) increasing hydrophobicity of the protein, (2) the extent of structural rearrangement of the protein, and (3) the number of positively charged groups which become located at the surface. Similarly, they indicate that the affinity of a negatively charged surface for a given protein increases with (1) increasing hydrophobicity of the surface, (2) a lowering of surface charge, and (3) screening of the surface charge by specifically adsorbed cations.

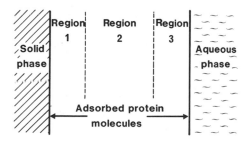

Figure 3. Model of a protein layer adsorbed at a solid-liquid interface showing the three regions described in the text. Adapted from Norde and Lyklema (1978).

3.2.1. Surface Charge

From the model in Fig. 3, it is obvious that the solid surface will assume a net surface charge characteristic of the outermost portion (region 3) of the protein. Norde and Lyklema (1978) note that this outer portion of region 3 extends to the hydrodynamic shear plane, that is, the outermost region of the surface plus adsorbed materials that move in an electrophoretic field. Neihof and Loeb (1972,1974) have elegantly demonstrated that the surface charge of particulates immersed in seawater is altered by adsorbed organic films. Using negatively, zero, and positively charged particulates, the authors noted a convergence of electrophoretic mobilities towards low negative values in natural seawater, but not in organic-free artificial seawater. Destruction of the organics in natural seawater by UV irradiation resulted in particulate mobilities comparable to those found in artificial seawater. Adsorbed organics led to a significant modification to the apparent surface charge of the particulates, and a similar modification can be expected with all substrata immersed into natural waters. Adsorption of organic and colloidal materials to the surfaces of soil bacteria also resulted in significant alterations to their electrophoretic mobilities (Lahav, 1962; Santoro and Stotzky, 1967; Marshall, 1968).

3.2.2. Surface Free Energy

By analogy with changes in surface charge, an immersed substratum should assume the surface free energy properties of the groups exposed at the outermost portion of region 3 of the conditioning film. Baier (1980,1981) has reviewed his extensive studies on the nature of conditioning films detected on surfaces immersed in a variety of diverse biological systems. Using substrata of similar texture (surface roughness) but of different surface free energy, Baier (1980) found spontaneous adsorption of proteinaceous material from calf serum on all surfaces, and a continuing buildup of this material in the same relative amounts and at the same rates. Similar observations were made using bacterial culture systems.

Based on analyses of dried films, Baier (1980,1981) has concluded that, although all surfaces spontaneously acquire adsorbed films dominated by glycoproteins soon after immersion, different strengths of adhesion are evident. Films acquired on high-surface-energy, high-polarity substrata are flattened (denatured) and adhere very strongly, whereas those acquired on low surface energy, nonpolar surfaces are more weakly bonded, but occasionally thicker. These low energy surfaces, dominated by closely packed methyl groups, bind macromolecules more loosely in more native, solutionlike configurations, even though the films are composed of essentially the same protein-dominated material. However, even lower energy surfaces dominated by fluorocarbon groups apparently denature and bind natural macromolecules as strongly as some higher energy surfaces.

Baier (1981) suggests that low-energy surfaces are more resistant to fouling accumulation or persistence and are easier to clean because of weaker binding at the solid-liquid interface. Differences observed in the quality of adhesion reveal that some features of the original substrata are still propagated through the conditioning films, probably resulting from differences in the configuration of each adsorbed film.

Resistance to fouling by low-energy substrata is consistent with the results of marine microbial fouling reported by Dexter et al. (1975) and Dexter (1979), who found maximum fouling on high-energy surfaces and minimal fouling on low-energy surfaces. However, the picture is not as clear as these results may indicate, since Fletcher and Loeb (1979) found that the marine *Pseudomonas* NCMB 2021 attached preferentially to low-energy surfaces. In fact, these authors reported a positive correlation between attached bacteria and increasing water contact angle (increased hydrophobicity) of the substrata. Fletcher (1980) has described a series of bacteria with a wide variety of preferences for high- and low-energy substrata. Obviously, a great deal remains to be learnt about conditioning films and their relationship to bacterial adhesion. Doubts must be raised as to the validity of studies on air-dried substrata, as the degree of denaturation, overall thickness, and molecular configuration must be affected by the drying process. Information on the nature of conditioning films in the presence of the aqueous phase is vital to understanding the role of such films in adhesion processes.

Using bubble contact angles in the aqueous phase, Fletcher and Marshall (1982) found consistent differences between angles observed on hydrophobic (PD) and hydrophilic (TCD) polystyrene substrata, as well as considerable differences between the conditioning effects of different proteins on the test substrata (Table III). As predicted by Norde and Lyklema (1978) (see Section 3.2), the somewhat hydrophobic, fatty-acid-free, bovine serum albumin appeared to form the most complete conditioning films on both substrata, whereas protamine sulphate did not alter the surface properties. Addition of a protease enzyme (pronase) following conditioning of substrata also revealed differences in protein behavior on the two substrata. No change in contact angle was observed on TCD, but the contact angles on PD changed from 64° to 84° for bovine glycoprotein and from <15° to 40° for fatty-acid-free bovine serum albumin. An increase in contact angle indicates a decrease in wettability. Such results suggest differences in the configurations of the proteins on each substratum resulting in different susceptibilities to protease digestion.

A culture supernatant of *Pseudomonas* NCMB 2021 gave an apparent increase in wettability of the TCD substratum (<15°), but no change in that of the PD substratum (90°). Similar results were obtained with extracted

Table III. Bubble Contact Angles Obtained Using Various Proteins as Conditioning Agents on the Polystyrene Tissue-Culture Dish (TCD) and Petri Dish (PD) Substrata.[a]

	Contact angle (deg) on	
Protein[b]	TCD	PD
No protein	29	90
Bovine glycoprotein	<15	64
Fatty acid free bovine serum albumin	<15	<15
Protamine sulphate	28	89

[a]A decrease in contact angle indicates an increase in wettability. Adapted from Fletcher and Marshall (1982).
[b]Conc. of 100 μg/ml.

Pseudomonas extracellular polymer, suggesting that conditioning film formation on substrata immersed in the culture filtrate may result from this polymer in solution.

3.3. Possible Nutritional Implications of Conditioning Films

The accumulation of macromolecules at solid surfaces suggests the possibility that these organic materials serve as carbon or energy sources, and in the case of proteins as nitrogen sources, for bacteria in the vicinity of the solid-liquid interface. Although the amounts of macromolecules on surfaces increase with time (Baier, 1980), bacterial colonization of the surface may result in utilization of the macromolecules almost as rapidly as they adsorb at the surface. Macromolecular utilization at surfaces has not been studied in any detail, but it almost certainly depends on the nutritional characteristics of the bacteria found at surfaces. This aspect will be considered in Section 5.

4. Adhesion Mechanisms

Most aquatic bacteria appear to attach to surfaces by means of extracellular surface polymers which act as adhesives (Costerton *et al.*, 1978). In some cases, more specialized structures, such as appendages, prosthecae, or blebs (Hirsch and Pankratz, 1970; Pertsovskaja *et al.*, 1972; Corpe *et al.*, 1976; Ellen *et al.*, 1978) are involved. For example, caulobacters attach by means of a substance produced at a special holdfast region at the end of the prosthecate stalk (Corpe *et al.*, 1976). Also, fimbriae are able to attach to a range of surfaces, but evidence of their significance to attachment in natural aquatic environments is scarce. Thus, this review will concentrate on attachment mechanisms utilizing cell-surface polymeric coats or adhesives.

4.1. Reversible Attraction

Frequently, when observed under a microscope, bacteria appear to be attached, but are, in fact, easily removed by washing, or, if motile, they may eventually swim away from the surface. Such weakly attached bacteria may exhibit Brownian motion, but the principal criterion for reversible attachment is that the bacteria are easily removed by washing (Marshall *et al.*, 1971a).

4.2. Irreversible Attraction

Reversible attachment may then be followed by irreversible attachment, when the bacterium-surface binding resists strong washing, although it is usually labile to more severe treatments, e.g., detergents, ultrasonication. This irreversible attachment is dependent upon the physicochemical adsorption of bacterial cell surface components onto the substratum surface, and there will be a concomitant reduction in free energy within the system (see Section 2.1.1).

Electron microscopy has demonstrated bacterial surface polymers which appear to be acting as adhesives (Fletcher and Floodgate, 1973; Marshall and Cruickshank, 1973; Costerton *et al.*, 1978), but this tells us little about the physicochemistry of the interaction or which cell-surface components elicit the adhesion process. A number of studies have suggested that cell-surface polysaccharides are the bacterial adhesives (Corpe, 1973; Fletcher and Floodgate, 1973; Costerton *et al.*, 1978). However, as some proteases are able to remove attached bacteria (Dannielsson *et al.*, 1977; Fletcher and Marshall, 1982), proteins are probably also involved.

The situation is complicated by an apparently wide range of bacterial surface polymers, which may alter in quantity or composition with growth or environmental conditions. For example, the bacterial adhesive may already be present on the cell surface when it encounters the substratum (Fletcher and Floodgate, 1973) or it may be produced during reversible attachment, so that it bridges the gap between the bacterial and substratum surfaces (ZoBell, 1943; Marshall *et al.*, 1971a). The firm attachment of oral streptococci is caused by the synthesis of glucans *in situ* (Clark and Gibbons, 1977), and similar attachment strengthening mechanisms could occur with slime-forming aquatic bacteria. Bacterial growth conditions which can affect adhesive production or efficiency are nutrient concentration (Marshall *et al.*, 1971a), dilution rate (Ellwood *et al.*, 1974; Rutter and Leech, 1980), whether the limiting nutrient is carbon or nitrogen (Brown *et al.*, 1977; Wardell *et al.*, 1980), temperature (Fletcher, 1977), cation concentration (Marshall *et al.*, 1971a,b;

Fletcher and Floodgate, 1976), and pH (Fletcher and Floodgate, 1973; Wood, 1980).

The three main types of bacterial cell surface polymer which are most likely to enter into adhesion interactions are (1) lipopolysaccharide, in gram-negative bacteria, (2) peptidoglycan, in gram-positive organisms, and (3) extracellular polymers and capsules, which occur on both types of cells. Lipopolysaccharide (LPS) is one of the major components of the outer membrane of gram-negative bacteria, and it comprises an "O" antigen polysaccharide side chain, an oligosaccharide core, and a glucosamine-containing lipid. Bacteria vary considerably in the amount of LPS produced. For example, *Escherichia coli* and *Salmonella typhimurium* may have LPS extending out to 150 nm from the cell (Shands, 1966), whereas rough mutants have reduced "O" side chains or oligosaccharide cores (Ward and Berkeley, 1980). Lipoproteins and proteins are also usually associated with LPS. The principal polymers occurring at the surface of gram-positive bacteria are peptidoglycan, containing glycan chains with peptide substituents, and teichoic and teichuronic acids (phosphate-containing wall polymers), and protein antigens may also be present. Capsules may be homopolysaccharides, e.g., glucans, but are usually heteropolysaccharides, which often contain uronic acids and/or pyruval ketal groups which confer an overall negative charge. In a survey of extracellular polysaccharides produced by freshwater and marine bacteria (Sutherland, 1980), no unusual components were found, and the principal constituents were usually D-mannose, D-glucose, D-galactose, and uronic acids (usually D-galacturonic).

Because of the presence of polysaccharides and proteins at the bacterial surface, it is equipped for a range of polar and electrostatic interactions, as well as hydrogen bonding. Moreover, molecules which have both polar and nonpolar regions (amphipathic molecules) may be involved in hydrophobic interactions, and these include lipoteichoic acids (Rolla, 1980) in gram-positive bacteria, and lipopolysaccharides (Wood, 1980) and lipoproteins in gram-negative bacteria.

Some attachment mechanisms may also depend on cations, which may bridge or screen negative charges on the bacterial and substratum surfaces, and thus reduce repulsion forces. For example, the addition of cations has been found to promote the attachment of marine pseudomonads (Marshall *et al.,* 1971a) and possibly oral streptococci (Olsson *et al.,* 1976).

4.3. Strength of Adhesion

Most studies of bacterial attachment have treated adhesion as an all-or-none phenomenon. This is, of course, misleading, as attachment is certain to vary in the strength of the adhesive bond. For example, the difference between reversible and irreversible attachment may be one of degree and depend upon the strength of adhesion of higher organisms, such as algal spores (Christie *et al.,* 1970) or tissue cells (Weiss, 1961a,b) by applying some kind of disruptive force, such as shear flow (Weiss, 1961a) or centrifugation (Weiss, 1961a,b). Measurement of adhesive strength has been more difficult with bacteria, however, because of their comparatively small size and the tenacious binding of irreversibly attached cells. Nevertheless, an attempt to measure bacterial adhesive strengths was made using centrifugal force to rupture the adhesive bond, and adhesive strengths of 4×10^{-12} to 4×10^{-9} N/cell were calculated (Zvyagintsev *et al.,* 1971). The danger in using such disruptive techniques, as pointed out by Weiss (1961b) and recognized by Zvyagintsev *et al.* (1971), is that the actual rupture may not be in the adhesive bond itself but in the bacterial or (less likely) substratum material. Thus, it is cohesion, rather than adhesion, which is being measured. Nevertheless, such methods are valuable in that they provide a minimum of adhesive strength.

Occasionally, what were once apparently irreversibly attached bacteria may subsequently become detached from substrata. This is particularly common with slime layers of bacteria, where sloughing of cells occurs after a period of buildup. The detachment, or desorption, of bacteria from monolayers may also occur, but the reasons for desorption are poorly understood. It may be associated with low physiological activity (J. J. Bright, unpublished data), a change or deterioration in surface adhesive polymers (Zvyagintsev *et al.,* 1977), or, in natural environments, by grazing and physical abrasion (see Section 2.2.2).

5. Bacterial Nutrition Relative to the Trophic Status of Waters

In considering the ecological significance of surfaces to aquatic bacteria, it is necessary to look closely at the nutritional groups of bacteria, especially since most aquatic habitats are nutritionally deficient (oligotrophic). Two major groups of bacteria are recognized in regard to low-nutrient conditions. These are the oligotrophs, capable of growth in low-nutrient environments, and the copiotrophs (Poindexter, 1981a,b), requiring relatively high levels of nutrients for growth. Both groups of bacteria are found in varying proportions in low nutrient waters, and it is obvious that each group must exhibit different strategies of survival.

5.1. Oligotrophic Bacteria

5.1.1. Properties of Oligotrophs

The concept of oligotrophy in bacteria has recently been reviewed by Kuznetzov *et al.* (1979), Hirsch (1979), and Poindexter (1981b). Poindexter (1981b) has concluded that oligotrophic bacteria may be defined as those whose survival in nature depends on their ability to multiply in habitats of low nutrient flux (from near zero to a fraction of a mg C/liter/day). This is in contrast to those organisms whose survival depends on the exploitation of a nutrient flux of at least 50-fold higher and does not fall to near zero for prolonged periods.

According to Hirsch (1979) and Poindexter (1981b), the characteristics of an ideal oligotroph are as follows: (1) it is an aerobic organism, (2) it is a small sphere, slender rod, or a prosthecate bacterium, (3) it exhibits a high affinity (low K_m) for the uptake of a variety of utilizable substrates, (4) its transport systems exhibit low substrate specificity and are repressible only in the presence of high substrate levels, (5) transport leads to reserve polymer formation and the enzymes associated with this process are constitutive, (6) its dissimilatory capacity is versatile, (7) its rate of "balanced growth" is low, and (8) regulation of net biosynthesis is effected by uptake rates rather than by intermediates of dissimilatory metabolism.

These proposed properties should maximize the ability of an oligotroph to gather nutrients over extended periods of time and across steep spatial gradients through conservative utilization of nutrients once they are transported into the cell (Poindexter, 1981b). These properties would be of no advantage under conditions of nutrient abundance. Poindexter (1981b) expressed the view that obligate oligotrophs may not exist, since oligotrophic habitats may be transiently enriched with nutrients. Successful oligotrophs must survive such enrichment, but may not be adequately prepared to exploit such a temporary situation. The ability of oligotrophs to grow, albeit slowly, in a nutrient deficient environment provides a competitive advantage over copiotrophs in the aqueous phase.

5.1.2. Oligotrophs and Surfaces

Accumulation at solid-liquid interfaces of macromolecules and nutrient ions constitutes the formation of sites of nutrient concentration in an otherwise nutrient deficient environment (Marshall, 1980). By definition, oligotrophic bacteria do not require such high levels of nutrients and would be expected to lose out in competition from copiotrophs at surfaces, at least initially. In most instances, this is precisely what has been observed in the first 24 hr following immersion of surfaces into marine habitats (Marshall et al., 1971b; Corpe, 1973) as initial colonizers are typical copiotrophs. Adhesion and growth of copiotrophs in the early stages of colonization may lead to a rapid utilization of nutrients accumulating at the interface, thereby creating a nutrient deficient status on the surface that would allow the oligotrophs a competitive advantage. Oligotrophs with a propensity for adhesion would then tend to colonize the surface. Again, this has been observed (Marshall et al., 1971b; Corpe, 1973) in marine habitats, where relatively large numbers of prosthecate, oligotrophic bacteria, such as Caulobacter and Hyphomicrobium (Hirsch, 1979), adhere after 24 or more hours of exposure of surfaces. These prosthecate bacteria possess highly effective holdfast mechanisms for attachment to surfaces (Tyler and Marshall, 1967; Hirsch and Pankratz, 1971; Marshall and Cruickshank, 1973). The subsequent development of extensive microbial films of 30 μm or more in thickness (Characklis, 1981a) indicates a complex system of nutrient entrapment and recycling occurring at such surfaces (see also Section 7.2).

5.2. Copiotrophic Bacteria

5.2.1. Properties of Copiotrophic Bacteria

Poindexter (1981a) introduced the term copiotrophic to describe those bacteria requiring an abundance of nutrients for normal growth; that is, they are dependent on a nutrient supply typically 100 times higher than that found in oligotrophic habitats. The vast majority of cultured bacteria, including those from aquatic sources, are copiotrophs and are incapable of growth under oligotrophic conditions (Hirsch, 1979). The physiological properties of copiotrophs are typified by those of *Escherichia coli* and other well-studied bacteria. In most instances, the properties are the opposite of those proposed for oligotrophs (see Section 5.1.1). Although it is possible to grow *E. coli* at very low growth rates in a chemostat (Koch, 1979), it is unlikely that this bacterium could withstand competition by oligotrophs under natural nutrient-poor conditions or even in mixed-culture chemostats (Jannasch, 1967).

Despite an inability to grow in the aqueous phase of low-nutrient waters, copiotrophic bacteria can be readily isolated from such sources. This suggests that they have evolved strategies for survival in low-nutrient conditions.

5.2.2. Survival Strategies of Copiotrophs in Low-Nutrient Waters

5.2.2.1. Response to Starvation. A number of recent studies on size distributions of bacteria in aquatic habitats have revealed the presence of significant numbers of small bacteria of less than 0.5 μm diameter (Anderson and Heffernan, 1965; Meyer-Reil, 1978; Zimmerman et al., 1978; Tabor et al., 1971; Fuhrman, 1981). Are these bacteria inherently small oligotrophs that are actively growing in low-nutrient waters or are they relatively dormant forms of copiotrophic bacteria? Current knowledge of these small bacteria is insufficient to provide an answer to these questions, but it is likely that both possibilities do exist. Stevenson (1978) has emphasized the state of exogenous dormancy in aquatic bacteria, wherein development is delayed because of the unfavorable physicochemical conditions of the habitat, as opposed to constitutive dormancy, which involves the formation of endospores. Fuhrman (1981), on the other hand, suggests that most of the small bacteria are in a physiologically active state, and this was supported by microautoradiographic evidence of substrate uptake by small marine bacteria (Hoppe, 1976).

Starvation of copiotrophic bacteria constitutes a stress that can influence both the size and activity of the organisms. A psychrophilic marine vibrio formed small cells on starvation (Novitsky and Morita, 1976), and these small bacteria retained a high level of viability in low nutrient habitats (Novitsky and Morita, 1977). In fact, Novitsky and Morita (1978) have suggested that the formation of small cells by the vibrio under oligotrophic conditions is a possible strategy for survival of this copiotrophic organism. The formation of small cells on starvation apparently is a regular feature of marine copiotrophs under oligotrophic conditions (Novitsky and Morita, 1976; Tabor et al., 1981; Dawson et al., 1981; Kjelleberg et al., 1982).

Although small starved bacteria exhibit low endogenous respiration rates, they show an instantaneous response to exogenous energy substrates (Novitsky and Morita, 1978; Kjelleberg et al., 1982). This finding indicates that the small starved forms rapidly respond to the high-level nutrient inputs that occur from time to time in typically oligotrophic habitats (Hirsch, 1979). In addition, this finding throws into doubt the suggestion that most small bacteria are actively metabolizing in nature (Fuhrman, 1981), particularly if this is tested by the addition of exogenous substrate. Once provided with adequate nutrients, small starved bacteria return to their normal size (approx. 1 μm) before cell division occurs.

   5.2.2.2. Adhesion of Small Starved Bacteria. Very small bacteria tend to be the primary colonizers of surfaces immersed in marine habitats (Marshall et al., 1971b), but they appear to be displaced on the surface by normal size bacteria after 12 or more hours. Marshall (1979) suggested that the small primary colonizing bacteria may be small starved cells and that the normal bacteria observed at later stages of the colonization process may represent the active growth form of the early colonizing cells. Subsequently, Dawson et al., (1981) proposed that adhesion of small starved forms to surfaces represents a tactic in the survival strategy of starved copiotrophic forms. These authors demonstrated that the starvation of the marine *Vibrio* DW1 resulted in small-cell formation and that these cells exhibited a greater propensity for adhesion to surfaces than did the unstarved, normal cells. Evidence was obtained by electron microscopy for a significant production of extracellular polymer in 5-hr and 5-day small cells that was not present in normal bacteria. It is feasible that this polymer enhanced the chances of adhesion to surfaces.

   Kjelleberg et al. (1982) have presented evidence for the actual growth of individual small starved cells at nutrient enriched solid-water and air-water interfaces. A small amount of nutrient was supported at an air-water interface by half a monolayer of oleic acid. Bacteria at the interface exhibited a rapid, but transient, increase in cell volume as the nutrient was utilized and rapidly exhausted (Table IV, 2 and 4 hr). Growth and division of small starved *Vibrio* DW1 were observed at a solid-water interface by microscopy using time-lapse videotape recording. In one sequence, the same attached cell was observed for four successive division cycles with times of 57, 63, and 54 min between each division following growth to normal size. Small starved cells did not grow in the nutrient-deficient bulk liquid in any of these experiments (Table IV). The appearance of exceptionally small cells at the air-water interface (Table IV, zero time) and at solid-water interfaces (Kjelleberg *et al.*, unpublished data) appears to be characteristic.

The reappearance of small starved forms at interfaces following growth in extremely nutrient-deficient conditions reinforces the suggestion made earlier (Section 5.1.2) that copiotrophs have a competitive advantage over oligotrophs in the initial colonization of nutrient-rich interfaces, but this advantage is probably less at later stages in the successional sequence.

6. Ecological Advantages of Attachment

There is a considerable range in the types of communities found on solid surfaces in aquatic environments. These include, on one hand, relatively simple communities of oligotrophic and copiotrophic bacteria on particulates in a very low-nutrient waters, as well as thick slime layer communities in eutrophic waters, where bacteria, algae, protozoa, and small invertebrates may reside together in a thick polymeric slime. Sessile communities exposed to the bulk water will also differ from those in the underlying sediments or in "specialist" habitats, such as salt marshes or mud flats. Thus, it is impossible to ascribe any ecological advantages to attached communities in general, as the possible advantages or disadvantages will depend upon individual conditions such as nutrient concentration, community diversity, and characteristics of the surface or particulate material.

6.1. Fast-Flowing Systems

Attached organisms have two quite clear advantages in lotic, or flowing waters. First, they do not have to expend energy searching for food, as water with fresh nutrients flows over them and removes waste products at the same

Table IV. Changes in Peak Cell Volume
(Coulter Counter) of 18-hr Small Starved Cells
at an Air-Water Interface Conditioned with
Nutrients and in the Bulk Aqueous Phase[a]

Time (hr)	Peak cell volume (μm^3)	
	Interface	Bulk aqueous phase
0	0.28	0.36
2	0.41	0.37
4	0.35	0.33

[a]Adapted from Kjelleberg *et al.*, 1982.

time. In nutrient-rich environments, there may be vast accumulations of attached growth, such as the sewage "fungus" *Sphaerotilus natans* (an aerobic bacterium), which is frequently found below sewage outfalls. The second advantage occurs in particularly fast-flowing waters where washout would occur without anchorage to a surface.

6.2. Attachment to Substrates

In some cases, the substratum is also the substrate, so that attachment, or at least close association, is a prerequisite for assimilation. For example, the bacterial utilization of organic detritus usually first requires hydrolytic digestion by extracellular enzymes. Thus, attachment to the substrate would allow bacteria to remain close to the nutrient source, and enzymes would be put to more efficient use if they remained at the substrate after production (perhaps through adsorption or through retention in a bacterial slime matrix), rather than diffusing into the surrounding medium. The production of enzymes is related to attachment to the substrate (Minato and Suto, 1976). Rumen bacteria attached to starch granules showed high specific amylase activity and little or no urease activity, whereas with nonattached bacteria, amylase and urease activities were comparatively low and high, respectively. Similarly, the dissolution of cortisone and hydrocortisone acetate crystals by *Mycobacterium globiforme* was considerably enhanced by direct contact between the bacteria and the crystals, as compared to experiments when the bacteria and steroids were separated by a semipermeable cellophane membrane (Zvyagintseva and Zvyagintsev, 1969). Attachment should also be an advantage to chemolithotrophic bacteria, which obtain energy by oxidizing reduced inorganic compounds, e.g., sulphide, ferrous iron. For the oxidation of elemental sulphur by *Thiobacillus thiooxidans,* the bacteria apparently must first attach to the sulphur by an energy-dependent process involving thiol groups in the cell envelope (Takakuwa *et al.,* 1979).

The attachment of gliding bacteria to solid substrates is achieved by the production of a viscous slime that promotes adhesion whilst allowing translational motion across the surface (Humphrey *et al.,* 1979). This mechanism is particularly useful in the case of *Cytophaga* species, which are capable of decomposing solid substrates such as cellulose, chitin, agar, or alginate in aquatic habitats.

6.3. The Adsorption of Nutrients at Surfaces

The significance of attachment for nutrient utilization by oligotrophic and copiotrophic bacteria has been explored in Sections 5.1.2 and 5.2.2, and there is convincing evidence that nutrient availability to bacteria at surfaces can, in certain situations, differ from that in the bulk liquid phase. For some years, it has been recognized that with experimental bacterial cultures of low-nutrient concentration, the addition of solid surfaces, e.g., beads or tubes, could facilitate bacterial growth. At peptone concentrations of <0.5 mg/liter, *E. coli* could not grow without the addition of glass beads (Heukelekian and Heller, 1940). With studies on marine bacteria, ZoBell (1943) described this effect in more detail, pointing out that the influence of solid surfaces upon bacterial growth depended upon both the concentration and type of organic matter present. No effect was observed with organic concentrations >5–10 mg/liter, and the increase in bacterial activity occurred with some organics, e.g., sodium caseinate, lignoprotein, or emulsified chitin, but not with others, e.g., glucose, glycerol, and lactate. Also, only certain surfaces promoted activity (ZoBell, 1943). It was not clear in these early experiments whether the attachment of the bacteria to the added solid surfaces was a prerequisite for the enhancement in activity or whether the association was superficial. However, more recent studies with river (Jannasch and Pritchard, 1972; Hendricks, 1974; Goulder, 1977) or salt marsh (Harvey and Young, 1980) bacteria have indicated that attached bacteria were more active than their free-living counterparts.

As this "surface effect" is frequently dependent upon nutrient concentration in the medium, it may be due to the adsorption, and hence concentration, of nutrients on the surfaces, so that the substrates are more accessible to attached bacteria. ZoBell (1943) found that with seawater containing glass beads, wool, or tubes to give 2–200 cm^2 glass surface/ml, that 2–27% of the organic material had been adsorbed on the glass. Another possible explanation is that extracellular enzyme activity may be enhanced. The solid surfaces may retard diffusion of exoenzymes and hydrolysates away from attached cells or they may affect enzyme orientation (ZoBell, 1943).

At this stage, it is very difficult to evaluate the effect of nutrient adsorption on accessibility to attached bacteria. Certainly, some processes requiring specific substrates may be facilitated, as the adsorption of ammonium ions on zeolite was shown to promote nitrification by attached bacteria (Sims and Little, 1973). However, the situation with low molecular weight organic compounds or macromolecules is more complex, making it difficult to generalize about surface effects on nutrient assimilation. For example, the comparative uptake of amino acids by free-living marine bacteria and those attached to a variety of plastic or glass surfaces depended on the substratum composition (Fletcher, 1979; Bright and Fletcher, unpublished data), and there was no general "surface effect" for all the substrata.

Glass has frequently been used in studies on the effects of surfaces on activity and has been found to increase bacterial (Heukelekian and Heller, 1940; ZoBell, 1943; Hendricks, 1974) and yeast (Navarro and Durrand, 1977) activity. However, a number of other types of surfaces have also been used for such studies. Clays, in particular, have received much attention because of their significance in soils and sediments. Results obtained with different bacteria or experimental systems have tended to vary so that it is difficult to generalize about the effects of clays on activity. Some clay minerals of the 2:1 type, particularly montmorillonite, have promoted bacterial respiration (Stotzky, 1966a,b; Stotzky and Rem, 1966), or substrate decomposition (Kunc and Stotzky, 1970), whereas other 2:1 clay minerals e.g., vermiculate-mica mixtures, or 1:1 minerals, e.g., kaolinite, have had little effect (Stotzky, 1966a). By contrast, in other experiments, the addition of kaolinite has led to increased bacterial growth (Conn and Conn, 1940), or, conversely, considerably retarded the initial stages of mineralization of peptone by bacteria (Nováková, 1970). Clearly, the influence of clay minerals on bacterial activity are complex, and a number of possible modes of action have been suggested. The clays may serve to maintain a favorable pH for growth, as they are able, because of their cation exchange capacity, to replace hydrogen ions produced during metabolism with cations (Stotzky, 1966a,b). For example, the activity of certain bacterial enzyme systems involved in substrate (e.g., aldehydes) breakdown may be increased through the buffering action of certain clay minerals (Kunc and Stotzky, 1970). Also, as with glass, clay may serve as an adsorption, hence concentration, site for nutrients; apparently, amino acids which have been adsorbed onto clays are still accessible to bacteria, and amino acid uptake may even be stimulated (Zvyagintsev and Velikanov, 1968).

Ion exchange resins, which share some of the properties of clays, have also been used in bacterial activity experiments, but conflicting results have been obtained. The addition of cation and anion exchange resins have resulted in reduced substrate (e.g., glucose, succinate) oxidation (Hattori and Furusaka, 1959,1960; Hattori and Hattori, 1963), whereas the addition of Amberlite CG-120, a cation exchange resin, to cultures of *Methylococcus capsulatus* stimulated growth (Harwood and Pirt, 1972).

Where surfaces are found to affect bacterial numbers or activity, factors other than nutrient availability or enzyme action may be involved. For example, surfaces may provide microenvironments which are more suitable for growth; they may have a more favorable pH (see above, Stotzky, 1966a,b), may allow the establishment of anoxic microenvironments for the growth of anaerobes (Douglas *et al.,* 1917), or may provide some protection from phage attack (Roper and Marshall, 1974). The bacterial-substratum interactions involved in attachment could also induce fundamental changes in bacterial physiology by affecting membrane charge densities or structure. Hence, membrane-associated enzyme activity, such as electron transport, or substrate transport mechanisms would be affected. The growth of suspended bacteria would also be promoted by the addition of surfaces if inhibitors are present which can be adsorbed and, thus, removed from the bacterial culture (Sutherland and Wilkinson, 1961; Harwood and Pirt, 1972). Finally, in mixed bacterial cultures, attachment by one or more bacterial types may affect their competitive ability and, thus, their relative numbers in the system (Maigetter and Pfister, 1975). For example, in a continuous culture containing two bacterial strains, one strain was dominant without the addition of kaolinite, whereas the other strain, which was apparently able to attach to kaolinite, became dominant when the clay was added (Jannasch and Pritchard, 1972).

Clearly, the possible advantages (or disadvantages) of attachment in aquatic environments are extremely complex. They depend on hydrographic factors and on the types of surfaces and nutrients being assimilated, on whether assimilation depends on simple uptake mechanisms or exoenzyme activity, on the influence of substratum proximity on envelope structure, and on the presence of acids, alkalis, or toxins. Equally important are the complexity of the bacterial community, the requirements and capabilities of the individual bacterial types (e.g., oligotrophs and copiotrophs), and the competitive relationships between community members.

7. Succession and Microbial Film Development

7.1. Film Development

The development of a microbial film on a surface exposed to high fluid shear rates (Characklis, 1981a) is adequately described by a sigmoid-shaped curve (Fig. 4). The phase of a steady increase in biomass with time is probably a function of growth of attached bacteria as well as continuing adhesion of bacteria to the film. The plateau represents a critical film thickness at a point where the film penetrates the boundary or viscous sublayer (Characklis, 1981a). The critical film thickness is dependent upon the magnitude of the fluid shear rate. For a shear stress of 6.5 to 7.9 N/m^2, the viscous sublayer is approximately 40 μm thick, and this corresponds with an observed critical film thickness of 30 to 35 μm at this shear stress range (Characklis, 1980). The protrusion of film irregularities through the viscous sublayer creates frictional flow resistance and, hence, turbulence in the water flowing past the film surface.

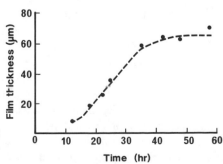

Figure 4. Development of a microbial film at a surface exposed to high fluid shear rates. Adapted from Characklis (1981a).

Characklis (1980) has reported the protrusion of microbial filaments into the viscous sublayer. These filaments were observed to resonate or "flutter," the frequency of the filament resonance being a direct function of the flow velocity past the surface. Frictional flow resistance was found to increase with increasing filament length. This frictional flow resistance creates problems in the free movement of vessels through water or of water through pipeline systems.

Rates of oxygen uptake by algal and bacterial communities attached to sand, pebble, rock, and detrital surfaces are directly related to the available surface area under constant conditions of temperature, pH, and oxygen consumption (Hargraves and Phillips, 1977). These authors report that microbial communities consume between 0.1 and 10×10^{-4} mg oxygen/cm^2/hr when expressed on an areal basis. During succession, as the attached community biomass increased, metabolism per unit organic weight decreased rapidly to a lower level, indicating some physicochemical limitations to diffusion. When the thickness of an attached community exceeds the limit of oxygen diffusion, about 21 μm in bacterial films, reduced oxygen supply may limit aerobic respiration despite the presence of an adequate nutrient supply (Sanders, 1967). This would account for the detection of anaerobic sulfate-reducing bacteria in films on metal surfaces suspended in well-aerated seawater (Gerchakov et al., 1977). It has been proposed that half-order kinetics best describe diffusion-limited biofilm situations (Saunders and Bazin, 1973; Harremoës, 1978).

7.2. Role of Prior Conditioning or Colonization on Succession in Microbial Films

Successions are characterized by changes in species composition and in relative abundance of any species within a community. Allogenic succession occurs when the succession is controlled by external environmental factors, whereas autogenic succession results from the resident population altering its own environment (Alexander, 1971). In considering the colonization of surfaces immersed in aquatic habitats, the external physicochemical conditions in the aqueous phase will usually remain relatively constant over the period taken for substantial microbial film development. Consequently, successional sequences in the microbial film community will be induced by changes induced by succeeding populations in the film.

As emphasized in Section 3, the spontaneous adsorption of macromolecules to surfaces immersed in aquatic habitats results in a concentration of useful nutrient materials at the surfaces. This situation favors colonization of the solid-liquid interface by copiotrophic bacteria, and a case has been presented in Section 5.2.2.2 for the selection of small starved copiotrophs as the initial colonizers of freshly immersed surfaces. Marshall et al. (1971b) reported that very small bacteria selectively colonized glass and formvar-coated electron microscope grid surfaces immersed in seawater. Further studies of the very early stages of colonization of surfaces immersed in nutrient deficient water are needed to confirm and extend these observations. Also, the suggestion that conditions at freshly immersed surfaces favor colonization by copiotrophs requires rigorous testing on a variety of different substrata.

Corpe (1973) reported that early colonizing bacteria were predominantly gram-negative rods, of which 50–90% were *Pseudomonas* species and 10–49% were pigmented *Flavobacterium* species or nonpigmented, nonmotile *Achromobacter* species. Gram-positive bacteria were rarely detected. Corpe noted that the gram-negative bacteria grew readily on surfaces in the presence of low nutrient levels (0.005% yeast extract) and suggested that concentration of nutrients at the surface was an important factor in colonization by these copiotrophic bacteria. Marshall *et al.* (1971b) and Corpe (1973) found that initial colonization by rod-shaped bacteria was followed after 24 to 72 hr by other species such as *Caulobacter, Hyphomicrobium,* and *Saprospira.* Corpe (1973) indicated the possibility that the secondary colonizers are dependent upon conditions created by primary colonizers, but emphasized that experimental evidence to support this idea is lacking. As suggested in Section 5.1.2, prior colonization by copiotrophs may result in a rapid utilization of nutrients accumulated at the surface, leading to a nutrient-deficient condition that would favor oligotrophs such as *Hyphomicrobium* and *Caulobacter.* The predatory gliding bacteria *(Saprospira)* would obtain their nutrients by scavenging bacteria from the colonized surface. Gliding bacteria are able to adhere to and move across surfaces as a result of the production of a viscous slime that acts as a Stefan adhesive (Humphrey *et al.,* 1979).

Jordan and Staley (1976) found that the biomass, numbers, and diversity of attached microorganisms increased as succession proceeded. The Shannon diversity index was shown to increase from 3.1 at day 1, to 4.2 at day 3, and to a maximum of 4.8 at day 10. Gerchakov *et al.* (1977) confirmed that colonization of surfaces began within several hours of immersion and demonstrated that bacterial numbers attaching to glass and metal surfaces increased over a 3-month period. Scanning electron microscope studies (Gerchakov *et al.,* 1977,1978; Marszalek *et al.,* 1979; Dempsey, 1981) again show a progression from simple rod-shaped primary colonizing bacteria to prosthecate forms, along with the formation of an extensive slime matrix that may be acidic polysaccharide (Corpe, 1970; Fletcher and Floodgate, 1973). The production rate of attached extracellular polysaccharide increases during film development, with attached polymer increasing nonlinearly, whereas attached biomass increases linearly (Bryers and Characklis, 1981).

The composition of the microbial film becomes increasingly complex with time and varies with the nature of the initial surface (Gerchakov *et al.,* 1977; Marszalek *et al.,* 1979; Dempsey, 1981). Normally, microalgae are not primary colonizers of surfaces (Marshall *et al.,* 1971b; Corpe, 1973; Jordan and Staley, 1976), but appear as a later stage in succession on surfaces. Scanning electron microscopy revealed intensive development of fungi, diatoms, and protozoa in later stages of succession on glass and stainless steel surfaces (Gerchakov *et al.*, 1977; Marszalek *et al.*, 1979).

7.3. Role of Original Surface Properties in Influencing Succession

Original substratum properties do influence the rate of microbial colonization of surfaces. The initial surface free energy of various substrata resulted in varying degrees of microbial colonization in the sea (Loeb, 1977; Dexter, 1979) and under laboratory conditions (Fletcher and Loeb, 1979). Gerchakov *et al.* (1978) and Marszalek *et al.* (1979) reported that substrata that are biologically inert (glass, stainless steel) fouled rapidly and produced a complex, two-tier microfouling layer. The first tier was in intimate contact with the substratum and consisted mainly of bacteria, fungi, and nonmotile diatoms. The second tier, appearing after about a 5-week exposure, consisted of large, colonial, motile diatoms, other diatoms, ciliates, and flagellates. Active substrata (brass, copper-nickel alloys) fouled at a slower rate, were selective for bacteria that secrete extracellular mucoid material, and were characterized by a less diverse fouling community. Similar results were reported for inert and active substrata by Dempsey (1981). Marszalek *et al.* (1979) suggested that the substratum characteristics influence microfouling at all stages, especially in the early stages of conditioning-film and primary-film formation.

The above results add weight to the suggestion by Baier (1980, 1981) that the original substratum properties do influence subsequent adhesion events despite the formation of surface-conditioning films that alter the surface properties of the substrata. Several questions may be posed in view of these results. Do all surfaces acquire the same conditioning films? Are the macromolecules of these conditioning films oriented in a similar fashion on different surfaces? Does the nature of conditioning films influence the types of microorganisms colonizing various substrata? Do the initial colonizing organisms modify the surface properties so that colonization by other particular organisms is favored?

7.4. Possible Control of Film Formation

It is desirable that some means of control of microbial film formation be devised for certain situations. Microbial fouling of boat and water pipeline surfaces leads to frictional flow resistance, and is costly in terms of cleaning procedures. Similarly, microbial fouling of heat transfer surfaces leads to losses in efficiency in such equipment (Characklis, 1981a). For example, Bell (1978) has calculated that a microbial film of 55 μm thickness would cause unacceptable reduction in the heat transfer coefficient of heat exchanger tubes in an Ocean Thermal Energy Conversion (OTEC) power system.

More detailed information is required as to the effects of substratum properties on conditioning film formation, on the rate of substratum colonization, on the successional sequence of microorganisms, and on the actual strength of microbial adhesion. Although promising results have been obtained, the wide variety of bacteria in aquatic habitats make for difficulties in proposing simple solutions to such problems. For instance, the wide differences in the observations of Fletcher and Loeb (1979) and of Dexter (1979) on the effects of substratum surface free energy on adhesion emphasize this difficulty.

8. Concluding Remarks

Although the data on bacterial attachment to solid surfaces in aquatic environments contain some inconsistencies, it is still possible to reach some conclusions about attachment processes and their significance. Over the last 10 years, substantial progress has been make in elucidating attachment mechanisms and demonstrating the importance of substratum properties in the adhesion process. Surface characteristics such as charge, surface energy, or hydrophobicity affect the numbers of bacteria able to attach, and these properties may even be expressed to some extent when conditioned by an adsorbed macromolecular film. However, adhesion mechanisms involve not only physicochemical adsorption processes, but also the physiological properties and possible responses of the bacteria themselves. Attachment is to a certain extent a selective process, whereby bacteria with certain adhesive and probably metabolic capabilities are preferentially adsorbed. Moreover, after attachment has occurred, the physiology of bacteria in the interfacial microenvironment will be conditioned by different environmental factors than those affecting their free-living counterparts. However, there is still much to be learned about bacterial attachment mechanisms, the dominant physiocochemical factors, the significance of conditioning films, the extent of physiological involvement in adhesion, and the response of organisms to the proximity of the surface.

There are also many questions to be answered concerning the nutritional status of aquatic bacteria, and although these are usually considered with respect to suspended bacteria, attached organisms are equally involved. The significance of oligotrophy and copiotrophy in aquatic environments is still not clear, and the relationship between these two nutritional types and surface microenvironments may give some clue to their growth and survival strategies. The properties of the interfacial microenvironment differ significantly from those of the bulk aqueous phase so that resident organisms in each will be presented with different environmental conditions. Overall, the surface microenvironment appears to be the more advantageous, as attached organisms do not expend energy searching for food, they resist washout in fast-flowing waters, and they can take advantage of nutrients adsorbed at the surface. On the other hand, the surface mode of growth loses some of these advantages as thick bacterial slime layers develop and growth conditions alter considerably. The formation of slime layers from attached monolayers creates a different type of microbial environment where interaction with the adjacent aqueous phase is considerably reduced for the majority of cells. Most of the bacteria are surrounded by a polymeric matrix, which tends to protect the cells from outside perturbations but introduces new limitations on mass transfer of nutrients and gases. In slime layers, cell-cell interactions are more significant than cell-water interactions, and these will influence or control the successional development of biofilms and be a major feature of cellular processes within mature biofilms. In natural environments, the relationships between attached organisms must be extremely complex, as slime layers contain a diverse assemblage of organisms, including not only bacteria, but also unicellular and filamentous algae, protozoa, and various invertebrates.

Although this review has been concerned with the significance of bacterial attachment in aquatic environments, essentially all environments inhabited by bacteria are "aquatic." Thus, it is likely that the problems, progress, and principles which apply to microbial attachment in natural waters will be relevant to other habitats where attachment plays a significant role. Such environments include terrestrial environments, where bacteria colonize soil particles and detritus, plant and animal environments, where attachment may be a prerequisite for disease (e.g. gastrointestinal disease, gonorrhea, dental caries), and industrial plants utilizing microbial processes (e.g., waste-water treatment in trickling filters).

References

Alexander, M., 1971, *Microbial Ecology*, John Wiley & Sons, New York.

Anderson, J. I. W., and Heffernan, W. P., 1965, Isolation and characterization of filterable marine bacteria, *J. Bacteriol.* **90:**1713–1718.

Baier, R. E., 1980, Substrata influences on adhesion of microorganisms and their resultant new surface properties, in: *Adsorption of Microorganisms to Surfaces* (G. Bitton and K. C. Marshall, eds.), pp. 59–104, Wiley-Interscience, New York.

Baier, R. E., 1981, Early events of micro-biofouling of all heat transfer equipment, in: *Fouling of Heat Transfer Equipment* (E. F. C. Somerscales and J. G. Knudsen, eds.), pp. 293–304, Hemisphere Publ. Co., Washington, D.C.

Baier, R. E., and Loeb, G. I., 1971, Multiple parameters characterizing interfacial films of a protein analogue, polymethyl glutamate, in: *Polymer Characterization: Interdisciplinary Approaches* (D. D. Craver, ed.), pp. 79–96, Plenum Press, New York.

Bell, K. J., 1978, The effect of fouling on OTEC heat exchanger design, construction and operation, in: *Proceedings OTEC Biofouling and Corrosion Symposium* (R. H. Gray, ed.), pp. 19–30, Pacific Northwest Laboratory, Richland, Wash.

Boyle, P. J., and Mitchell, R., 1980, Interactions between microorganisms and wood-boring crustaceans, in: *Biodeterioration* (T. A. Oxley, G. Becker, and D. Allsopp, eds.), pp. 179–186, Pitman Publ., London, and Biodeterioration Society.

Brown, C. M., Ellwood, D. C., and Hunter, J. R., 1977, Growth of bacteria at surfaces: influence of nutrient limitation, *FEMS Microbiol. Lett.* **1**:163–166.

Bryers, J. D., and Characklis, W. G., 1981, Kinetics of initial biofilm formation within a turbulent flow system, in: *Fouling of Heat Transfer Equipment* (E. F. C. Somerscales and J. G. Knudsen, eds.), pp. 313–333, Hemisphere Publ. Co., Washington, D.C.

Characklis, W. G., 1980, Biofilm development and destruction, Electric Power Research Institute Report 902-1 (September, 1980).

Characklis, W. G., 1981a, Microbial fouling: a process analysis, in: *Fouling of Heat Transfer Equipment* (E. F. C. Somerscales and J. G. Knudsen, eds.), pp. 251–291, Hemisphere Publ. Co., Washington, D.C.

Characklis, W. G., 1981b, Bioengineering report: fouling biofilm development: a process analysis, *Biotech. Bioeng.* **23**:1923–1960.

Chet, I., Asketh, P., and Mitchell, R., 1975, Repulsion of bacteria from marine surfaces, *Appl. Microbiol.* **30**:1043–1045.

Christie, A. O., Evans, L. V., and Shaw, M., 1970, Studies on the ship-fouling algae *Enteromorpha:* II The effect of certain enzymes on the adhesion of zoospores. *Ann. Bot.* **34**:467–482.

Clark, W. B., and Gibbons, R. J., 1977, Influence of salivary components and extracellular polysaccharide synthesis from sucrose on the attachment of *Streptococcus mutans* 6715 to hydroxyapatite surfaces, *Infect. Immunity* **18**:514–523.

Conn, H. J., and Conn, J. E., 1940, The stimulating effect of colloids upon the growth of certain bacteria, *J. Bacteriol.* **39**:99–100.

Corpe, W. A., 1970, An acid polysaccharide produced by primary film-forming bacteria, *Develop. Ind. Microbiol.* **11**:402–412.

Corpe, W. A., 1973, Microfouling: the role of primary film-forming bacteria, in: *Proc. 3rd Int. Congr. Mar. Corrosion Fouling* (R. F. Acker, B. F. Brown, J. R. DePalma, and W. P. Iverson, eds.), pp. 598–609, Northwestern Univ. Press, Evanston, Ill.

Corpe, W. A., Matsuuchi, L., and Armbruster, B., 1976, Secretion of adhesive polymers and attachment of marine bacteria to surfaces, in: *Proc. 3rd Int. Biodegradation Symp.* (J. M. Sharpley and A. M. Kaplan, eds.). pp. 433–442, Applied Science Publ., London.

Costerton, J. W., Geesey, G. G., and Cheng, K-J., 1978, How bacteria stick, *Sci. Am.* **238**:86–95.

Cundell, A. M., Sleeter, T. D., and Mitchell, R., 1977, Microbial populations associated with the surface of the brown alga *Ascophyllum nodosum, Microb. Ecol.* **4**:81–91.

Curtis, A. S. G., 1979, Summing-up, in: *Adhesion of Microorganisms to Surfaces* (D. C. Ellwood, J. Melling, and P. Rutter, eds.), pp. 199–208, Academic Press, London.

Dahlbäck, B., Hermansson, M., Kjelleberg, S., and Norkrans, B., 1981, The hydrophobicity of bacteria—an important factor in their initial adhesion at the air-water interface, *Arch. Microbiol.* **128**:267–270.

Danielsson, A., Norkrans, B., and Björnsson, A., 1977, On bacterial adhesion—the effect of certain enzymes on adhered cells of a marine *Pseudomonas* sp, *Botan. Mar.* **20**:13–17.

Dawson, M. P., Humphrey, B. A., and Marshall, K. C., 1981, Adhesion, a tactic in the survival strategy of a marine vibrio during starvation. *Curr. Microbiol.* **6**:195–198.

Dazzo, F. B., 1980, Microbial adhesion to plant surfaces, in: *Microbial Adhesion to Surfaces* (R. C. W. Berkeley, J. M. Lynch, J. Melling, P. R. Rutter, and B. Vincent, eds.), pp. 311–328, Ellis Horwood Publ., Chichester.

Dempsey, M. J., 1981, Marine bacterial fouling: a scanning electron microscope study, *Mar. Biol.* **61**:305–315.

Dexter, S. C., 1979, Influence of substratum critical surface tension on bacterial adhesion—*in situ* studies, *J. Coll. Interface Sci.* **70:**346–354.

Dexter, S. C., Sullivan, J. D. Jr., Williams, J., and Watson, S. W., 1975, Influence of substratum wettability on the attachment of marine bacteria to various surfaces, *Appl. Microbiol.* **30:**298–308.

Douglas, S. R., Fleming, A., and Colebrook, M. B., 1917, On the growth of anaerobic bacilli in fluid media under apparently aerobic conditions, *Lancet* **2:**530–532.

Ellen, R. P., Walker, D. L., and Chan, K. H., 1978, Association of long surface appendages with adherence-related functions of the Gram-positive species *Actinomyces naeslundii, J. Bacteriol.* **134:**1171–1175.

Ellwood, D. C., Hunter, J. R., and Longyear, V. M. C., 1974, Growth of *Streptococcus mutans* in a chemostat, *Arch. Oral Biol.* **19:**659–664.

Fletcher, M., 1976, The effects of proteins on bacterial attachment to polystyrene, *J. Gen. Microbiol.* **94:**400–404.

Fletcher, M., 1977, The effects of culture concentration and age, time and temperature on bacterial attachment to polystyrene, *Can. J. Microbiol.* **23:**1–6.

Fletcher, M., 1979, A microautoradiographic study of the activity of attached and free-living bacteria, *Arch. Microbiol.* **122:**271–274.

Fletcher, M., 1980, The question of passive versus active attachment mechanisms in nonspecific bacterial adhesion, in: *Microbial Adhesion to Surfaces* (R. C. W. Berkeley, J. M. Lynch, J. Melling, P. R. Rutter, and B. Vincent, eds.), pp. 197–210, Ellis Horwood Publ., Chichester.

Fletcher, M., and Floodgate, G. D., 1973, An electron-microscopic demonstration of an acidic polysaccharide involved in adhesion of a marine bacterium to solid surfaces, *J. Gen. Microbiol.* **74:**325–334.

Fletcher, M., and Floodgate, G. D., 1976, The adhesion of bacteria to solid surfaces, in: *Microbial Ultrastructure: the Use of the Electron Microscope* (R. Fuller and D. W. Lovelock, eds.), pp. 101–107, Academic Press, London.

Fletcher, M., and Loeb, G. I., 1979, The influence of substratum characteristics on the attachment of a marine pseudomonad to solid surfaces, *Appl. Environ. Microbiol.* **37:**67–72.

Fletcher, M., and Marshall, K. C., 1982, A bubble contact angle method for evaluating substratum interfacial characteristics and its relevance to bacterial attachment *Appl. Environ. Microbiol.* (in press).

Fuhrman, J. A., 1981, Influence of method on the apparent size distribution of bacterioplankton cells: epifluorescence microscopy compared to scanning electron microscopy, *Mar. Ecol. Prog. Ser.* **5:**103–106.

Gerchakov, S. M., Marszalek, D. S., Roth, F. J., and Udey, L. R., 1977, Succession of periphytic microorganisms on metal and glass surfaces in: *Proc. 4th Int. Congr. Mar. Corrosion Fouling* (V. Romanovsky, ed.), pp. 203–211, Centre de Recherches et d'Etudes Oceanographiques, Boulogne, France.

Gerchakov, S. M., Marszalek, D. S., Roth, F. J., Sallman, B., and Udey, L. R., 1978, Observations on microfouling applicable to OTEC systems, in: *Proceedings OTEC Biofouling and Corrosion Symposium* (R. H. Gray, ed.), pp. 63–75, Pacific Northwest Laboratory, Richland, Wash.

Gerson, D. F., 1980, Cell surface energy, contact angles and phase partition I. Lymphocytic cell lines in biphasic aqueous mixtures, *Biochim. Biophys. Acta* **602:**269–280.

Gerson, D. F., and Akit, J., 1980, Cell surface energy, contact angles and phase partition II. Bacterial cells in biphasic aqueous mixtures, *Biochim. Biophys. Acta* **602:**281–284.

Gerson, D. F., and Scheer, D., 1980, Cell surface energy, contact angles and phase partition III. Adhesion of bacterial cells to hydrophobic surfaces. *Biochim. Biophys. Acta* **602:**506–510.

Goulder, R., 1977, Attached and free bacteria in an estuary with abundant suspended solids, *J. Appl. Bacteriol.* **43**:399–405.

Harden, V. P., and Harris, J. O., 1953, The isoelectric point of bacterial cells, *J. Bacteriol.* **65**:198–202.

Hargraves, B. T., and Phillips, G. A., 1977, Oxygen uptake of microbial communities on solid surfaces, in: *Aquatic Microbial Communities* (J. Cairns, eds.), pp. 545–587, Garland Publ., New York.

Harremoës, P., 1978, Biofilm kinetics, in: *Water Pollution Microbiology, Vol. 2* (R. Mitchell, ed.), John Wiley & Sons, New York.

Harvey, R. W., and Young, L. Y., 1980, Enumeration of particle-bound and unattached respiring bacteria in the salt marsh environment, *Appl. Environ. Microbiol.* **40**:156–160.

Harwood, J. H., and Pirt, S. J., 1972, Quantitative aspects of growth of the methane oxidizing bacterium *Methylococcus capsulatus* on methane in shake flask and continuous chemostat culture, *J. Appl. Bacteriol.* **35**:597–607.

Hattori, T., and Furusaka, C., 1959, Chemical activities of *Escherichia coli* adsorbed on a resin, *Biochim. Biophys. Acta* **31**:581–582.

Hattori, T., and Furusaka, C., 1960, Chemical activities of *Escherichia coli* adsorbed on a resin, *J. Biochem.* **48**:831–837.

Hattori, R., and Hattori, T., 1963, Effect of a liquid-solid interface on the life of microorganisms, *Ecol. Rev.* **16**:64–70.

Hendricks, C. W., 1974, Sorption of heterotrophic and enteric bacteria to glass surfaces in the continuous culture of river water, *Appl. Microbiol.* **28**:572–578.

Heukelekian, H., and Heller, A., 1940, Relation between food concentration and surface for bacterial growth, *J. Bacteriol.* **40**:547–558.

Hirsch, P., 1979, Life under conditions of low nutrient concentrations, in: *Strategies of Microbial Life in Extreme Environments* (M. Shilo, ed.), pp. 357–372, Dahlem Konfr. Life Sci. Res. Report 13, Verlag Chemie, Weinheim.

Hirsch, P., and Pankratz, S. H., 1971, Studies of bacterial populations in natural environments by use of submerged electron microscope grids, *Z. Allg. Mikrobiol.* **10**:589–605.

Hoppe, H.-G., 1976, Determination and properties of actively metabolizing heterotrophic bacteria in the sea, investigated by means of microautoradiography, *Mar. Biol.* **36**:291–302.

Hossell, J. C., and Baker, J. H., 1979, A note on the enumeration of epiphytic bacteria by microscopic methods with particular reference to two freshwater plants, *J. Appl. Bacteriol.* **46**:87–92.

Humphrey, B. A., Dickson, M. R., and Marshall, K. C., 1979, Physicochemical and in situ observations on the adhesion of gliding bacteria to surfaces, *Arch. Microbiol.* **120**:231–238.

Jannasch, H. W., 1967, Enrichments of aquatic bacteria in continuous culture, *Arch. Mikrobiol.* **59**:165–173.

Jannasch, H. W., and Pritchard, P. H., 1972, The role of inert particulate matter in the activity of aquatic microorganisms, *Mem. Ist. Ital. Idrobiol. 29 Suppl:* 289–308.

Jordan, T. L., and Staley, J. T., 1976, Electron microscopic study of succession in the periphyton community of Lake Washington, *Microb. Ecol.* **2**:241–251.

Kjelleberg, S., Lagercrantz, C., and Larsson, T., 1980, Quantitative analysis of bacterial hydrophobicity studied by the binding of dodecanoic acid, *FEMS Microbiol. Lett.* **7**:41–44.

Kjelleberg, S., Humphrey, B. A., and Marshall, K. C., 1982, The effects of interfaces on small starved marine bacteria, *Appl. Environ. Microbiol.* **43**:1166–1172.

Koch, A. L., 1979, Microbial growth in low concentrations of nutrients, in: *Strategies of Microbial Life in Extreme Environments* (M. Shilo, ed.), pp. 261–279, Dahlem Konfr. Life Sci. Res. Report 13, Verlag Chemie, Weinheim.

Kunc, F., and Stotzky, G., 1970, Breakdown of some aldehydes in soils with different amounts of montmorillonite and kaolinite, *Folia Microbiol.* **15**:216.

Kunznetzov, S. I., Dubinina, G. A., and Lapteva, N. A., 1979, Biology of oligotrophic bacteria, *Annu. Rev. Microbiol.* **33:**377–387.

Lahav, N., 1962, Adsorption of sodium bentonite particles on *Bacillus subtilis, Plant and Soil* **17:**191–208.

Leech, R., and Hefford, R. J. W., 1980, The observation of bacterial deposition from a flowing suspension, in: *Microbial Adhesion to Surfaces* (R. C. W. Berkeley, J. M. Lynch, J. Melling, P. R. Rutter, B. Vincent, eds.), pp. 544–545, Ellis Horwood Publ., Chichester.

Loeb, G. I., 1977, The settlement of fouling organisms on hydrophobic surfaces, Naval Research Laboratory Memorandum Report 3665, Washington, D.C.

Lupton, F. S., and Marshall, K. C., 1981, Specific adhesion of bacteria to heterocysts of *Anabaena* spp. and its ecological significance, *Appl. Environ. Microbiol.* **42:**1085–1092.

Maigetter, R. Z., and Pfister, R. M., 1975, A mixed bacterial population in a continuous culture with and without kaolinite, *Can. J. Microbiol.* **21:**173–180.

Marshall, K. C., 1968, Interaction between colloidal montmorillonite and cells of *Rhizobium* species with different ionogenic surfaces, *Biochim. Biophys. Acta* **156:**179–186.

Marshall, K. C., 1979, Growth at interfaces, in: *Strategies of Microbial Life in Extreme Environments* (M. Shilo, ed.), pp. 281–290, Dahlem Konfr. Life Sci. Res. Report 13, Verlag Chemie, Weinheim.

Marshall, K. C., 1980, Reactions of microorganisms, ions and macromolecules at interfaces, in: *Contemporary Microbial Ecology* (D. C. Ellwood, J. N. Hedger, M. J. Latham, J. M. Lynch, and J. H. Slater, eds.), pp. 93–106, Academic Press, London.

Marshall, K. C., and Cruickshank, R. H., 1973, Cell surface hydrophobicity and the orientation of certain bacteria at interfaces, *Arch. Mikrobiol.* **91:**29–40.

Marshall, K. C., Stout, R., and Mitchell, R., 1971a, Mechanism of the initial events in the sorption of marine bacteria to surfaces, *J. Gen. Microbiol.* **68:**337–348.

Marshall, K. C., Stout, R., and Mitchell, R., 1971b, Selective sorption of bacteria from seawater, *Can J. Microbiol.* **17:**1413–1416.

Marszalek, D. S., Gerchakov, S. M., and Udey, L. R., 1979, Influence of substrate composition on marine microfouling, *Appl. Environ. Microbiol.* **38:**987–995.

Meyer-Reil, L-A, 1978, Autoradiograpny and epifluorescnece microscopy combined for the determination of number and spectrum of actively metabolizing bacteria in natural waters, *Appl. Environ. Microbiol.* **36:**506–512.

Minato, H., and Suto, T., 1976, Technique for fractionation of bacteria in rumen microbial ecosystem. I. Attachment of rumen bacteria to starch granules and elution of bacteria attached to them, *J. Gen. Appl. Microbiol.* **22:**259–276.

Moss, B., 1980, *Ecology of Fresh Waters,* Blackwell, Oxford.

Navarro, J. M., and Durand, G., 1977, Modification of yeast metabolism by immobilisation onto porous glass, *Eur. J. Appl. Microbiol.* **4:**243–254.

Neihof, R. A., and Loeb, G. I., 1972, The surface charge of particulate matter in seawater, *Limnol. Oceanogr.* **17:**7–16.

Neihof, R., and Loeb, G., 1974, Dissolved organic matter in seawater and the electric charge of immersed surfaces, *J. Mar. Res.* **32:**5–12.

Neumann, A. W., Good, R. J., Hope, C. J., and Sejpal, M., 1974. An equation-of-state approach to determine surface tensions of low energy solids from contact angles, *J. Coll. Interface Sci.* **49:**291–304.

Neumann, A. W., Absolom, D. R., Van Oss, C. J., and Zingg, W., 1979, Surface thermodynamics of leukocyte and platelet adhesion to polymer surfaces, *Cell. Biophys.* **1:**79–92.

Norde, W., and Lyklema, J., 1978, Adsorption of proteins from aqueous solution on negatively charged polystyrene surfaces, in: *Ions in Macromolecular and Biological Systems* (D. H. Everett and B. Vincent, eds.), pp. 11–35, Scientechnica, Bristol.

Nováková, J., 1970, Effect of clay minerals on the mineralisation of peptone in liquid medium, *Folia Microbiol.* **15:**217.

Novitsky, J. A., and Morita, R. Y., 1976, Morphological characterization of small cells resulting from nutrient starvation of a psychrophilic marine vibrio, *Appl. Environ. Microbiol.* **32**:617–622.

Novitsky, J. A., and Morita, R. Y., 1977, Survival of a psychrophilic marine vibrio under long term nutrient starvation, *Appl. Environ. Microbiol.* **33**:635–641.

Novitsky, J. A., and Morita, R. Y., 1978, Possible strategy for the survival of marine bacteria under starvation conditions, *Mar. Biol.* **48**:289–295.

Olsson, J., Glantz, P.-O., and Krasse, B., 1976, Surface potential and adherence of oral streptococci to solid surfaces, *Scand. J. Dent. Res.* **84**:240–242.

Ørstavik, D., 1977, Sorption of streptococci to glass: effects of macromolecular solutes, *Acta. Pathol. Microbiol. Scand.* **85**:47–53.

Owens, D. K., and Wendt, R. C., 1969, Estimation of the surface free energy of polymers, *J. Appl. Polym. Sci.* **13**:1741–1747.

Paerl, H. W., 1976, Specific association of the blue-green algae *Anabaena* and *Aphanizomenon* with bacteria in freshwater blooms, *J. Phycol.* **12**:431–435.

Pertsovskaya, A. F., Duda, V. I., and Zvyagintsev, D. G., 1972, Surface ultrastructure of adsorbed microorganisms, *Soviet Soil Sci.* **4**:684–689.

Poindexter, J. S., 1981a, The caulobacters: ubiquitous unusual bacteria, *Microbiol. Rev.* **45**:123–179.

Poindexter, J. S., 1981b, Oligotrophy: fast and famine existence, in: *Advances in Microbial Ecology, Vol. 5* (M. Alexander, ed.), Plenum Press, New York, 1981.

Rheinheimer, G., 1980, *Aquatic Microbiology* (2nd ed.), John Wiley & Sons, Chichester.

Rolla, G., 1980, On the chemistry of the matrix of dental plaque, in: *Microbial Adhesion to Surfaces* (R. C. W. Berkeley, J. M. Lynch, J. Melling, P. R. Rutter, and B. Vincent, eds.), pp. 425–439, Ellis Horwood Publ., Chichester.

Roper, M. M., and Marshall, K. C., 1974, Modification of the interaction between *Escherichia coli* and bacteriophage in saline sediments, *Microb. Ecol.* **1**:1–13.

Rutter, P., and Leech, R., 1980, The deposition of *Streptococcus sanguis* NCTC 7868 from a flowing suspension, *J. Gen. Microbiol.* **120**:301–307.

Rutter, P., and Vincent, B., 1980, The adhesion of microorganisms to surfaces: physico-chemical aspects, in: *Microbial Adhesion to Surfaces* (R. C. W. Berkeley, J. M. Lynch, J. Melling, P. R. Rutter and B. Vincent, eds.), pp. 79–92, Ellis Horwood Publ., Chichester.

Sanders, M. W., III, 1967, The growth and development of attached stream bacteria, Part I. Theoretical growth kinetics of attached stream bacteria, *Water Resources Res.* **3**:81–87.

Santoro, T., and Stotzky, G., 1967, Effect of cations and pH on the electrophoretic mobility of microbial cells and clay minerals, *Bacteriol. Proc.* **1967**:A15.

Saunders, P. T., and Bazin, M. J., 1973, Nonsteady state studies of nitrification in soil: theoretical considerations, *Soil Biol. Biochem.* **5**:545–557.

Shands, J. W., 1966, Localization of somatic antigen on gram-negative bacteria using ferritin antibody conjugates, *Ann. N.Y. Acad. Sci.* **133**:292–298.

Sieburth, J. McN., 1968, The influence of algal antibiosis on the ecology of marine microorganisms, *Adv. Microbiol. Sea* **1**:63–94.

Sieburth, J. McN., 1975, *Microbial Seascapes*, University Park Press, Baltimore.

Sims, R. C., and Little, L., 1973, Enhanced nitrification by addition of clinoptilolite to tertiary activated sludge units, *Environ. Lett.* **4**:27–34.

Stevenson, L. H., 1978, A case for dormancy in aquatic systems, *Microb. Ecol.* **4**:127–133.

Stotzky, G., 1966a, Influence of clay minerals on microorganisms II. Effect of various clay species, homionic clays, and other particles on bacteria, *Can. J. Microbiol.* **12**:831–848.

Stotzky, G., 1966b, Influence of clay minerals on microorganisms III. Effect of particle size, cation exchange capacity and surface area on bacteria, *Can. J. Microbiol.* **12**:1235–1246.

Stotzky, G., and Rem, L. T., 1966, Influence of clay minerals on microorganisms, I. Montmorillonite and kaolinite on bacteria, *Can. J. Microbiol.* **12**:547–563.

Sutherland, I. W., 1980, Polysaccharides in the adhesion of marine and freshwater bacteria, in: *Microbial Adhesion to Surfaces* (R. C. W. Berkeley, J. M. Lynch, J. Melling, P. R. Rutter, and B. Vincent, eds.), pp. 329–338. Ellis Horwood Publ., Chichester.

Sutherland, I. W., and Wilkinson, J. F., 1961, A new growth medium for virulent *Bordetella pertussis, J. Path. Bact.* **82**:431–438.

Tabor, P. S., Ohwada, K., and Colwell, R. R., 1981, Filterable marine bacteria found in the deep sea: distribution, taxonomy, and response to starvation, *Microb. Ecol.* **7**:67–83.

Tadros, T. F., 1980, Particle-surface adhesion, in: *Microbial Adhesion to Surfaces* (R. C. W. Berkeley, J. M. Lynch, J. Melling, P. R. Rutter, and B. Vincent, eds.), pp. 93–116, Ellis Horwood Publ., Chichester.

Takakuwa, S., Fujimori, T., and Iwasaki, H., 1979, Some properties of cell-sulfur adhesion in *Thiobacillus thiooxidans, J. Gen. Appl. Microbiol.* **25**:21–29.

Tyler, P. A., and Marshall, K. C., 1967, Form and function in manganese-oxidizing bacteria, *Arch. Mikrobiol.* **56**:344–353.

Ward, J. B., and Berkeley, R. C. W., 1980, The microbial cell surface and adhesion, in: *Microbial Adhesion to Surfaces* (R. C. W. Berkeley, J. M. Lynch, J. Melling, P. R. Rutter, and B. Vincent, eds.), pp. 47–66, Ellis Horwood Publ., Chichester.

Wardell, J. N., Brown, C. M., and Ellwood, D. C., 1980, A continuous culture study of the attachment of bacteria to surfaces, in: *Microbial Adhesion to Surfaces* (R. C. W. Berkeley, J. M. Lynch, J. Melling, P. R. Rutter, and B. Vincent, eds.), pp. 221–230, Ellis Horwood Publ., Chichester.

Weise, W., and Rheinheimer, G., 1978, Scanning electron microscopy and epifluorescence investigation of bacterial colonization of marine sand sediments, *Microb. Ecol.* **4**:175–188.

Weiss, L., 1961a, The measurement of cell adhesion, *Exp. Cell Res.* (Suppl.) **8**:141–153.

Weiss, L., 1961b, Studies on cellular adhesion in tissue culture—IV. The alteration of substrata by cell surfaces. *Exp. Cell Res.* **25**:504–517.

Wood, J. M., 1980, The interaction of microorganisms with ion exchange resins, in: *Microbial Adhesion to Surfaces* (R. C. W. Berkeley, J. M. Lynch, J. Melling, P. R. Rutter, and B. Vincent, eds.), pp. 163–185, Ellis Horwood Publ., Chichester.

Wu, S., 1980, Surface tension of solids: generalisation and reinterpretation of critical surface tension, in: *Adhesion and Adsorption of Polymers* (L.-H. Lee, ed.), pp. 53–65, Plenum Press, New York.

Zimmermann, R., Iturriaga, R., and Becker-Birck, J., 1978, Simultaneous determination of the total number of aquatic bacteria and the number thereof involved in respiration, *Appl. Environ. Microbiol.* **36**:926–935.

Zisman, W. A., 1964, Relation of the equilibrium contact angle to liquid and solid constitution, *Adv. Chem.* **43**:1–51.

ZoBell, C. E., 1943, The effect of solid surfaces upon bacterial activity, *J. Bacteriol.* **46**:39–56.

Zvyagintsev, D. G., and Velikanov, L. L., 1968, Influence of adsorbents on the activity of bacteria growing on media with amino acids, *Microbiology* **37**:861–866.

Zvyagintsev, D. G., Pertsovskaya, A. F., Yakhnin, E. D., and Averbakh, E. I., 1971, Adhesion value of microorganism cells to solid surfaces, *Microbiology* **40**:889–893.

Zvyagintsev, D. G., Guzev, V. S., and Guzeva, I. S., 1977, Relationship between adsorption of microorganisms and the stage of their development, *Microbiology* **46**:245–249.

Zvagintseva, I. S., and Zvyagintsev, D. G., 1969, Effect of microbial cell adsorption onto steroid crystals on the transformation of the steroid, *Microbiology* **38**:691–694.

Index